Encyclopaedia of Mathematical Sciences

Volume 48

Editor-in-Chief: R.V. Gamkrelidze

Yu. D. Burago   V. A. Zalgaller (Eds.)

Geometry III

Theory of Surfaces

With 80 Figures

Springer-Verlag

Berlin Heidelberg New York
London Paris Tokyo
Hong Kong Barcelona
Budapest

Consulting Editors of the Series:
A.A. Agrachev, A.A. Gonchar, E.F. Mishchenko,
N.M. Ostianu, V.P. Sakharova, A.B. Zhishchenko

Title of the Russian edition:
Itogi nauki i tekhniki, Sovremennye problemy matematiki,
Fundamental'nye napravleniya, Vol. 48, Geometriya 3
Publisher VINITI, Moscow 1989

Mathematics Subject Classification (1991): 53A05

ISBN 3-540-53377-X Springer-Verlag Berlin Heidelberg New York
ISBN 0-387-53377-X Springer-Verlag New York Berlin Heidelberg

Library of Congress Cataloging-in-Publication Data
Geometriía 3. English Geometry III:
theory of surfaces / Yu. D. Burago, V. A. Zalgaller, eds.
p. cm.—(Encyclopaedia of mathematical sciences; v. 48)
Includes bibliographical references and index.
ISBN 0-387-53377-X (U.S.: alk. paper)
1. Surfaces. I. Burago, IU. D. (IUrii Dmitrievich) II. Zalgaller, V. A. (Viktor A.) III. Title.
IV. Title: Geometry three. V. Title: Geometry 3. VI. Series.
QA571.G36 1992 516.3'63—dc20 92-16958

List of Editors, Authors and Translators

Editor-in-Chief

R.V. Gamkrelidze, Russian Academy of Sciences, Steklov Mathematical Institute, ul. Vavilova 42, 117966 Moscow, Institute for Scientific Information (VINITI), ul. Usievicha 20a, 125219 Moscow, Russia

Consulting Editors

Yu. D. Burago, LOMI, Fontanka 27, 191011 St. Petersburg, Russia
V. A. Zalgaller, LOMI, Fontanka 27, 191011 St. Petersburg, Russia

Authors

Yu. D. Burago, LOMI, Fontanka 27, 191011 St. Petersburg, Russia
E. R. Rozendorn, Department of Mathematics and Mechanics, Moscow University, 119899 Moscow, Russia
I. Kh. Sabitov, Department of Mathematics and Mechanics, Moscow University, 119899 Moscow, Russia
S. Z. Shefel'†

Translator

E. Primrose, 12 Ring Road, Leicester LE2 3RR, England

Contents

I. The Geometry of Surfaces in Euclidean Spaces

Yu.D. Burago, S.Z. Shefel'

Translated from the Russian
by E. Primrose

Contents

Preface

The original version of this article was written more than five years ago with S.Z. Shefel', a profound and original mathematician who died in 1984. Since then the geometry of surfaces has continued to be enriched with ideas and results. This has required changes and additions, but has not influenced the character of the article, the design of which originated with Shefel'. Without knowing to what extent Shefel' would have approved the changes, I should nevertheless like to dedicate this article to his memory. (Yu.D. Burago)

We are trying to state the qualitative questions of the theory of surfaces in Euclidean spaces in the form in which they appear to the authors at present. This description does not entirely correspond to the historical development of the subject. The theory of surfaces was developed in the first place mainly as the theory of surfaces in three-dimensional Euclidean space E^3; however, it makes sense to begin by considering surfaces F in Euclidean spaces of any dimension $n \geqslant 3$. This approach enables us, in particular, to put in a new light some unsolved problems of this developed (and in the case of surfaces in E^3 fairly complete) theory, and in many cases to refer to the connections with the present stage of development of the theory of multidimensional submanifolds.

The leading question of the article is the problem of the connection between classes of metrics and classes of surfaces in E^n. The first chapter is a brief survey of general questions in the theory of surfaces from this point of view. Chapters 2 and 3 are devoted to a more detailed consideration of convex and saddle surfaces respectively. The subject of Chapter 4 consists of classes of metrics not associated directly with the condition that the Gaussian curvature has a definite sign, and G-stable immersions of them.

A whole series of important questions in the theory of surfaces remain outside the framework of the article. We only touch on questions of the purely extrinsic geometry of surfaces. This applies above all to the most developed and complete theory of convex surfaces. Thus, the geometric theory of equations (basically of Monge-Ampère type) is only recalled, and there is no description of existence and uniqueness theorems for surfaces with given conditional curvatures. The reader can become acquainted with these questions from the monographs Bakel'man, Verner and Kantor (1973), Pogorelov (1969), Pogorelov (1975). We do not consider boundary-value problems of the theory of bending of convex surfaces, infinitesimal bendings of high orders, or subtle questions of the bending of surfaces in a neighbourhood of an isolated zero of the curvature. For these questions see Part III of the present book.

Chapter 1
The Geometry of Two-Dimensional Manifolds and Surfaces in E^n

§ 1. Statement of the Problem

As the title itself emphasizes, in our article we consider only questions in the theory of surfaces in E^n, although many of the results recalled carry over automatically to surfaces in spaces of constant curvature, and sometimes in Riemannian manifolds. Of course, there are aspects that are specific for such spaces; we shall not dwell on them, see Pogorelov (1969), Milka (1980), for example.

1.1. Classes of Metrics and Classes of Surfaces. Geometric Groups and Geometric Properties. It is well known that every (for simplicity, sufficiently smooth) surface in E^n, considered from the viewpoint of its intrinsic metric, uniquely determines a Riemannian manifold. On the other hand, an abstractly defined Riemannian manifold can always be isometrically immersed in some E^n, but such an immersion is not unique, and generally speaking the properties of the Riemannian metric do not have an appreciable influence on the geometry of the immersed surface. In the natural problem of the connection between properties of a surface and properties of its intrinsic metric we shall be mainly interested in the following two aspects.

Firstly, we have the question of which of the intrinsic properties of a surface can be guaranteed by some completely determined extrinsic geometrical properties of it. (Of course, the answer to this question depends on what one understands by a "geometric" property of a surface.) Secondly, there is the question of the restriction of the class of admissible immersions to "regular" ones, that is, immersions for which the properties of the metric have an appreciable influence on the extrinsic properties of the surface. The following definition of a geometric property of a surface is basic for our later arguments.

A property of a surface is said to be *geometric* if it is preserved by transformations of E^n that belong to some group G. We always assume that G contains the group of similarities and is distinct from it. Such groups are called *geometric*. A classification of geometric groups was obtained in G.S. Shefel' (1984), G.S. Shefel' (1985). Leaving a detailed discussion of this question to 2.2 of Ch. 4, we note that it is meaningful to consider only the group of affine transformations[1], the pseudogroup of Möbius transformations (generated by similarities and inver-

---

[1] Since the dimension n of the ambient space is not fixed, it is a question, strictly speaking, of an infinite choice of groups A_n of affine transformations of E^n for all $n > 2$ and similarly in the other cases.

sions when $n > 2$) and, to rather different ends, the group of all diffeomorphisms of fixed smoothness.

The given definition of a geometric property makes more precise the first of the questions posed above and suggests an answer to the second. Namely by the "regularity" of an immersion we shall understand its G-stability.

Definition. A surface F in E^n is called a G-*stable* immersion of the metric of some class $\mathscr{K}$ if any transformation belonging to the group G takes F into a surface whose intrinsic metric also belongs to the class $\mathscr{K}$.

Here it is assumed that G is a geometric group (or pseudogroup) of transformations in E^n. Since the identity transformation id belongs to G, it is obvious that the intrinsic metric of the surface F itself belongs to $\mathscr{K}$. In this definition the class of metrics $\mathscr{K}$ is not necessarily exhausted by Riemannian metrics. Correspondingly, by an immersion of a metric here we understand a C^0-smooth (topological) immersion which is an isometry.

It is essential that the requirement of G-stability of a surface does not impose any a priori restrictions on the dimension n of the ambient space. We note that G-stable immersions of metrics of some class $\mathscr{K}$ (not exhausting all admissible metrics) always have a certain general geometric property. Transition from any immersions to G-stable ones enables us to establish a dual connection between extrinsic and intrinsic properties of surfaces.

The naturalness of the concept of G-stability is illustrated by the following assertions, proved in the most general form in S.Z. Shefel' (1969), S.Z. Shefel' (1970), Sabitov and S.Z. Shefel' (1976). The only affine-stable immersions in E^n, $n \geqslant 3$, for the class of two-dimensional Riemannian metrics of positive curvature are locally convex surfaces in some $E^3 \subset E^n$. The class of affine stable immersions for two-dimensional Riemannian metrics of negative curvature is by no means exhausted by surfaces in E^3, but all such immersions belong to the class of so-called saddle surfaces, that is, surfaces that locally do not admit strictly supporting hyperplanes; for the details see 3.1 of Ch. 3. Now suppose that G is the group of diffeomorphisms in E^n of smoothness C^∞. Then the only G-stable immersions for the class of Riemannian metrics of smoothness $C^{l,\alpha}$, $l \geqslant 2$, $0 < \alpha < 1$, are surfaces of the same smoothness.

The most attractive situation is that in which the class of metrics $\mathscr{K}$, the group G and the class of surfaces $\mathscr{M}$ have the following relations.

$1°$. The class of surfaces $\mathscr{M}$ coincides with the class of all G-stable immersions of metrics of the corresponding class of metrics $\mathscr{K}$.

$2°$. Every metric of the class $\mathscr{K}$ admits an immersion in the form of a surface of class $\mathscr{M}$.

In this case the class of surfaces $\mathscr{M}$ and the class of metrics $\mathscr{K}$ are said to be G-*connected*.

Later we shall also use the concept of G-connectedness "in the small" and G-connectedness "in the large"; for details see the next section.

The given definition admits gradations depending on how we understand the terms surface, metric, and immersion of a metric. For example, affine-stable immersions in E^n of a one-element class of plane metrics on E^2 contain all cylinders (with rectifiable directrix) or consists only of smooth cylinders, depending on whether we understand by a surface any C^0-immersion or only a smooth one. We must take into account that the fact that a surface and all its images under affine transformations have a smooth intrinsic metric does not imply, generally speaking, that the surface itself is smooth[2]. Therefore in §2 all metrics, surfaces and immersions are a priori assumed to be smooth. In the examination of non-regular surfaces and metrics, by isometric immersions we understand topological (of smoothness C^0) immersions that are isometries.

Otherwise it is a question of immersions that are stable with respect to the group of diffeomorphisms; see §5 of Ch. 4.

§2. Smooth Surfaces

2.1. Types of Points. We assume that F is a smooth surface, that is, an immersion of smoothness C^l, $l \geqslant 3$, of a two-dimensional manifold M in E^n, $n \geqslant 3$. In differential geometry it is usual to describe surfaces by means of the first and second fundamental forms. The first fundamental form specifies the intrinsic (induced) metric of the surface – a metric where the distance between points is equal to the greatest lower bound of lengths of curves joining these points on the surface. The second fundamental form determines at each point of the surface a family of osculating paraboloids. Let us explain this.

Let B be the second fundamental form of a surface F at a fixed point p. If F is specified by a vector-valued function $r(u^1, u^2)$, then

$$B(X, Y) = \sum_{i,j=1}^{2} X^i Y^j (r_{ij})^N.$$

Here X^i and Y^j are the coordinates of vectors X and Y tangent to F in the basis (r_1, r_2), where $r_i = \partial r/\partial u^i$, $r_{ij} = \partial^2 r/\partial u^i \partial u^j$, and the index N denotes projection into the normal (that is, orthogonal to $T_p F$) subspace.

Every projection of the graph Γ of the map $X \mapsto B(X, X)$ onto the three-dimensional space spanned by $T_p F$ and some normal v is a paraboloid (or degenerates into a cylinder) and is called the *osculating paraboloid*. In the case of degeneracy to a cylinder we shall call the latter a parabolic paraboloid by analogy with elliptic and hyperbolic paraboloids.

We note that the subspace spanned by Γ is said to *osculate* F at the point p. Its dimension is at most five. For it is spanned in E^n by the two-dimensional subspace $T_p F$ and the vectors $(r_{11})^N$, $(r_{12})^N$, $(r_{22})^N$.

In the case of a surface in E^3 the family of osculating paraboloids consists of one paraboloid. According to the type of osculating paraboloid the points of a

---

[2] A remarkable exception consists of smooth metrics of positive curvature under locally convex immersions; see §3 of Ch. 2.

surface in E^3 are traditionally divided into elliptic, hyperbolic and parabolic (in particular, flat points), which forms the only possible affine classification of points of a surface in E^3 up to infinitesimals of the second order. The affine classification of points coincides with the classification according to the sign of the Gaussian curvature.

When $n > 3$ the affine classification of points of a smooth surface in E^n is also determined by the affine-invariant properties of the family of osculating paraboloids at a point p and leads to eight different types of points (S.Z. Shefel' (1985)). Without giving the classification itself here, we note that for two of these types the Gaussian curvature[3] of the surface at p is zero. Points of these two types are called *parabolic*. For another type of point the Gaussian curvature is positive (*elliptic point*). For three other types of point the Gaussian curvature is negative (*hyperbolic point*), and in two cases the sign of the Gaussian curvature is not determined by the type of point (such points are said to be *movable*). The first two of these types – parabolic points – have a common property: among the osculating paraboloids there are no elliptic or hyperbolic ones. One type – elliptic point – is characterized by the fact that among the osculating paraboloids there are elliptic but no hyperbolic or non-degenerate parabolic ones. Three more types are characterized by the fact that there are hyperbolic paraboloids but no elliptic ones (hyperbolic point). Finally, the two remaining types are characterized by the fact that at a point there are elliptic, hyperbolic and parabolic paraboloids.

2.2. Classes of Surfaces. The classification of points enables us to distinguish six classes of smooth surfaces. Surfaces of the first three classes M^+, M^-, M_0 consist, respectively, of only elliptic, hyperbolic or parabolic points. Surfaces of class M_0^+ consist only of elliptic and parabolic points, and surfaces of class M_0^- consist only of hyperbolic and parabolic points. Finally, the class M is formed by all smooth surfaces.

Surfaces of the class M_0^+ are called *normal* surfaces of non-negative curvature, and surfaces of the classes M_0^- and M^- are called *saddle* surfaces and *strictly saddle* surfaces respectively.

Theorem 2.2.1 (S.Z. Shefel' (1970)). *The class M^+ in E^n consists of locally convex surfaces each lying in some $E^3 \subset E^n$. A complete surface of class M^+ is a complete convex surface (the boundary of a convex body in E^3). Normal surfaces of non-negative curvature (of class M_0^+) are characterized by the fact that either every point of such a surface has a neighbourhood in the form of a convex surface or through this point there passes a rectilinear generator with its ends on the boundary of the surface, and the tangent plane along this rectilinear generator is stationary. A complete surface of class M_0^+ is either a convex surface in E^3 or a cylinder in E^n.*

---

[3] By the Gaussian curvature K of a smooth surface in E^n we always have in mind the Gaussian (that is, sectional) curvature of its intrinsic metric. By the generalized Gauss theorem $K = B(X, X)B(Y, Y) - B(X, Y)^2$ when $\|X \wedge Y\| = X^2 Y^2 - \langle X, Y \rangle^2 = 1$.

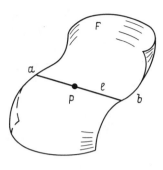

Fig. 1

The class M_0 consists of developable surfaces. The complete surfaces of this class are cylinders.

Saddle surfaces F (the class M_0^-) can be characterized by the property that no hyperplane cuts out from F a *crust*, that is, a region whose closure is compact and does not go out to the boundary of F.

Fig. 1 shows the case when a surface of class M_0^+ in a neighbourhood of a point p is neither locally convex nor developable (ab is a rectlinear generator). We should emphasize that, in contrast to the class M^+, surfaces of the class M_0^+, like all the subsequent classes, can be essentially n-dimensional for any $n > 3$, that is, they do not lie in any proper subspace of E^n.

Thus, the theory of convex surfaces is, by necessity, the theory of surfaces in E^3, while surfaces of all the remaining classes are naturally regarded as surfaces in E^n for all $n \geqslant 3$.

The reason for such an exceptional position of convex surfaces has a simple algebraic nature. Let B be the second fundamental form of a surface F at some point p. Consider a linear map L of the normal space to F at p into $\mathbb{R}^3$ according to the following rule: we fix a basis in $T_p F$ and associate with each normal v an ordered triple of numbers (a, b, c), the coefficients of the quadratic form $B^v(X, X) := \langle B(X, X), v \rangle$, where $\langle \ , \ \rangle$ is the scalar product. The type of osculating paraboloid corresponding to the normal v (and vectors parallel to it) is determined by the sign of the discriminant $ac - b^2$. In particular, every direction for which the osculating paraboloid is elliptic is mapped inside the cone $ac - b^2 > 0$, Fig. 2. Therefore all osculating paraboloids can be elliptic or degenerate only if $q = \dim$ image $L \leqslant 1$. Similarly at a hyperbolic point, where there are no elliptic paraboloids, we certainly have $q = \dim$ image $L \leqslant 2$.

If $q = 3$ at all points, then the immersion (surface) is said to be *free*. Surfaces consisting only of variable points form the closure of the set of free immersions in the corresponding topology. In the class of saddle surfaces it is natural to regard the situation of general position as that in which $q = 2$ everywhere (the osculating space is four-dimensional), and in the class of convex surfaces $q = 1$ (the osculating space is three-dimensional). For convex surfaces the condition $q = 1$ (that is, $q \neq 0$) means that the Gaussian curvature does not vanish.

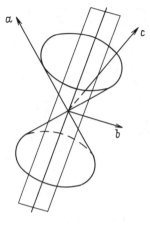

Fig. 2

2.3. Classes of Metrics. According to the sign of the Gaussian curvature it is natural to distinguish the following classes of two-dimensional Riemannian metrics: the classes K^+, K^-, K_0 of Riemannian metrics of positive, negative and zero curvature, the classes K_0^+, K_0^- of metrics of non-negative and non-positive curvature, and the class K of all Riemannian metrics. The classes of surfaces and metrics marked with the same indices will be called corresponding.

2.4. G-Connectedness. Local properties of smooth surfaces and metrics usually reduce to conditions on the surface (or metric) at each point of it. As a rule, these conditions describe the behaviour of the surface (metric) in a neighbourhood of a point up to the second order of smallness. Henceforth a geometric property of a surface will be called *local* if it is a property of a point of the surface, and its fulfilment at some point p of the surface F implies its fulfilment at p for any other surface that coincides with F in a neighbourhood of p up to infinitesimals of the second order.

For classes of surfaces and metrics distinguished on the basis of their local properties we shall distinguish G-connectedness in the small and G-connectedness in the large and correspondingly formulate two problems: in the small and in the large.

The class of surfaces $\mathcal{M}$ and the class of metrics $\mathcal{K}$ are said to be *G-connected in the small* if 1) the class of surfaces $\mathcal{M}$ coincides with the class of G-stable immersions of metrics of $\mathcal{K}$, 2) every metric of $\mathcal{K}$ admits a local immersion in the form of a surface of $\mathcal{M}$. The problem in the small consists in looking for classes of surfaces and metrics that are G-connected in the small.

The class $\tilde{\mathcal{M}}$ of complete surfaces and the class $\tilde{\mathcal{K}}$ of complete metrics are said to be *G-connected in the large* if 1) the class of surfaces $\tilde{\mathcal{M}}$ coincides with the class of G-stable immersions of metrics of $\tilde{\mathcal{K}}$, 2) every metric of $\tilde{\mathcal{K}}$ admits an immersion (in the large) in the form of a surface of $\tilde{\mathcal{M}}$. The problem in the large consists in looking for classes of surfaces and metrics that are G-connected in the large.

In contrast to the problem in the small, here even in those cases when local properties are fundamental for the distinction of classes, we need to impose a priori conditions of non-local character on classes of complete surfaces and metrics that are G-connected in the large.

This is because the local conditions that distinguish classes of metrics and surfaces that are G-connected in the small may lead to topological restrictions that are different for surfaces and metrics. For example, on the projective plane there are metrics of positive curvature, but none of them admits affine stable immersions in E^n. Moreover, in the case of classes of surfaces and metrics defined by local conditions that are G-connected in the small there may exist non-local obstructions for G-stable isometric immersions that have not only topological but also mixed topological-metric character. Thus, on a sphere with three punctures there are complete Riemannian metrics of non-positive curvature that are immersible in E^3 and non-immersible as a saddle surface in any E^n; see 1.4 of Ch. 3.

In the case of complete metrics of positive curvature, and correspondingly complete convex surfaces, the only (purely topological) obstruction is non-connectedness. The matter is simple in the case of zero curvature. However, finding all obstructions to immersibility of complete metrics of non-positive (negative) curvature in the form of complete saddle (strictly saddle) surfaces in at least one E^n is a difficult problem. (The case of simply-connected surfaces is simpler; for them it may be that all obstructions are trivial; see 1.3 of Ch. 3 and 4.3 of Ch. 4.)

2.5. Results and Conjectures. In this chapter a fundamental question is that of the correspondence of surfaces and metrics in the case of smooth surfaces[4] and for the affine transformation group, as in the general case, it consists of the problem in the small and the problem in the large. The problem in the small for the classes K^+, K^-, K_0, K has been solved completely; we have the following two theorems.

Theorem 2.5.1. *The classes M^+, M^-, M_0, M of smooth surfaces and the corresponding classes of metrics are pairwise affine connected in the small.*

Theorem 2.5.2. *If we restrict ourselves to those classes of smooth surfaces, each of which is defined by a local geometric property, then there are no pairs that are affine connected in the small other than those listed in Theorem 2.5.1 and possibly the pairs K_0^-, M_0^-.*

Theorem 2.5.1 combines the following assertions.

$1°$. Each of the classes of surfaces mentioned above is affine-invariant.

---

[4] We recall that a smooth surface is always understood to be an immersion of class C^l, $l \geqslant 3$. Special cases, such as C^0-smoothness (topological immersion) or $C^{l,\alpha}$-smoothness, will be treated specially.

2°. The intrinsic metric of a surface of any of these classes belongs to the corresponding class of metrics.

3°. An affine-stable immersion in E^n of a metric of any class belongs to the corresponding class of surfaces.

4°. Every metric of any of the classes admits a local immersion in the form of a surface of the corresponding class in E^n.

Assertions 1°–3° hold for all six classes. The first of them is obvious. The second follows from the many-dimensional generalization of Gauss's theorem. The third assertion is proved in S.Z. Shefel' (1970). The fourth assertion has been proved (see Pogorelov (1969), Poznyak and Shikin (1974)) only for the classes listed in the theorem.

Let us proceed to complete metrics and surfaces.

Theorem 2.5.3. *The classes $\tilde{M}^+$, $\tilde{M}_0$, $\tilde{M}$ of smooth complete simply-connected surfaces and the corresponding classes of Riemannian metrics are affine connected in the large[5].*

Like Theorem 2.5.1, this theorem combines four assertions. The first three of them are the same as in Theorem 2.5.1, and are therefore proved. The fourth assertion is as follows: every complete simply-connected Riemannian manifold of any of the classes $\tilde{K}^+$, $\tilde{K}_0$, $\tilde{K}$ admits an immersion in the form of a (complete) surface of the corresponding class. In the case of $\tilde{K}_0$ this is obvious. Also, every complete Riemannian metric of positive curvature, defined on a sphere or plane, admits an immersion in E^3 in the form of a smooth complete convex surface. This is the solution of Weyl's famous problem and its analogue for non-compact surfaces; for details see Ch. 2. Therefore in the case of the classes $\tilde{K}^+$ and $\tilde{K}_0$ Theorem 2.5.3 is true. It is also true for the class $\tilde{K}$ (even without the requirement of simply-connectedness) by a general theorem of Nash on isometric immersions (Nash (1956)).

Let us state the proposition that the classes $\tilde{K}^-$ and $\tilde{M}^-$ of smooth simply-connected surfaces and metrics are affine-connected in the large. This proposition combines four parts, of which the first three are the same as in Theorems 2.5.1 and 2.5.3, and are automatically true. The fourth part can be stated as follows.

Conjecture A[6]. *A complete simply-connected Riemannian metric of negative curvature admits an isometric immersion in some E^n in the form of a saddle surface.*

Together with Theorem 2.5.3, Conjecture A, when it is true, can be regarded as a generalization of Weyl's problem. In any case all the results about non-immersibility, in the first place Hilbert's classical theorem and the well-known

---

[5] Here and later a tilde over a letter implies the completeness of the metric or surface.

[6] This conjecture was made in S.Z. Shefel' (1978), S.Z. Shefel' (1979), but with superfluous generality, without the assumption of simply-connectedness; as we mentioned above, such a generalized conjecture is false.

more general theorem of Efimov (see §1.1 of Ch. 3), do not contradict our conjecture, since here the class of immersions is restricted not by the dimension of the space but by a geometric property, the saddle form.

The class $\tilde{K}^+$ of metrics and the corresponding class of surfaces do not form an affine connected pair. It is true that a complete simply-connected Riemannian manifold of non-negative curvature admits an immersion in E^3 in the form of a convex surface, but the smoothness of this surface may turn out to be substantially lower than the smoothness of the metric at the zeros of the curvature; see the example in 1.1 of Ch. 2. Such a lowering of the smoothness also takes place when considered locally; it is easy to verify this on the basis of an example from Pogorelov (1971). The authors do not have corresponding examples for the classes K_0^- and $\tilde{K}_0^-$. We observe that in the case of analytic metrics and surfaces the classes $^aK_0^-$ and $^a\tilde{K}_0^-$ of analytic metrics are affine connected in the small with the corresponding classes of surfaces; see Poznyak (1973).

The fact that not all the classes of surfaces under consideration are affine connected with the corresponding classes of metrics is probably stipulated by the eclectic character of these classes: they are distinguished simultaneously by geometric properties (convexity, saddle form, and so on) and the a priori requirement of smoothness. However, as we mentioned at the end of §1, smoothness is not an affine stable property in general; for details see §5 of Ch. 4. We can therefore hope that in the case of not necessarily smooth surfaces distinguished on the basis of just geometric properties there arise only classes that are affine connected with the corresponding classes of metrics; see §3 below.

2.6. The Conformal Group. Let us now dwell on the conformal group of transformations. At each point of any smooth surface either 1) all the osculating paraboloids are paraboloids of rotation or degenerate, or 2) by a conformal transformation we can arrange that the Gaussian curvature of the surface at this point takes any value. Hence it follows easily that apart from the class of all surfaces and the class of all metrics the only ones that are conformally connected in the small are the class of surfaces in E^3 locally congruent to a sphere or a plane, and the class of metrics of constant curvature.

If a group of diffeomorphisms that preserves the subgroup of similarities is not affine or conformal, then by the action of this group we can achieve any value of the Gaussian curvature at some point of the surface (G.S. Shefel' (1985)). Therefore all other groups distinguish only the class of all metrics and the class of all surfaces, and consideration of them from these positions is not meaningful.

The principle of correspondence between classes of surfaces and metrics distinguishes classes of surfaces and metrics that play a central role in the theory of surfaces and in Riemannian geometry, and this is one of the basic forms of connection between intrinsic and extrinsic geometry. Only metrics of constant negative curvature have not found their natural place in this scheme. It is possible that a similar approach in the case of a pseudo-Euclidean space could distinguish such metrics instead of metrics of constant positive curvature.

§3. Convex, Saddle and Developable Surfaces with No Smoothness Requirement

3.1. Classes of Non-Smooth Surfaces and Metrics. The classes of surfaces considered above, apart from the general class M, admit synthetic definitions (that is, purely geometric, not requiring any analytic apparatus). These definitions, but without any a priori assumption of smoothness, distinguish the wider classes $\mathcal{M}_0^+$, $\mathcal{M}_0^-$, $\mathcal{M}_0$, $\mathcal{M}^+$, $\mathcal{M}^-$ of generally speaking non-regular surfaces. Complete surfaces of the first three classes are complete convex surfaces, complete saddle surfaces, and cylinders.

These classes, apart from possibly non-simply-connected saddle surfaces, have the compactness property: if compact surfaces F_i, lying in E^n, of one of the classes have the same topology and their boundaries form a compact family, then we can pick out from them a convergent subsequence (it is a question of Fréchet convergence); see Aleksandrov (1939), G.S. Shefel' (1984). The classes $\mathcal{M}_0^+$, $\mathcal{M}_0^-$, $\mathcal{M}_0$ and the corresponding classes $\tilde{\mathcal{M}}_0^+$, $\tilde{\mathcal{M}}_0^-$, $\tilde{\mathcal{M}}_0$ are closed[7] in the sense that a convergent subsequence of surfaces of one class converges to a surface of the same class. The classes $\mathcal{M}^+$ and $\mathcal{M}^-$ are not closed and in this connection they play a minor role.

What we have said here about surfaces can largely be repeated for metrics. The classes of Riemannian metrics considered above admit a simple synthetic description. The classes K_0^+, K_0^-, K_0 are characterized by the fact that the excess (that is, the difference between the sum of the angles and π) of any simply-connected triangle of shortest curves is respectively non-negative, non-positive and equal to zero. (For the classes K^+ and K^- we need to compare the excess with the area of the triangle.) Let us now give up the fact that the metric is Riemannian, that is, we shall consider a two-dimensional manifold with an intrinsic metric (given directly by distances, and not by means of a quadratic form). For precise definitions of a triangle, an angle, and other concepts in such a space, we refer the reader to Aleksandrov and Zalgaller (1962). Then, depending on the sign of the excesses of triangles, we distinguish five classes of generally speaking non-Riemannian metrics. These are the classes $\mathcal{K}_0^+$, $\mathcal{K}_0^-$, $\mathcal{K}_0$ of metrics of non-negative, non-positive and zero curvature (the last class consists merely of flat Riemannian metrics), and two more classes $\mathcal{K}^+$, $\mathcal{K}^-$ of metrics of strictly positive and strictly negative curvature. The classes $\mathcal{K}_0^+$, $\mathcal{K}_0^-$, $\mathcal{K}_0$ are closed, but $\mathcal{K}^+$, $\mathcal{K}^-$ are not closed. Criteria for compactness of these classes are apparently not known.

3.2. Questions of Approximation. Another approach, which leads to non-regular surfaces and metrics, is as follows. We complete the classes M_0^+, M_0^-, M_0

---

[7] Surfaces in a Euclidean space of fixed dimension form a metric space T with a Fréchet metric. This space is complete. The fact that a class Ψ is closed means that the set $T \cap \Psi$ is closed in T. If we regard $T \cap \Psi$ as a metric space, it is a question of its completeness.

by adjoining to them surfaces admitting an approximation (in the sense of Fréchet convergence) by smooth surfaces of the corresponding classes. We denote the new classes, closed with respect to.Fréchet convergence, by $\overline{M}_0^+$, $\overline{M}_0^-$, $\overline{M}_0$.

Similarly we complete the classes K_0^+, K_0^-, K_0 of Riemannian metrics by limiting elements. The new classes $\overline{K}_0^+$, $\overline{K}_0^-$, $\overline{K}_0$ consist of two-dimensional manifolds M with intrinsic metrics ρ that admit an approximation (in the sense of uniform convergence) by Riemannian metrics defined on M.

Along with approximation by smooth objects, we can consider the classes $\overline{P}_0^+$, $\overline{P}_0^-$, $\overline{P}_0$ of surfaces and classes $\overline{\Pi}_0^+$, $\overline{\Pi}_0^-$, $\overline{\Pi}_0$ of metrics that are the closures of the corresponding classes P_0^+, P_0^-, P_0 of polyhedra in E^n and the classes Π_0^+, Π_0^-, Π_0 of polyhedral metrics.

The synthetic approach is more natural from the general geometrical point of view and enables us to use direct geometrical constructions. Approximation by smooth surfaces and smooth metrics promotes the use of analytical apparatus, and the consideration of polyhedra enables us to simplify the objects of consideration and construction. Hence the combination of the synthetic and approximative approaches has turned out to be very successful.

Let us compare the classes of surfaces and metrics obtained by these approaches. There is a conjecture that all three approaches lead to the same classes.

For metrics this has been completely proved; see Aleksandrov and Zalgaller (1962), Reshetnyak (1960b). It is easy to see that the classes of surfaces obtained approximatively are contained in the corresponding classes constructed synthetically. Hence the conjecture we have stated can be reformulated as follows.

Conjecture B. *Each surface of the classes $\mathcal{M}_0^+$, $\mathcal{M}_0^-$, $\mathcal{M}_0$ can be approximated by smooth surfaces (or polyhedra) of the corresponding class.*

For surfaces of classes $\mathcal{M}_0^+$ and $\mathcal{M}_0^-$ this has been proved in the most important cases (see Pogorelov (1956b), S.Z. Shefel' (1974), Aleksandrov (1948)); the methods used in these papers can probably be applied in the general case.

For saddle surfaces the question is still open, though for approximation by polyhedra a partial result was obtained in S.Z. Shefel' (1964). We need to bear in mind that although any smooth saddle surface in E^3 can be approximated by saddle polyhedra, the converse is not all obvious[8]. However, the question of the coincidence of $\overline{M}_0^-$ and $\overline{P}_0^-$ is now not so acute.

Along with questions of approximation of general convex and saddle surfaces by smooth surfaces or polyhedra of the corresponding class, we mention the question of approximation of a (smooth) saddle surface by smooth surfaces that are strictly saddle in the sense that all their osculating paraboloids are hyperbolic. Strictly speaking, the interesting case is $n > 3$, which has apparently not

---

[8] It is also not clear whether a saddle surface in E^n, $n > 3$, can be approximated by saddle polyhedra. If such an approximation exists, it is not at all "good"; see 3.3 of Chapter 3 and 1.4 of Chapter 4.

been solved even on the assumption that the Gaussian curvature of the surface is negative everywhere. For convex surfaces the possibility of a similar approximation is well known.

3.3. Results and Conjectures. Let us turn to the central question of this chapter about connections between classes of surfaces and metrics.

Conjecture C. *The problem in the large (for the affine group) has the following solution.*

The classes $\tilde{\mathscr{M}}_0^+$, $\tilde{\mathscr{M}}_0^-$, $\tilde{\mathscr{M}}_0$ of complete simply-connected surfaces and the classes $\tilde{\mathscr{K}}_0^+$, $\tilde{\mathscr{K}}_0^-$, $\tilde{\mathscr{K}}_0$ of complete simply-connected metrics are pairwise affine connected in the large.

A similar proposition can be stated about the problem in the small; from the viewpoint of the difficulties in the path of the solution it apparently differs little from the problem in the large.

For the classes $\tilde{\mathscr{M}}_0^+$, $\tilde{\mathscr{M}}_0$ of surfaces and the corresponding classes of metrics this conjecture has been completely proved. Since the proof is comparatively simple for complete developable surfaces and metrics of zero curvature, we dwell in more detail on the case of complete metrics of non-negative curvature. In this case the problem reduces to the proof of three assertions: 1. With respect to its intrinsic geometry a convex surface is a manifold of non-negative curvature. 2. Every complete metric of non-negative curvature specified on a sphere or a plane is the metric of a complete convex surface. 3. An affine-stable immersion of a complete non-flat metric of non-negative curvature is a convex surface. The answer to the first two questions lies in the sources of non-regular geometry and forms the main content of a classic book of A.D. Aleksandrov (Aleksandrov (1948)). The third assertion was proved in S.Z. Shefel' (1970).

Let us turn to saddle surfaces. The question of the affine connectedness of the classes $\tilde{\mathscr{M}}_0^-$ and $\tilde{\mathscr{K}}_0^-$ reduces to two problems. Firstly, does a general saddle surface (of class $\tilde{\mathscr{M}}_0^-$) have an intrinsic metric of non-positive curvature? Here it would be sufficient to give a positive answer to the conjecture about approximation (even locally and in either the smooth or polyhedral version). Secondly, can every simply-connected manifold of non-positive curvature be isometrically mapped into some E^n as a saddle surface? In contrast to the case of a smooth surface, this question has not been solved (and is hardly any simpler) even in the local formulation.

In conclusion, a few words about non-regular surfaces generally. Firstly, in the question we are considering of connections between natural classes of two-dimensional surfaces and two-dimensional metrics, non-regular surfaces and metrics appear as objects of investigation having the same rights as smooth surfaces and metrics. Moreover, in the case when non-regular surfaces are considered along with smooth ones, the answers to the questions sound very simple. Secondly, the use of non-regular surfaces not only enables us to extend the methods of proof, but also leads to new formulations of the problems.

§4. Surfaces and Metrics of Bounded Curvature

4.1. Manifolds of Bounded Curvature. Convex and saddle surfaces serve as examples of non-regular objects having, in view of their geometrical properties, intrinsic metrics that are not Riemannian but preserve the essential features of the latter and can be regarded as generalizations of a Riemannian metric of non-positive or non-negative curvature. Since the natural formulations sometimes lead to non-regular surfaces other than convex and saddle surfaces, the need for further generalization of Riemannian geometry arises. Such a generalization consists of two-dimensional manifolds of negative curvature in the sense of Aleksandrov (see Aleksandrov and Zalgaller (1962)). The fact that this generalization is not only successful but apparently the only possible one is emphasized by the result that the three independent approaches, axiomatic, approximative and analytic, lead to the same generalization of a two-dimensional Riemannian space – a manifold of bounded curvature.

Let us recall the definition of a manifold of bounded curvature, referring the reader for the details to Aleksandrov (1950b), Aleksandrov and Zalgaller (1962). In a metric space with an intrinsic metric it is natural to define shortest curves and upper angles between curves. A two-dimensional manifold with an intrinsic metric ρ is called a *manifold of bounded curvature* if for some neighbourhood U_p of any point p of it the sum of the positive excesses of pairwise non-overlapping triangles does not exceed a number $C < \infty$ that depends only on U_p. We understand a triangle as a domain homeomorphic to a disc bounded by three shortest curves with pairs of ends in common, and the excess of a triangle T is the expression $\delta(T) = \alpha + \beta + \gamma - \pi$, where α, β, γ are the upper angles of the triangle. Triangles are assumed to be non-overlapping if their interiors are disjoint.

A characteristic property of manifolds of bounded curvature is the possibility of defining in them the concept of the *curvature of a set*. This is a locally finite completely additive Borel set function ω such that for any triangle T satisfying the condition $\omega^-(\partial T) = 0$ its value is equal to the excess of the triangle: $\omega(T) = \delta(T)$.

The necessity of the condition $\omega^-(\partial T) = 0$ is illustrated in Fig. 3, which shows a cylinder with base in the form of a "concave" triangle T; $\omega(\text{int } T) = 0$, and the excess $\delta(T)$ is negative. We denote the class of manifolds of bounded curvature

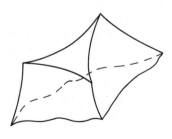

Fig. 3

by $\mathcal{K}$. Two-dimensional Riemannian manifolds belong to $\mathcal{K}$, and the curvature $\omega(E) = \int_E K \, dS$, where K is the Gaussian curvature and dS is the element of area. Polyhedral metrics also belong to $\mathcal{K}$; in this case the curvature is concentrated on the discrete set of vertices and at each vertex it is equal to $2\pi - \theta$, where θ is the total angle around the vertex.

A two-dimensional manifold with an intrinsic metric is a two-dimensional manifold of bounded curvature (that is, it belongs to the class $\mathcal{K}$) if and only if it admits an approximation (even if just locally) by polyhedral or Riemannian metrics whose positive curvatures are bounded in aggregate. (Here we have weak convergence of curvatures as set functions; see Aleksandrov and Zalgaller (1962).

Finally, the analytic approach is based on the application of generalized isothermal coordinates. The crux of the matter is that the metric of a manifold of bounded curvature can be specified by means of a (non-regular) line element $ds^2 = \lambda(du^2 + dv^2)$, where $\ln \lambda(u, v)$ is the difference between two subharmonic functions. Conversely, the metric introduced in this way belongs to $\mathcal{K}$; see Reshetnyak (1960b).

4.2. Surfaces of Bounded Extrinsic Curvature. The question of which extrinsic geometrical properties can distinguish the class of surfaces corresponding to the intrinsic concept of the class of manifolds of bounded curvature has been discussed by a number of authors, but the character of the correspondence has not been clearly expressed. A number of classes of surfaces supporting metrics of class $\mathcal{K}$ have been suggested, and also some specific connections have been traced between extrinsic and intrinsic geometries; see Aleksandrov (1949), Aleksandrov (1950a), Bakel'man (1956), Borisov (1958–1960), Pogorelov (1956b), Reshetnyak (1956), Yu.D. Burago (1968b).

From the point of view of the criteria discussed above, so long as we understand a geometrical property as an invariant of the transformation group, this question is part of the general question of this chapter; it is formulated as follows. Is there a class of surfaces G-connected with the class $\mathcal{K}$ of metrics, and what kind is it? (Here the problems in the small and in the large hardly differ in essence.)

We begin with the affine group G; we discuss other possibilities below. We assume that the class of surfaces affine-connected with the class $\mathcal{K}$ of metrics is *the class $\mathcal{M}$ of surfaces of bounded extrinsic positive curvature* defined in this section. Let $q(v)$ be the number of points of a surface F in E^n at which the surface has locally strictly supporting hyperplanes with "outward" normal v. We put $\mu^+(F) = \int_{S^{n-1}} q(v) \, d\sigma_v$, where the integration is carried out over the unit sphere. To the class $\mathcal{M}$ we refer all surfaces with $\mu^+ < \infty$ that locally have finite area; for the details see Ch. 4.

In contrast to the cases of convex and saddle surfaces the conjecture about affine connectedness in the large of the classes $\mathcal{K}$ and $\mathcal{M}$ may not depend on a priori topological or non-local topological-metric assumptions. The fact is that the conditions for a surface to be convex or saddle-shaped have the character of an equality: the extrinsic curvature is equal to the intrinsic, while the class $\mathcal{M}$ is distinguished by an inequality, and for surfaces of this class, for example, the extrinsic positive curvature μ^+ may be unequal to (greater than) the positive part ω^+ of the curvature of the intrinsic metric.

In the scheme for confirming the last conjecture not much has been achieved so far. This is possibly because it has not proved possible to apply any developed apparatus to the solution of similar problems; in this connection any progress requires a new original construction; evidently the fact that up to now the main attention has been paid to immersions in E^3 has played a part. There are no publications on the possibility of immersing a metric of class $\mathcal{K}$ in the class $\mathcal{M}$. It is only known that any metric of class $\mathcal{K}$ can be realized on some surface; see Yu.D. Burago (1960), Yu.D. Burago (1970). The fact that a surface of class $\mathcal{M}$ has a metric of class $\mathcal{K}$ has been proved only under additional assumptions; see Yu.D. Burago (1968b).

In contrast to the previous considerations, in the case of manifolds of bounded curvature the affine transformation group is not the only possible one. It may be that the class of surfaces connected with $\mathcal{K}$ relative to the whole group of diffeomorphisms consists of surfaces of class $\mathcal{M}$ having finite integral mean curvature (understood in the well-known generalized sense; see 2.1 of Ch. 4). The fact that a surface preserves a metric of class $\mathcal{K}$ under inversions if and only if its integral mean curvature is finite (S.Z. Shefel' (1970)) supports this conjecture; see 2.2 of Ch. 4.

To conclude this chapter we should like to dwell on two points. All the contents of the chapter testify to the fact that in the main question of connections between the theory of surfaces and the theory of Riemannian manifolds essentially only two specific problems remain unsolved; the question of an isometric immersion of a metric of negative curvature by a saddle surface, and the question of approximation of saddle surfaces.

The general questions considered here for two-dimensional surfaces also arise in the multidimensional case. Despite the differences and difficulties, the same ideas and concepts can apparently be used successfully in the general case. This has played a definite role in the choice of the character of the presentation on the basis of the ideas developed in S.Z. Shefel' (1970), S.Z. Shefel' (1978), S.Z. Shefel' (1979).

Chapter 2
Convex Surfaces

§ 1. Weyl's Problem

1.1. Statement of the Problem. In its original formulation, *Weyl's problem* is as follows. Suppose we are given a Riemannian metric of positive curvature on a sphere. Is there a convex surface (unique up to a motion) with this metric in

Euclidean space E^3? Recently this problem has been completely solved (both as to existence and as to uniqueness and smoothness).

Theorem 1.1.1. *A $C^{l,\alpha}$-smooth $(l \geqslant 2, 0 < \alpha < 1)$ two-dimensional Riemannian manifold of positive curvature, homeomorphic to a sphere, admits a (unique up to a motion) $C^{l,\alpha}$-smooth isometric immersion in E^3 as a convex surface. If the Riemannian metric is analytic, then the immersion is also analytic.*

Here the condition that the curvature is strictly positive cannot be discarded, as the following example shows: the surface $z = (x^2 + y^2)^{3/2}, x^2 + y^2 \leqslant 1$, can be completed to a closed convex surface that is C^∞-smooth and has positive Gaussian curvature everywhere except at the origin 0. The metric of the resulting surface is C^∞-smooth (and even analytic in a neighbourhood of 0), but the surface is not even C^3-smooth at the point 0. By Theorem 5.1.1 on uniqueness, this metric cannot be immersed in E^3 with greater smoothness[1].

Two approaches to the proof of this theorem are known. In one of them the original problem is reduced to the question of the solubility of a non-linear partial differential equation of Monge-Ampère type (the so-called *Darboux equation*)[2]. The solubility of the latter has been proved in the standard way, by continuation with respect to the parameter, but it has not been possible to prove Theorem 1.1.1 in this way merely by the methods of the theory of equations: the decisive step – obtaining a priori estimates – makes essential use of geometrical considerations. Below we dwell in a little more detail on the basic steps of this approach.

In the other approach Theorem 1.1.1 is obtained as a consequence of two fundamental results, Theorem 2.2.2 of A.D. Aleksandrov about the existence of a convex surface with any metric of non-negative curvature defined on the sphere, and Theorem 3.1.1 of Pogorelov on the smoothness of a convex surface with Riemannian intrinsic metric of positive curvature. We should emphasize that in the first theorem the metric may not be Riemannian, and the second theorem has a local character – completeness of the surface is not assumed.

Such a method, in which we first prove the existence of a generalized solution, and then establish its smoothness, is widely used in the theory of equations. However, historically one of its main sources has consisted of problems of the existence of convex surfaces with preassigned properties, principally the problems of Minkowski (see 7.2 below) and Weyl; see Aleksandrov and Pogorelov (1963).

---

[1] Pogorelov (Pogorelov 1971)) constructed an example of a $C^{2,1}$-Riemannian manifold, homeomorphic to a disc, with $K \geqslant 0$ and with strictly convex boundary, that does not admit $C^{2,0}$-isometric immersions in E^3 (however, the zeros of the curvature fill a domain). For C^n-smooth metrics with $K \geqslant 0$ and $n \geqslant 10$, local immersibility in the form of a C^{n-6}-smooth convex surface has been proved; see Lin (1985).

[2] This equation made its appearance with Weierstrass in 1884.

1.2. Historical Remarks. Weyl (Weyl (1916)) formulated the analytic case of Theorem 1.1.1, suggested a method of solution, and realized it in the case of analytic metrics sufficiently close to the metric of a sphere. A complete solution, under assumptions of analyticity, was obtained only after 20 years by Lewy; see Lewy (1938). As essentially new step was taken by A.D. Aleksandrov; see § 2. From his results it followed in particular that every Riemannian metric of positive curvature on a sphere can be "realized" as the intrinsic metric of a convex surface, a priori not necessarily smooth – a "generalized solution". The complete solution of the problem (for classes of smoothness C^k) was obtained independently and almost at the same time by Nirenberg (Nirenberg (1953)) and Pogorelov (Pogorelov (1949b), Pogorelov (1949c)). Nirenberg's proof followed the first of the approaches we have mentioned, and Pogorelov's proof followed the second. Thus, in Pogorelov's works not only is a solution of Weyl's problem given, but an independent result of fundamental character is obtained – the smoothness theorem (see § 3).

Pogorelov's proof of the theorem on smoothness relies on a theorem about the uniqueness of a convex surface with given metric. This theorem is very difficult, but for the theorem on smoothness a special case of it is sufficient: the uniqueness of a convex cap with regular metric. In this case uniqueness is proved much more easily; see Pogorelov (1969). A proof of the theorem on smoothness, independent of the uniqueness theorm, was obtained much later in Nikolaev and Shefel' (1982), Nikolaev and Shefel' (1985). Originally the proofs of Nirenberg and Pogorelov required increased smoothness assumptions. Later in a number of papers these assumptions were weakened; see Heinz (1959), Heinz (1962), Pogorelov (1969). For Hölder classes the solution in the form of Theorem 1.1.1 was obtained in Sabitov (1976).

As for the uniqueness of a closed convex surface with a given metric, here the first result was Cauchy's theorem about the non-bendability of a convex closed polyhedron. Liebmann and Minkowski in 1899 proved that a sphere is not bendable, and later Liebmann proved the impossibility of a continuous bending of a closed convex surface. The theorem about the uniqueness of a closed convex surface with a given metric was first proved by Cohn-Vossen, originally for piecewise-analytic surfaces, and then for C^3-smooth surfaces of positive curvature. His proof was based on an estimate of the sum of the indices of singular points of a specially constructed vector field. A simpler proof of the uniqueness theorem, based on an integral formula, was proposed by Herglotz (Herglotz (1943)) and completed by a number of authors; it was possible to lower the smoothness requirements to $C^{1,1}$ (a result of Aleksandrov in Efimov (1948). For general convex surfaces without any smoothness assumptions the uniqueness theorem was proved by Pogorelov; for more details see § 5. As well as this, Olovyanishnikov and Pogorelov completely investigated bendings of complete infinite convex surfaces, see § 6.

1.3. Outline of One of the Proofs. Let us dwell in detail on the basic steps of the first approach to Weyl's problem. The set of $C^{1,\alpha}$-smooth Riemannian metrics on a sphere forms a topological space (with the weak topology induced by C^0-smoothness). To prove the existence theorem it is sufficient to establish that the set of $C^{l,\alpha}$-smoothly immersible metrics of this space is (a) open, (b) closed, and (c) connected. This approach, which goes back to the research of S.N. Bernstein, was proposed by Weyl (Weyl (1916)); at present it is one of the basic methods in the theory of partial differential equations.

The main difficulty in the proof of Theorem 1.1.1, as usual, is part (b); in the theory of equations it is justified by means of so-called a priori estimates; in the given case the well-known ways of obtaining them require geometrical arguments.

Parts (a) and (c) were essentially proved by Weyl. Part (c) can be verified by means of Koebe's uniformization theorem (see Ahlfors and Sario (1960)), which asserts that every Riemannian metric ds^2 on a sphere S^2 is conformally equivalent to the standard metric ds_0^2 of constant curvature 1, that is, it can be represented in the form $ds^2 = e^\varphi ds_0^2$. In our case $\varphi \in C^{l,\alpha}$. A simple calculation shows that all metrics $ds_t^2 = e^{t\varphi} ds_0^2$, $0 \leqslant t \leqslant 1$, have positive curvature.

Assertion (a) is a consequence of a suitable inverse function theorem. To make this clearer, let us reformulate our question in the language of differential equations. It is well know that Theorem 1.1.1 reduces to the question of the solubility of the Darboux equation mentioned above. This is a second-order equation of Monge-Ampère type. It can be obtained as follows.

Suppose that in E^3 we have introduced polar coordinates r, φ, θ, and that the equations $r = r(w)$, $\varphi = \varphi(w)$, $\theta = \theta(w)$, where $w \in S^2$, specify a surface with metric ds^2. It is not difficult to see (see Kagan (1947–1948), for example) that if we know the function $r(w)$, then from it and ds^2 the remaining functions $\varphi(w)$ and $\theta(w)$ can be found uniquely up to a motion on the sphere. In turn, the function $r(w)$ must satisfy the Darboux equation, which we obtain if we observe that $(1/r^2)(ds^2 - dr^2)$ is the metric of the unit sphere, and equate the Gaussian curvature of this metric to one. The converse is also true: if $r(w)$ is a solution of the Darboux equation, and $\varphi(w)$, $\theta(w)$ are found from $r(w)$, as we mentioned above, then such functions specify a surface with metric ds^2.

If we put $\rho = \frac{1}{2}r^2$ and make certain transformations, the Darboux equation reduces to the form

$$(EG - F^2)^{-1}(\rho_{11}\rho_{22} - \rho_{12}^2) - \Delta\rho + K(|\text{grad }\rho|^2 - 2\rho) + 1 = 0. \qquad (1)$$

Here K is the Gaussian curvature of the original metric $ds^2 = Edu^2 + 2Fdudv + Gdv^2$, ρ_{ij} are the second covariant derivatives of ρ, and Δ and grad are respectively the Laplace operator and the gradient with respect to the metric ds^2. Thus the problem of isometric immersion has reduced to the question of the solubility of equation (1). Now part (a) – the openness of the set of immersible metrics – is obtained by applying the inverse function theorem to equation (1). Of course, this needs preparatory investigation of the linearized equation. There is a detailed account of these questions in Nirenberg (1953).

To prove part (b), and thereby complete the proof of the theorem, it would be sufficient to have a priori estimates of the solution of (1) in the norm $C^{l,\alpha}$ in terms of suitable norms of its coefficients. Such estimates are still not known[3], but a simple method enables us to prove that the set of $C^{l,\alpha}$ immersible metrics is closed, using well-known (Heinz (1959), Nikalaev and Shefel' (1985)) estimates for the simpler equation

$$z_{uu} z_{vv} - z_{uv}^2 = (1 + z_u^2 + z_v^2) K(u, v).$$

In fact, we shall assume for simplicity that an atlas of finitely many charts is fixed on the sphere and the metrics are specified by quadratic forms of class $C^{l,\alpha}$ in the corresponding local coordinates. Suppose that the $C^{l,\alpha}$-smooth metrics ρ_i are the intrinsic metrics of $C^{l,\alpha}$-smooth convex surfaces F_i in E^3 and converge in $C^{l,\alpha}$ to some metric ρ. Then the Gaussian curvatures K_i of the metrics ρ_i are uniformly bounded, $0 < a^2 \leqslant K(u, v) \leqslant b^2 < \infty$, and by Bonnet's classical theorem there are numbers $0 < r < R < \infty$ such that every surface F_i contains a ball of radius r and is contained in a ball of radius R. We may assume that these balls have a common centre Q. Let $r = r_i(\varphi, \theta)$ be the equations of the surfaces F_i in polar coordinates with centre O. It follows easily from Sabitov and Shefel' (1976) that the Gaussian curvatures of the surfaces F_i, as functions of the parameters φ, θ, belong to $C^{l-2,\alpha}$, and the functions r_i that specify the F_i belong to $C^{l,\alpha}$, and their $C^{l,\alpha}$-norms are uniformly bounded. Thus there is a subsequence of surfaces F_i converging in $C^{l,\alpha}$ to some convex surface that is a $C^{l,\alpha}$-smooth immersion of the metric ρ.

§2. The Intrinsic Geometry of Convex Surfaces. The Generalized Weyl Problem

It is well known from differential geometry that a sufficiently smooth convex surface is a Riemannian manifold of non-negative Gaussian curvature with respect to its intrinsic geometry; if the surface is strictly convex (in the special sense that at each point of it the osculating paraboloid is non-degenerate), then it supports a Riemannian metric of positive Gaussian curvature. By Theorem 1.1.1 positivity of the Gaussian curvature of a Riemannian metric defined on the sphere is a necessary and sufficient condition that this metric is the intrinsic metric of some (sufficiently smooth) strictly convex surface. Since convexity is a natural geometrical requirement, by its nature not connected with a smoothness assumption, Aleksandrov considered the following problem: to characterize the intrinsic metrics of general convex surfaces. To this end he constructed a generalization of the concept of a Riemannian manifold of non-negative curvature to the case of non-smooth metrics.

---

[3] It would be interesting to determine whether the required estimates could be obtained in the same way as Theorems 1' and 2' of Nikolaev and Shefel' (1985).

2.1. Manifolds of Non-Negative Curvature in the Sense of Aleksandrov. Let M be a two-dimensional manifold with an intrinsic metric, that is, a metric space satisfying two conditions: 1) every point of it has a neighbourhood homeomorphic to a two-dimensional disc; 2) the distance between any two points of M is equal to the greatest lower bound of the lengths of curves joining these points. The metric space M is locally compact, so any two sufficiently close points of it can be joined by a shortest curve. If σ_1, σ_2 are curves in M with common origin p, then the *upper angle* $\bar{\alpha}$ between them is defined as follows. Suppose that $x \in \sigma_1$, $y \in \sigma_2$, and that $\gamma(x, y)$ is the angle at the vertex corresponding to p of the flat triangle with sides respectively equal to the sides of the triangle pxy. Then

$$\bar{\alpha} := \overline{\lim_{x,y \to p}} \gamma(x, y).$$

As we mentioned above, the excess $\delta(T)$ of a triangle T is the difference between the sum of its upper angles and π. A two-dimensional manifold M with an intrinsic metric is called a *manifold of non-negative curvature* if 1) the excess of every (sufficiently small) triangle is non-negative; 2) every point $x \in M$ has a neighbourhood G with compact closure such that the sum of the excesses of any set of pairwise non-overlapping[4] triangles contained in G is bounded by some constant $C(G) < \infty$.

Remark 1. Condition 2) cannot be discarded, as the example of the Minkowski plane shows. However, we can give it up if instead of the upper angle we use the so-called lower angle, defined not as $\lim_{x,y \to p} \gamma(x, y)$, but in a much more complicated way. Originally the construction of the theory of manifolds of non-negative curvature in Aleksandrov (1948) relied on the concept of lower angle; the approach in terms of the upper angle was developed in Aleksandrov and Zalgaller (1962).

A polyhedral metric is a metric of non-negative curvature if and only if the total angle around each vertex (that is, the sum of the angles of the triangles of the triangulation adjoining this vertex) is not greater than 2π.

Theorem 2.2.1 (Aleksandrov (1948)). *A convex surface is a manifold of non-negative curvature with respect to its intrinsic geometry.*

This theorem is a consequence of three facts. 1) A convex polyhedron has an intrinsic metric of non-negative curvature. 2) If convex polyhedra converge to a convex surface F, then their intrinsic metrics converge to the intrinsic metric of F. 3) If a two-dimensional manifold with an intrinsic metric admits an approximation by polyhedral metrics of non-negative curvature, then it is itself a manifold of non-negative curvature.

The first two assertions are quite simple. In the proof of the third assertion an important role is played by the following condition for convexity of a metric: for

---

[4] We have in mind that G is homeomorphic to a disc, so for a triangle contained in G it is natural to define its interior. Two triangles are assumed to be non-overlapping if they do not have essential intersections (that is, not removable by a small movement).

any two shortest curves σ_1, σ_2 with common origin p and points $x \in \sigma_1$, $y \in \sigma_2$, $y' \in \sigma_2$ such that y lies on σ_2 between p and y', we have $\gamma(x, y') \leqslant \gamma(x, y)$, that is, the angle γ does not increase as the points x and y move away from p. For the proof it is essential that the convexity condition easily implies that the curvature of the metric is non-negative (in fact, these conditions are equivalent, but to prove this is not trivial); on the other hand, we can verify that convex polyhedra satisfy the convexity condition. Now part 3 (and with it Theorem 2.2.1 also) follows from the fact that the convexity condition is easily verified in the case of a polyhedral metric of non-negative curvature and, like the property in the large, it is preserved on transition to the limit.

2.2. Solution of the Generalized Weyl Problem

Theorem 2.2.2 (Aleksandrov (1948)). *Every two-dimensional manifold of non-negative curvature homeomorphic to the sphere is isometric to some convex surface.*

It is understood that among convex surfaces we include surfaces that degenerate into a doubly covered convex domain in the plane, for example, two flat discs placed one on top of the other and glued along the boundary.

Theorems 2.2.1 and 2.2.2 establish a one-to-one correspondence between the class of metrics of non-negative curvature defined on the sphere and the class of all closed convex surfaces. It is essential that these classes are distinguished on the basis of simple and natural geometrical criteria, and in the proofs, instead of using the theory of partial differential equations, we use direct geometrical methods.

Theorem 2.2.2 of Aleksandrov differs fundamentally from Theorem 1.1.1 in two respects: firstly, as a criterion for the "regularity" of an immersion we take not its degree of smoothness, but a geometrical property – convexity. Secondly, the problem is considered for the widest class of objects. Such a widening of the class made it possible, in particular, to include in it convex polyhedra, which are a convenient instrument in the proof.

The proofs of Theorem 2.2.2 proposed by Aleksandrov in general outline consist in the following. By approximating an arbitrary metric of non-negative curvature by polyhedra the general case is reduced to the analogous theorem for polyhedra. This is proved by means of one of the modifications of the method of continuation with respect to the parameter. Since both polyhedral metrics and also polyhedra in E^3 with a fixed number of vertices can be regarded as elements of a finite-dimensional space, the closedness of the set of realizable metrics becomes almost obvious, and the centre of gravity of the difficulties (in contrast to the situation in the theory of equations) is transferred to the openness of this set. Here we make essential use of Cauchy's theorem on the uniqueness of a convex manifold with a given metric, or a rigidity theorem that replaces it.

2.3. The Gluing Theorem. Consider domains $G_1, \ldots, G_m$ in a manifold M of non-negative curvature, having compact closures and each bounded by finitely many rectifiable curves. Suppose also that each boundary curve is divided by finitely many points (vertices) into parts (we shall call them sides). Suppose that each domain G_i is chosen from the manifold M, that is, G_i is considered with the metric induced by the inclusion $G_i \subset M$, and that the boundaries of the domains G_i are glued to each other by identifying the sides in pairs in such a way that the gluing maps are isometries, and as a result we obtain a two-dimensional manifold. It is natural to define an intrinsic metric in this manifold Q.

Theorem 2.3.1 (Aleksandrov (1948)). *In order that the manifold Q constructed above should be a manifold of non-negative curvature it is necessary and sufficient that the following two conditions should be satisfied:*

1) The sum of the turns[5] of any two identifiable parts of the sides is non-negative.

2) The sum of the angles of the domains G_i that meet at one vertex does not exceed 2π.

One of the main methods of investigating bendings of convex surfaces is based on the combined application of Theorem 2.2.2 on immersibility and Theorem 2.3.1 on gluing. Together with the smoothness theorems (see § 3) it has become a method of differential geometry that is very effective both in problems of the existence of local isometric immersions and in questions of bending of surfaces (see 4.5). The details can be found in Pogorelov (1969).

§3. Smoothness of Convex Surfaces

3.1. Smoothness of Convex Immersions. Convex surfaces with an intrinsic metric of positive Gaussian curvature have the remarkable property that the smoothness of their intrinsic metric implies the smoothness of the surface itself. The first research in this direction was done by Aleksandrov, who proved that a convex surface with bounded positive *specific curvature*[6] is C^1-smooth.

Further progress was made, mainly by Pogorelov. The following theorem contains his basic result with some improvements due to Sabitov (Sabitov (1976)).

---

[5] The *right turn* τ_1 (respectively, *left turn* τ_2) *of a simple open polygon* of shortest curves is defined as $\sum_i (\pi - \alpha_i)$, where α_i is the angle of the right (left) sector between the sides of the open polygon starting from the vertex i; the sum is taken over all vertices. For a simple arc L by definition $\tau_1(L) = \lim_{i \to \infty} \tau_1(L_i)$, $\tau_2(L) = \lim_{i \to \infty} \tau_2(L_i)$, where the open polygons L_i lie to the right (left) of L, they have ends in common with L, and they converge to L in such a way that the angles between L_i and L at the end-points tend to zero. A turn exists if L has definite directions at its ends. For a smooth curve its turns are equal to $\pm \int k_g \, dl$, where k_g is the geodesic curvature, and dl is the element of length.

[6] That is, a surface such that for any Borel subset E of it such that $S(E) > 0$ we have $0 < \omega(E)/S(E) \leqslant C < \infty$, where ω and S are its integral curvature and area respectively.

Theorem 3.1.1 (Pogorelov (1969), Sabitov (1976)). *A convex surface with $C^{l,\alpha}$-smooth metric, $l \geqslant 2$, $0 < \alpha < 1$, and positive Gaussian curvature is $C^{l,\alpha}$-smooth[7]. If the metric is analytic, then the surface is also analytic.*

We should emphasize that this theorem has a local character, that is, it holds for any domain on a convex surface. The assertion of the theorem is best possible: the intrinsic metric of a $C^{l,\alpha}$-smooth surface is $C^{l,\alpha}$-smooth (Sabitov and Shefel' (1976)). The condition that the curvature is positive cannot be dropped, as the example of the surface $z = (x^2 + y^2)^{3/2}$ shows; compare with the example of 1.1.

3.2. The Advantage of Isothermal Coordinates. As we know, every two-dimensional manifold of bounded (in particular, non-negative) curvature admits an atlas of isothermal coordinates, which defines a canonical analytic structure on it (see Reshetnyak (1959)). In these coordinates the metric has maximal smoothness, that is, if in some coordinates the components of the metric tensor have smoothness $C^{l,\alpha}$, then they have at least this smoothness in isothermal coordinates, and so its Gaussian curvature is a $C^{l-2,\alpha}$-smooth function of the isothermal coordinates. The converse is also true: if the curvature is $C^{l-2,\alpha}$-smooth with respect to isothermal coordinates, then the metric is $C^{l,\alpha}$-smooth; see S.Z. Shefel' (1970). Hence Theorem 3.1.1 is equivalent to the following assertion: a convex surface with $C^{l-2,\alpha}$-smooth positive Gaussian curvature is $C^{l,\alpha}$-smooth. If the curvature is analytic, then the surface is also analytic.

We mentioned above that Theorem 3.1.1 is local; in fact, even its "point" version holds.

Definition. A function f defined in a domain $\Omega \subset \mathbb{R}^n$ has an *approximative differential* of order (l, α) at a point x_0 if there is a polynomial $P_l(x)$ of degree l such that

$$|f(x) - P_l(x)| \leqslant C|x - x_0|^{l+\alpha}.$$

Theorem 3.2.1 (Nikolaev and Shefel' (1985)). *If the Gaussian curvature of a convex surface F, regarded as a function of the isothermal coordinates, is positive at some point p and has an approximative differential of order $(l-2, \alpha)$, $l \geqslant 2$, $0 < \alpha < 1$, then close to p the surface F is the graph of a function that has an approximative differential of order (l, α) at p.*

Since a function that has an approximative differential of order (l, α) everywhere is $C^{l,\alpha}$-smooth, Theorem 3.1.1 follows from Theorem 3.2.1.

Theorem 3.2.2 (Nikolaev and Shefel' (1985)). *If the specific curvature of a convex surface F satisfies the inequalities*

$$0 < a \leqslant \omega(E)/S(E) \leqslant C < \infty,$$

---

[7] In addition, it was established in Pogorelov (1953) that if the metric is C^2-smooth, then the convex surface is $C^{1,\alpha}$-smooth.

then F is $C^{1,\alpha}$-smooth, where α depends on a and C, $0 < \alpha < 1$, and $\alpha \to 1$ as $a/C \to 1$. This result covers the assertion mentioned in the footnote to Theorem 3.1.1.

3.3. Consequences of the Smoothness Theorems. As we have already mentioned, from Theorems 2.2.2 and 3.1.1 (or 3.2.1, 3.2.2 instead of 3.1.1) there follows a solution of Weyl's problem in the form stated in Theorem 1.1.1.

The next theorem can serve as a typical example of the application of the assertions under consideration.

Theorem 3.2.3. *Every point p of a two-dimensional $C^{l,\alpha}$-smooth ($l \geqslant 2$, $0 < \alpha < 1$) Riemannian manifold of positive curvature has a neighbourhood that admits a $C^{l,\alpha}$-smooth isometric embedding in E^3.*

In fact, by Theorem 2.3.1 on gluing it is not difficult to complement some neighbourhood of p to a (possibly not smooth) manifold of negative curvature, homeomorphic to a sphere. By Theorem 2.2.2 the latter is isometric to a convex surface. The domain on the surface corresponding to the chosen neighbourhood of p is a $C^{l,\alpha}$-smooth surface, by Theorem 3.1.1.

§4. Bendings of Convex Surfaces

4.1. Basic Concepts. Let $\mathcal{M}$ be a class of surfaces. We say that a surface $F \in \mathcal{M}$ is *uniquely determined* by its metric (in the class $\mathcal{M}$) if every surface $F' \in \mathcal{M}$ isometric to F is congruent to F. Here surfaces are assumed to be congruent if we can make them coincide by a motion, possibly including a reflection.

A *bending* of a surface F is a continuous family F_t of surfaces isometric to it, where $F_0 = F$. If not all the F_t are congruent to F, the bending is said to be *non-trivial*, and the surface F is *bendable*. If all the surfaces F_t belong to some smoothness class, we talk about a bending in this smoothness class. A bending is said to be *smooth* (*analytic*) if surfaces of the family F_t depend smoothly (respectively, analytically) on t. These two properties are independent of one another, so for example it makes sense to talk about an analytic (with respect to the parameter) bending in the class of general convex surfaces.

Remark. In a number of books and articles (see Pogorelov (1969), for example) a different terminology is used; what we have called a bending they call a *continuous bending*, and they call any isometry simply a *bending*. Then unique determination implies the absence of non-trivial bendings (other than a congruence).

Suppose that a surface F is specified by a vector-function $r(x)$; consider the family of surfaces F_t specified by the equation $r(t, x) = r(x) + t\rho(x)$, where ρ is a continuous vector-function. To a curve γ in the domain of the parameters there corresponds on each surface F_t a curve γ_t: $(r + t\rho) \circ \gamma$. Let $s(t)$ denote the length of γ_t. A homotopy F_t, $0 \leqslant t \leqslant \varepsilon$, is called an *infinitesimal bending* of a surface if for any rectifiable curve γ we have $\left.\dfrac{ds}{dt}\right|_{t=0} = 0$, that is, if the lengths of all curves are stationary when $t = 0$. The corresponding vector field ρ is called the *bending field*. The *order of smoothness* of an infinitesimal bending is defined as the order of smoothness of the field ρ. An infinitesimal bending is taken as *trivial* if ρ is the velocity field for some motion of the surface F as a rigid body, that is, $\rho(x) = a + b \times r(x)$ for some (constant) vectors a, b; here the $\times$ sign denotes the vector product. In this case the bending field ρ is also said to be trivial.

In cases when it is a question of bendings and infinitesimal bendings simultaneously, in order to emphasize the difference we shall call the first ones *finite* bendings.

A surface that does not admit non-trivial infinitesimal bendings is said to be *rigid*. The concept of rigidity of a surface is the mathematical expression of the idea of stability of the construction. In this connection the theory of infinitesimal bendings has numerous applications in mechanics, in the first place in the theory of thin *shells*; see Vekua (1959), Vekua (1982), Rozendorn (1989).

Let F be a smooth surface with parametrization $r(u, v)$. A smooth vector field $\rho(u, v)$ is a bending field for F if and only if it satisfies the system of equations

$$r_u \rho_u = 0, \quad r_v \rho_v = 0, \quad r_u \rho_v + r_v \rho_u = 0 \qquad (2)$$

or in differentials $dr\,d\rho = 0$.

Generally speaking, it is also allowable to consider non-smooth bending fields. In the case of a convex surface, even one not subject to any smoothness conditions, a bending field is necessarily locally Lipschitz. In addition, if a locally Lipschitz field ρ satisfies (2) almost everywhere, then it is a bending field for a convex surface F with parametrization $r(u, v)$; see Aleksandrov (1942). We recall that a convex surface is Lipschitz everywhere, moreover, the second differential for it exists almost everywhere; see Aleksandrov (1939).

Along with infinitesimal bendings (of the first order) infinitesimal bendings of higher orders have been studied; however, we shall not dwell on this question, referring the reader to the survey Efimov (1949), and to Part III of the present book.

4.2. Smoothness of Bendings. From Theorem 3.1.1 it obviously follows that if a surface F is $C^{l,\alpha}$-smooth, $l \geqslant 2$, $0 < \alpha < 1$, (or analytic), and has positive Gaussian curvature, then every convex surface isometric to it is also $C^{l,\alpha}$-smooth (analytic); in particular, every bending of F in the class of convex surfaces is a bending in the class of $C^{l,\alpha}$-smooth (analytic) surfaces. A similar assertion holds also for infinitesimal bendings: the bending field of a $C^{l,\alpha}$-smooth (analytic) convex surface with positive curvature is $C^{l,\alpha}$-smooth (analytic); see Pogorelov (1969).

4.3. The Existence of Bendings. It is usual to distinguish the following three basic questions.

(1) For a given surface F of a certain class (in our case, convex) is there a surface of this class that is isometric to it but not congruent to it?

(2) In the class under consideration, are there continuous bendings of the surface F?

(3) Are there infinitesimal bendings of the surface F?

These three questions are independent; the connection between them will be discussed in the next section.

Considered locally, that is, when it is a question of a sufficiently small (without stating the dimensions) neighbourhood of a fixed point p on a smooth surface, these questions can be solved comparatively easily and positively so long as the curvature $K(p)$ of the surface at the point p is not equal to zero – it is positive in the case of interest to us. (However, there are analytic surfaces that do not admit continuous bendings in any neighbourhood of a flat point; this is a non-trivial result of Efimov (1949), see Part III of the present book.)

All three problems reduce to questions about solutions in the small for differential equations and can be solved on the basis of general theorems. For the first two problems this is the Darboux equation (elliptic if $K(p) > 0$ and hyperbolic if $K(p) < 0$). For infinitesimal bendings this is a linear equation of the second order.

In the case of complete convex surfaces the solution of these equations consists of fundamental results in the theory of convex surfaces (basically these are theorems about the absence of bendings). They will be presented below in §§ 5–6. Here we dwell on compact convex surfaces with non-empty boundary and with total curvature less than 4π. The restriction on the curvature is caused by the fact that in this case the first two questions are easily solved by a simple but elegant application of Aleksandrov's gluing theorem.

In fact, let us consider the convex hull of the surface F. Its boundary is a closed convex surface consisting of F and a developable surface Q. We shall consider F and Q from the viewpoint of their intrinsic metric (for Q it is locally Euclidean). Since the total curvature of F is less than 4π, we can choose instead of Q a different locally Euclidean domain Q' so that, as before, the conditions of the gluing theorem are satisfied for F and Q'. By Theorem 2.2.2 there is a convex surface in E^3 isometric to $F \cup Q'$. It is not difficult to see that the part F' of this surface corresponding to F is not congruent to F. If instead of Q' we construct a family of "subgluings" Q_t that depend continuously on t, then as before we obtain a family of surfaces F_t isometric to F. By Theorem 5.1.1 (see below) this family is continuous with respect to t.

A complete answer to the third question is apparently still not known, but sufficiently smooth surfaces, homeomorphic to a closed disc, with strictly positive curvature (less than 4π) are always non-rigid. This follows, for example, from the previous work and 4.4. Concerning the existence of infinitesimal bendings of general (non-smooth) convex surfaces, the following result is known.

Theorem 4.3.1 (Pogorelov (1969)). *If F is the graph of a function over a strictly convex domain $\Omega \subset \mathbb{R}^2$ and does not have "vertical" planes of support,* *then it admits infinitesimal bendings with great arbitrariness. Namely, for any Lipschitz function f, defined on $\partial\Omega$, there is a bending field whose vertical component on $\partial\Omega$ coincides with f. If F does not contain flat domains, the bending field is unique.*

Vekua in his monograph Vekua (1959) developed another method for investigating infinitesimal bendings of convex surfaces. This method is connected with the reduction of the original problem to a system of linear partial differential equations of special form. For the investigation of such systems Vekua developed a theory of so-called generalized analytic functions. This approach enabled him to study by a purely analytic method the possible infinitesimal bendings of convex surfaces for different methods of fixing the boundary of the surface.

Further development of Vekua's method proceeded in two directions. Firstly, Sabitov, Fomenko and his students, and a number of other mathematicians investigated various boundary-value problems in the theory of infinitesimal bendings. For a more detailed account, see Part III of the present book.

Another direction was that of applying Vekua's methods to *non-linear* systems that describe finite bendings. The first results of this kind were obtained by Fomenko; see Fomenko (1964), Fomenko (1965). Although these papers contained an error, noticed later by Klimentov, they clearly stimulated further research into non-linear systems of elliptic type that describe finite bendings of surfaces of positive curvature. (It should not be thought that such systems reduce to one second-order equation of elliptic type!) More complete results (partially formulated below) here are due to Klimentov, see Klimentov (1982), Klimentov (1948); they were achieved by introducing new approaches to Vekua's theory, connected with wide use of the methods of functional analysis and the theory of infinite-dimensional manifolds.

The conditions for applicability of Vekua's methods are not to do with the surface being convex, but with its Gaussian curvature being positive. For infinitesimal bendings this leads only to an extension of the class of objects under consideration. However, in the case of finite bendings this means that we are concerned only with bending in the class of surfaces of positive Gaussian curvature (locally convex, but not necessarily convex in the large), even if the original surface was convex.

Without dwelling on numerous results concerning boundary-value (and some other) problems in the theory of bending, we formulate here just the following theorem of fundamental character.

Theorem 4.3.2 (Klimentov (1982)). *Suppose that the surfaces F, F', both homeomorphic to a closed disc, are isometric and $C^{l,\alpha}$-smooth (up to the boundary), $l \geqslant 4, 0 < \alpha < 1$, and have strictly positive Gaussian curvature. Then F is bendable (analytically with respect to the parameter) in the class of $C^{l,\alpha}$-smooth surfaces either to F' or to a surface that is the mirror image of it.*

The requirement $l \geqslant 4$ is apparently associated only with the method of proof.

We emphasize that the conditions of smoothness and positivity of the curvature are assumed to be satisfied up to the boundary, so a surface can be regarded as a smooth submanifold with boundary of a "large" $C^{l,\alpha}$-smooth surface with positive curvature. This requirement is essential, since there are actually examples of isometric (and arbitrarily close to each other) convex surfaces that are not bendable continuously into each other in the class of convex surfaces; see Shor (1969). However, in these examples the surfaces do not have even minimal smoothness and, which is more important, their specific curvature vanishes on the boundary, and the boundary itself has breaks.

4.4. Connection Between Different Forms of Bendings. There is a general result for all two-dimensional surfaces in E^3 (Efimov (1948)): if a surface admits a non-trival finite bending, analytic with respect to the parameter, then it also has a non-trivial infinitesimal bending (it is not rigid). We do not assert that such an infinitesimal bending is the starting point for the original continuous bending.

For convex surfaces the converse is also true. Namely, we have the following result.

Theorem 4.4.1 (Isanov (1979a), Isanov (1979b), Klimentov (1984)). *Every infinitesimal bending of a $C^{l,\alpha}$-smooth (analytic) surface of positive Gaussian curvature can be extended to a (finite) $C^{l,\alpha}$-smooth (analytic) bending that is analytic with respect to the parameter.*

In Isanov (1979a), Isanov (1979b) this was proved for surfaces that serve as the graph of a function when $l \geqslant 2, 0 < \alpha < 1$. In Klimentov (1984) it was proved for an arbitrary surface when $l \geqslant 3, 0 < \alpha < 1$.

Despite the fact that this theorem refers to fundamental questions of the theory of bendings, it was obtained comparatively recently.

The smoothness $C^{3,\alpha}$ is probably associated with the method of proof in Klimentov (1984), and nothing to do with the essence of the matter. The paper Klimentov (1987) also contains similar results for bendings of higher orders. Similar assertions under boundary conditions were obtained in Klimentov (1986).

There is also an unexpectedly simple connection between the absence of unique determination and infinitesimal bendings.

Lemma 4.4.2. *If the surfaces F_1 and F_2 are isometric, and $r_1(u, v)$ and $r_2(u, v)$ are their parametrizations, under which points corresponding to each other under the isometry correspond to the same parameters u, v, then the vector field $\rho = r_1 - r_2$ is a bending field for the mean surface $r(u, v) = \frac{1}{2}(r_1(u, v) + r_2(u, v))$.*
In fact,

$$0 = dr_1^2 - dr_2^2 = (dr_1 + dr_2)(dr_1 - dr_2) = 2drd\rho.$$

Remark. Generally speaking, the mean surface can have singularities (it is not regular), and in the case of convex F_1, F_2 it may not be convex. However, if

F_1 and F_2 are convex surfaces sufficiently close to each other, then their mean surface is also convex. Thus, if F is a convex surface, and if there are isometric convex surfaces sufficiently close to it but not congruent to it, then F admits non-trivial infinitesimal bendings.

§ 5. Unbendability of Closed Convex Surfaces

5.1. Unique Determination. We have the following theorem, which is remarkable in its generality.

Theorem 5.1.1 (Pogorelov (1952a)). *A closed convex surface in E^3 is uniquely determined in the class of convex surfaces.*

In other words: if two closed convex surfaces are isometric, then they can be made to coincide by a motion in E^3 (including a mirror reflection)[8].

Despite the simplicity of the formulation, its proof is very complicated. At present three approaches are known. First there is a direct proof; see Pogorelov (1952a). Next, Theorem 5.1.1 can be obtained by means of the following theorem on rigidity, which is also difficult to prove.

Theorem 5.1.2 (Pogorelov (1969)). *A closed convex surface is rigid outside flat domains.*

Let us explain the connection between Theorems 5.1.1 and 5.1.2. The following lemma is not too difficult.

Lemma 5.1.3. *Let F be a closed convex surface. If there is a convex surface isometric but not congruent to it, then there are convex surfaces arbitrarily close to F that are isometric but not congruent to it.*

This lemma has no independent interest, because it deals with a non-existent object. However, together with Lemma 4.4.2 and the remark following it, Lemma 5.1.3 shows that Theorem 5.1.1 follows from Theorem 5.1.2.

5.2. Stability in Weyl's Problem. The third approach consists in obtaining Theorem 5.1.1 as a direct consequence of a theorem of Yu.A. Volkov on stability in Weyl's problem. Let us state this theorem.

Consider two closed convex surfaces F, F' and a homeomorphism $f: F \to F'$. Let ρ, ρ' be the intrinsic metrics of these surfaces, and d the distance in E^3. For

---

[8] It is obvious that if we give up convexity in Theorem 5.1.1, then it ceases to be true. However, apparently up to now there are no examples of closed non-convex surfaces embedded in E^3 that admit bendings with the smoothness C^l preserved, $l \geqslant 2$. In the case of polyhedra things are quite different: Connelly (Connelly (1978)) constructed an example of a polyhedron in E^3 homeomorphic to a sphere (and not having self-intersections!) that is bendable in the class of polyhedra in such a way that no breaking of the faces occurs, that is, every face is moved as a rigid body. Bendings of a closed surface in the class of C^1-smooth surfaces are always possible; see 3.1 of Ch. 4.

any $x, y \in F$ we put $\Delta d(x, y) = |d(f(x), f(y)) - d(x, y)|$,

$$\Delta \rho(x, y) = |\rho'(f(x), f(y)) - \rho(x, y)|.$$

Theorem 5.2.1 (Volkov (1968)). *For any homeomorphism $f : F \to F'$ of closed convex surfaces we have the inequality*

$$\Delta d \leqslant C(\Delta \rho)^{\alpha}.$$

Here $\alpha > 0$ is an absolute constant; the constant $C > 0$ depends only on the diameters of the surfaces F, F'. For α we can take 0.04, for example. The largest possible value of α is not known, but it cannot exceed $1/2$; this is shown by comparing a doubly covered disc F and the surface F' obtained from F if one copy of the disc is replaced by a conical surface whose vertex projects into the centre of the disc and whose height is small.

A similar theorem is true for convex caps, that is, convex surfaces with boundary lying in a plane; in fact, the corresponding assertion for caps (with constant $\alpha \geqslant 0.08$) was first proved in Verner (1970a), and Theorem 5.2.1 can be derived from it.

Theorem 5.2.1 gives a positive answer to a question posed by Cohn-Vossen as long ago as 1936 (Klimentov (1984)). Clearly, it is sufficient to prove this theorem for any class that is dense in the space of all closed convex surfaces, for example, only for polyhedra or only for analytic surfaces. In Volkov (1968) the proof was carried out in the class of polyhedra. It is interesting that this proof makes essential use of the theory of mixed volumes of convex bodies. The possible geometrical applications of this theory have apparently still not been exhausted.

5.3. Use of the Bending Field. Let us turn to Theorems 5.1.1 and 5.1.2 on the unique determination and rigidity of convex surfaces. Geometrical observations, which serve as the starting point for the proofs of these theorems, are quite simple but not trivial. In the case of sufficiently smooth objects they lead quite quickly to the proofs. We cannot present all the key features of the proofs, particularly in the non-regular case, so we dwell on certain assertions that suggest the path along which these proofs were obtained.

Lemma 5.3.1 (Pogorelov (1969)). *Suppose that a convex surface F, not containing flat domains, is given by an equation $z = f(x, y)$, and that ζ is the component along the z-axis of the bending field of F. Then the equation $z = \zeta(x, y)$ specifies a saddle surface.*

From Lemma 5.3.1 it follows that the bending field of a surface is determined uniquely (up to a trivial term) by specifying its component ζ along the z-axis on the boundary of the surface.

The proof of Theorem 5.1.2 on the basis of Lemma 5.3.1 can be obtained comparatively easily.

The difficult proof of this lemma is based on a subtle approximation of general objects by smooth ones. If the surface and the bending field are C^2-

smooth, then the proof becomes very simple. In fact, it is not difficult to show that ζ satisfies the equation

$$f_{yy}\zeta_{xx} - 2f_{xy}\zeta_{xy} + f_{xx}\zeta_{yy} = 0. \tag{3}$$

For this we need to turn to the equations (2) of infinitesimal bendings, and putting $u = x, v = y, \rho = (\xi, \eta, \zeta)$, differentiate them with respect to x and y, and then eliminate the derivatives of ξ and η from the resulting system.

Now suppose that a smooth surface $z = \zeta(x, y)$ is not a saddle surface. Then in a neighbourhood of some point we have $\zeta_{xx}\zeta_{yy} - \zeta_{xy}^2 > 0$. Since $d^2f \not\equiv 0$ in this neighbourhood, there is a point at which as well as $\zeta_{xx}\zeta_{yy} - \zeta_{xy}^2 > 0$ we have $d^2f \neq 0$. By a rotation of the axes we may assume that at this point $\zeta_{xx} > 0$, $\zeta_{yy} > 0$, $\zeta_{xy} = 0$. Then equation (3) takes the form $f_{yy}\zeta_{xx} + f_{xx}\zeta_{yy} = 0$. Hence either $f_{xx}f_{yy} < 0$, which contradicts the convexity of F, or $f_{xx} = f_{yy} = 0$. But then, again from the convexity of F, we have $f_{xy} = 0$, which contradicts $d^2f \not\equiv 0$.

§6. Infinite Convex Surfaces

6.1. Non-Compact Surfaces. By an *infinite* convex surface we mean a complete non-compact surface without boundary; such a surface is either homeomorphic to a plane or isometric to a cylinder. The total curvature of an infinite convex surface is always at most 2π. There is an essential difference between surfaces with curvature 2π and surfaces with curvature strictly less than 2π.

Theorem 6.1.1 (Pogorelov (1969)). *For any complete non-compact two-dimensional manifold of non-negative curvature with total curvature $\omega = 2\pi$ there is a unique (up to congruence) complete convex surface in E^3 isometric to it.*

6.2. Description of Bendings. Infinite surfaces F with total curvature ω less than 2π are bendable. To give a complete description of possible bendings of such a surface we need the concept of a limiting cone.

Let $h(p, \lambda)$ be a homothety with centre p and coefficient λ. If F is an infinite convex surface, then the surfaces $h(p, 1/n)(F)$ converge as $n \to \infty$ to a cone Q, which is called the *limiting cone of the surface F*.

The cone Q is defined up to a parallel displacement and may degenerate to a doubly covered angle, a ray, or a line (the last only if F is a cylinder).

Now consider on F a ray γ, that is, an infinite geodesic extended on one side that is a shortest curve on any finite part of it. As $n \to \infty$ the curves $h(p, 1/n)(\gamma)$ converge to some generator of the limiting cone Q, which we denote by $R(\gamma)$.

Theorem 6.2.1 (Olovyanishnikov (1946)). *Let M be a complete non-compact oriented two-dimensional manifold of non-negative curvature with total curvature less than 2π. We fix a ray γ in M and let Q be an infinite convex conical surface with the same total curvature as M, and l a generator of Q. Then there is an infinite*

*convex surface F in E^3 and an isometry ψ of M onto F preserving the orientation[9]
such that Q is the limiting cone of F and $R(\psi(\gamma)) = l$.*

Theorem 6.2.2 (Pogorelov (1969)). *The surface F satisfying the conditions of
Theorem 6.2.1 is unique.*

§7. Convex Surfaces with Given Curvatures

7.1. Hypersurfaces. As a rule, the problems considered above are specific to
two-dimensional surfaces. In fact, for surfaces in E^n with $n > 3$ even an arbitrarily small neighbourhood of any point at which the Gauss-Kronecker curvature is non-zero is unbendable. One the other hand, the conditions under which
a Riemannian manifold M^{n-1} at least locally admits an isometric immersion in
E^n have at present been obtained only in a form that excludes their geometrical
applications; see Rozenson (1940–1943). As for the geometry of submanifolds of
high codimension, we recall that within the bounds of the criteria adopted here
there is no natural concept of convexity for them; see S.Z. Shefel' (1969).

In this section we dwell essentially on questions of extrinsic geometry. Many
of them can be generalized in dimension. We shall therefore describe them in the
natural n-dimensional formulation. The inclusion of these questions in the present article is justified by the generality of Aleksandrov's school and the methods
of obtaining the results.

7.2. Minkowski's Problem. We recall that the *Gauss-Kronecker curvature $\tilde{K}$*
(or more briefly Gaussian curvature) at a point p of an oriented C^2-smooth
hypersurface F in E^n is the product of the principal curvatures. Suppose that a
convex surface F in E^n has positive Gaussian curvature $\tilde{K}$. We denote by K the
Gaussian curvature "carried over" to the sphere S^{n-1} by means of the *spherical
map v*; in other words, $K(\xi) = \tilde{K}(v^{-1}(\xi))$ for $\xi \in S^{n-1}$.

According to Gauss's theorem,

$$K(\xi) = \lim_{E \to \xi} \frac{\sigma(E)}{S(v^{-1}(E))}, \tag{4}$$

where E are Borel subsets of S^{n-1}, and S and σ denote the $(n-1)$-dimensional
areas on F and S^{n-1} respectively. The equality (4) can serve as a definition of K
that preserves the sense and does not assume that F is C^2-smooth.

Theorem 7.2.1 (Minkowski (1903)). *If a positive continuous function K on the
sphere S^{n-1} satisfies the condition*

$$\int_{S^{n-1}} \frac{\xi d\sigma(\xi)}{K(\xi)} = 0, \tag{5}$$

---

[9] The space E^3 is assumed to be oriented.

where $d\sigma$ is the area $(n-1)$-form on S^{n-1}, then there is a unique (up to parallel displacement) closed convex hypersurface F for which K is the Gaussian curvature transferred to the sphere.

The condition (5) is necessary; it expresses the fact that the vector area $\int_F v(x)\, dS_x$ of a closed hypersurface is equal to zero. Since smoothness of F is not assumed a priori, the Gaussian curvature here is understood in the sense of definition (4), which goes back to Gauss.

The assumption in Theorem 7.2.1 that the function K is continuous was substantially weakened by Aleksandrov (Aleksandrov (1937–1938)) and Fenchel and Jessen (Fenchel and Jessen (1938)). For this instead of the point function K we introduce a (not normalized) measure on S^{n-1} – the so-called surface function ω defined by $\omega(E) = S(v^{-1}(E))$. For a C^2-smooth hypersurface we have $\omega(E) = \displaystyle\int_E \frac{d\sigma\,(\xi)}{K(\xi)}$. In the case of a polyhedron, ω is concentrated on a finite set of points $\xi_i \in S^{n-1}$, and $\omega(\xi_i)$ is the area of the face with normal ξ_i.

Theorem 7.2.2 (Aleksandrov (1937–1938)). *If the non-negative Borel measure ω on S^{n-1} satisfies the conditions*
1) $\int_{S^{n-1}} \xi d\omega\,(\xi) = 0$,
2) *for any unit vector e*

$$\int_{S^{n-1}} |e\xi|\, d\omega\,(\xi) \geqslant a > 0,$$

then there is a unique (up to parallel displacement) convex surface F whose surface function coincides with ω.

In Theorems 7.2.1 and 7.2.2 the uniqueness of the hypersurface F is a consequence of Minkowski's well-known inequality from the theory of mixed volumes. To prove the existence of a surface with given curvature K a similar problem is solved first in the discrete version, for polyhedra, and then a limiting process is carried out.

Minkowski's problem is equivalent to the question of solutions of a certain non-linear second-order equation on the sphere. From this point of view Theorems 7.2.1 and 7.2.2 guarantee the existence and uniqueness of the generalized solution. However, in these theorems there is no mention of the degree of smoothness of the solution if the function K is sufficiently smooth.

Theorem 7.2.3 (Pogorelov (1975)). *If under the conditions of Theorem 7.2.1 the function K is C^m-smooth, $m \geqslant 3$, then the surface F itself is $C^{m+1,\alpha}$-smooth, $0 < \alpha < 1$[10].*

---

[10] This result was re-proved by Cheng and Yau (Cheng and Yau (1976), Cheng and Yau (1977)), who, not knowing about the detailed publication Pogorelov (1975), suspected the presence of gaps in the proofs announced.

For two-dimensional surfaces this result was obtained much earlier than in the general case, Pogorelov (1952b), Nirenberg (1953). In addition, for $n = 2$ Theorem 7.2.3 can be made precise as follows.

Theorem 7.2.4 (S.Z. Shefel' (1977)). *If, under the conditions of Theorem 7.2.1, $n = 2$ and $K \in C^{m,\alpha}$, $m \geqslant 0$, $0 < \alpha < 1$, then the surface F is $C^{m+2,\alpha}$-smooth. If the Gaussian curvature K of the surface is bounded (but not necessarily continuous), then $F \in C^{1,\alpha}$.*

The assertions of Theorems 7.2.3 and 7.2.4 have a local character, that is, they hold also for an incomplete surface. The basis of the proof of Theorem 7.2.3 consists of a priori estimates for the normal curvatures of a hypersurface and their derivatives. These estimates make it possible in the case of a smooth function K to obtain the proof of Theorem 7.2.1 by the usual method of continuation with respect to the parameter, independently of its proof in the general case.

7.3. Stability. Although we do not intend to describe the whole cycle of questions connected with Minkowski's problem, let us state a theorem of Volkov on stability.

Theorem 7.3.1 (Volkov (1963)). *If the surface functions ω_A, ω_B of convex bodies A, B in E^n are ε-close in the sense that $|\omega_A(E) - \omega_B(E)| \leqslant \varepsilon\omega_A(E)$ for any Borel set $E \subset S^{n-1}$, then the distance $\delta(A, B)$ between these bodies satisfies the inequality*

$$\delta(A, B) \leqslant C_1 \varepsilon^{1/(n+2)}(1 + C_2(\varepsilon)).$$

Here $\delta(A, B)$ is the smallest of those numbers τ such that each of the bodies A, B can be moved by a parallel displacement into the τ-neighbourhood of another, the constants C_1, $C_2(\varepsilon)$ depend on n and the radii of the circumscribed and inscribed balls, and $C_2(\varepsilon) \to 0$ as $\varepsilon \to 0$.

This result was strengthened by Diskant to the estimate

$$\delta(A, B) \leqslant C\varepsilon^{1/n}$$

for sufficiently small $\varepsilon > 0$. The proof relies on a refinement of the analogues of the isoperimetric inequality associated with the coefficients of embeddability of one convex body in another; see Diskant (1988).

7.4. Curvature Functions and Analogues of the Minkowski Problem. The i-th order *curvature function* W_i of a convex hypersurface F in E^n is the i-th elementary symmetric function of the principal radii of curvature S_i $(R_1, \ldots, R_{n-1})$, regarded as a function of the normal. In other words, $S_i(R_1(v^{-1}(\xi)), \ldots, R_{n-1}(v^{-1}(\xi)))$, where v is the spherical mapping, and $\xi \in S^{n-1}$. We naturally assume that the hypersurface F is C^2-smooth and has positive Gaussian curvature. Obviously, $W_{n-1}(\xi) = K(\xi)^{-1}$.

There arises the question of what functions on the sphere S^{n-1} can be curvature functions of convex hypersurfaces and whether the curvature function W_i determines a hypersurface uniquely (up to parallelism). When $i = n - 1$, that is, in the case of the Minkowski problem, these questions, as we have seen, have a definite answer.

The question of uniqueness has been finally solved in the general case, while a complete solution of the existence problem is still not known, except for the cases $i = 1$ and $i = n - 1$.

Theorem 7.4.1 (Aleksandrov (1937–1938)). *If for some i, $0 < i < n$, the i-th order curvature functions of two convex hypersurfaces coincide, then these hypersurfaces can be obtained from each other by a parallel displacement.*

This theorem is a special case of a more general result (Aleksandrov (1937–1938)) concerned with the theory of mixed volumes and consisting in the following. With a convex compactum $A \subset E^n$ it is natural to associate the Borel measures ω_i, $i = 1, \ldots, n - 1$, on S^{n-1}, which are called *curvature measures* and characterized by the following properties:

(a) if the boundary A is a C^2-smooth hypersurface, then

$$\omega_i(E) = \int_E W_i(\xi) \, d\sigma(\xi),$$

(b) if the convex compacta $A_m \to A$ as $m \to \infty$, then the curvature measures ω_i^m corresponding to them converge weakly to ω_i.

Theorem 7.4.2 (Aleksandrov (1937–1938)). *If the convex compacta A, B in E^n are at least $(i + 1)$-dimensional and their i-th order curvature measures are equal, then the compacta themselves are equal up to a parallel displacement.*

For the last theorem, only in certain special cases have the corresponding stability theorems been established; see Diskant (1985).

For two-dimensional surfaces Theorem 7.4.1 can also be generalized in another direction. It is a question of the conditions for equality of surfaces such that at points with the same normals certain (not necessarily symmetric) functions of the principal radii of curvature, and possibly of the normal itself, coincide; see Aleksandrov (1956–1958), Pogorelov (1969), Pogorelov (1975). As an example let us state the following theorem.

Theorem 7.4.3. *Let $f(x, y, \xi)$ be a C^1-smooth function of the variables $x, y \in \mathbb{R}$, $\xi \in S^{n-1}$, $0 < x < y$, and suppose that $\dfrac{\partial f}{\partial x} \dfrac{\partial f}{\partial y} > 0$. If for C^3-smooth closed strictly convex surfaces F_1, F_2 we have $f(R_1^1, R_2^1, \xi) = f(R_1^2, R_2^2, \xi)$ for all ξ, where R_1^1, R_2^1 and R_1^2, R_2^2 are the principal radii of curvature of the surfaces F_1, F_2 respectively (where we assume that $R_1^1 \leqslant R_2^1$, $R_1^2 \leqslant R_2^2$), then the surfaces F_1 and F_2 are equal and parallel to each other.*

Let us turn to existence theorems. For all i, $1 \leqslant i \leqslant n - 1$, the condition

$$\int_{S^{n-1}} \xi W_i(\xi) \, d\sigma(\xi) = 0 \tag{6}$$

is necessary for the existence of a convex closed hypersurface with given curvature function W_i.

However, when $i > 1$ condition (6) is not sufficient; see Aleksandrov (1937–1938). If $i = 1$ (the Christoffel problem), then (6) is sufficient for the existence of a (unique) hypersurface with given curvature function $W_1(\xi) = R_1 + \cdots + R_{n-1}$, but this hypersurface need not be convex. Necessary and sufficient conditions for solubility of the Christoffel problem in the class of convex surfaces have been obtained by Firey (Firey (1968)). These conditions turn out to be very cumbersome. When $1 < i < n - 1$ such conditions are not known. The most general sufficient (but not necessary) conditions for the existence of a convex hypersurface with a given curvature function of order i, $1 < i < n - 1$, have been found by Pogorelov (Pogorelov (1975)).

7.5. Connection with the Monge-Ampère Equations.
As we have already mentioned, the theory of convex surfaces is closely connected with the theory of elliptic equations of Monge-Ampère type. In its most explicit form this connection appears in the case of the problem of "restoring" a convex surface given by the explicit equation $z = f(x, y)$ from its Gaussian curvature, transferred to the (x, y)-plane by orthogonal projection. In its analytic formulation this is a question of solutions of the equation

$$f_{xx}f_{yy} - (f_{xy})^2 = K(x, y)(1 + (f_x)^2 + (f_y)^2)^{3/2}. \tag{7}$$

Under a natural substitution the function $K(x, y)$ is defined either in a strictly convex bounded domain Ω or on the whole plane. In the first case it is a question of Dirichlet's problem, when the boundary of the surface is fixed: $f|_{\partial\Omega} = h$, where h is a given function. We must observe that this problem is not always soluble; since the total curvature of a surface is at most 2π, we necessarily have $\int_\Omega K \, dx \, dy \leqslant \int K \, dS \leqslant 2\pi$. The condition $\int_\Omega K \, dx \, dy < 2\pi$ is already sufficient for the solubility of (7). In the second case, that is, if $\Omega = E^2$, it is natural to specify the limiting cone of the surface.

Questions of this kind have now been well studied. As an example, the Dirichlet problem for equation (7) in the case of a strictly convex bounded domain Ω and a continuous function h on $\partial\Omega$ has (on condition that $\int_\Omega K \, dx \, dy < 2\pi$) exactly two solutions f with $f|_{\partial\Omega} = h$: one is convex upwards, the other convex downwards, and these solutions are symmetrical if the boundary lies in a plane. If K is positive and $C^{l,\alpha}$-smooth, $l \geqslant 0$, $0 < \alpha < 1$, then $f \in C^{l+2,\alpha}$.

It is essential that the same geometrical methods that enable us to obtain generalized solutions of (7), and establish their smoothness in the case when K is positive and smooth, can be applied to the much wider class of equations of

Monge-Ampère type. The fact is that the Dirichlet problem for such equations can be reformulated as a question of the "construction" of a convex surface with given conditional curvatures; in the simplest case the conditional curvature is the product of K and a weight function. One can become acquainted with these questions and multidimensional generalizations of them in the books Pogorelov (1969), Pogorelov (1975), Bakel'man, Verner and Kantor (1973).

§8. Individual Questions of the Connection Between the Intrinsic and Extrinsic Geometry of Convex Surfaces[11]

8.1. Properties of Surfaces. The extrinsic properties of convex surfaces (for example, the existence of supporting planes, the types of tangent cones, the existence almost everywhere of the second differential (Aleksandrov (1939)) and so on) are essentially concerned with the theory of convex bodies.

The generalized Gauss theorem due to Aleksandrov refers to the connections between extrinsic and intrinsic properties of a surface: the intrinsic curvature of a Borel set on a convex surface is equal to the area of its spherical image; see Aleksandrov (1948). The questions considered above are: the unique determination of a closed convex surface by its metric; an estimate of the extrinsic deformation of such a surface in terms of the change of the intrinsic metric; the inevitable smoothness of a convex surface of non-zero curvature when its metric is smooth also refers to connections between the intrinsic and extrinsic geometry.

8.2. Properties of Curves. A useful technique of the theory of convex surfaces consists of theorems on the connection between extrinsic and intrinsic properties of curves on a surface.

A point p lying outside a convex body Φ with boundary F has a unique closest point on F; we assume that it is the projection of p on F. As Busemann and Feller showed (see Busemann and Feller (1935)), the following result is true.

Lemma 8.2.1. *If two points on a convex surface F are joined in space by a curve of length l going outside Φ, then the length of the projection of this curve on F does not exceed l.*

Using this lemma, Liberman (Liberman (1941)) established the following result, which plays an important role.

Theorem 8.2.2. *A shortest curve on a convex surface has one-sided half-tangents in space at each of its points.*

The existence of a *direction in space* for a curve L at an initial point of it means the existence of a half-tangent, and the existence of a *direction in the intrinsic*

---

[11] This section was written by V.A. Zalgaller.

geometry means the property of forming a non-zero angle with itself (in the sense of Definition 2.2).

Theorem 8.2.3. *An arc L on a convex surface F always either has or does not have a direction both in the intrinsic and the extrinsic geometry simultaneously.*

For an arc L on F that has directions at the endpoints, in the footnote to 2.3 we defined the concept of a right turn τ_1 and a left turn τ_2. We naturally define the variations var τ_1 and var τ_2 of these turns; they may be infinite. On the other hand, in space by means of a limiting transition from inscribed open polygons we can define the *spatial turn* $\varkappa$; it may also be infinite. Obviously, var $\tau_i \leqslant \varkappa$, $i = 1, 2$.

The following qualitative connection was proved independently by Zalgaller (Zalgaller (1950)) and Pogorelov (Pogorelov (1969)).

Theorem 8.2.4. *For an arc L lying on a convex surface F to have restricted variations of turns* var τ_1, var τ_2 *it is necessary and sufficient that L should have finite turn $\varkappa < \infty$ in space.*

The new proof of this theorem in Usov's papers (Usov (1976b), Usov (1977)) is accompanied by an exact numerical estimate:

Theorem 8.2.5. *If the spherical image v(L) is contained in a disc of radius $R < \pi/2$ on S^2, then $\varkappa \leqslant (2 + \text{var } \tau_i) \tan R + \text{var } \tau_i$, $i = 1, 2$.*

8.3. The Spherical Image of a Shortest Curve. The ends of a shortest curve l on a convex surface F may be conical points, but an open arc l cannot be extended through a conical point. In this connection Pogorelov (Pogorelov (1969)) discussed the following question: is the image $v(l)$ of an open arc of a shortest curve a curve on the sphere, and is $v(l)$ rectifiable?

The answer to the second part of the question turned out to be negative. Suitable examples were constructed by Usov in Usov (1976a) and in a number of papers by Milka; see Milka (1977).

An obstacle to answering the first part of the question is the fact that l can be extended through a ridge point whose spherical image is an arc. Milka (Milka (1974a) and Dubrovin (Dubrovin (1974)) proved the following result.

Theorem 8.3.1. *An open arc l of a shortest curve on a convex surface can be extended through a ridge point only transversally to the direction of the ridge, except for the case when l is a rectilinear part of the ridge.*

From this we can easily obtain the following result.

Theorem 8.3.2. *The spherical image v(l) of an open arc l of a shortest curve is always a (continuous) curve on the sphere.*

8.4. The Possibility of Certain Singularities Vanishing Under Bendings. Milka (Milka (1977)) showed that non-rectifiability of $v(l)$ can hold even in a neighbourhood of each point of the curve $v(l)$. Non-rectifiability of $v(l)$ may vanish under bendings of F.

An isolated ridge point on a convex surface has not only an extrinsic but also an intrinsic singularity: at such a point the specific curvature is automatically infinite. Hence the following result of Pogorelov (Pogorelov (1953)) was rather unexpected: an isolated ridge point can lose its ridge property under bendings in the class of convex surfaces.

Chapter 3
Saddle Surfaces

§ 1. Efimov's Theorem and Conjectures Associated with It

1.1. Sufficient Criteria for Non-Immersibility in E^3. In the first two sections of this chapter we are concerned almost exclusively with smooth surfaces (immersions of smoothness C^m, $m \geq 2$). We recall (see 2.2 of Ch. 1) that a smooth surface in E^n is called a saddle surface if among its osculating paraboloids there are no elliptic ones, and a strictly saddle surface if in addition at each point of it there is at least one hyperbolic osculating paraboloid. The Gaussian curvature of a saddle surface is non-positive. A surface in E^3 is a saddle (strictly saddle) surface if and only if its Gaussian curvature is everywhere non-positive (negative). For a general definition of a saddle surface that is not connected with smoothness assumptions, see 3.1 below. Roughly speaking, a surface is a saddle surface if locally it does not have strictly supporting hyperplanes.

The central problems of the theory of saddle surfaces are the same as for convex surfaces. Basically this is the problem of the connection between intrinsic and extrinsic geometry for a surface with a Riemannian metric of negative (or non-positive) curvature; in the first place it is a question of the possibility of (isometrically) immersing such a metric in Euclidean space by a saddle surface. Another aspect is the problem of studying the purely extrinsic geometry of saddle surfaces. However, in contrast to the theory of convex surfaces, the results here are in many respects far from complete. It seems to us that this is connected with the fact already mentioned that saddle surfaces have been studied primarily in E^3. This is due to long tradition, rather than the internal logic of the subject. Thus, by no means all simply-connected two-dimensional Riemannian manifolds of negative curvature can be isometrically immersed in E^3. The first result of this kind is Hilbert's well-known theorem that in E^3 there is no regular surface isometric to the whole Lobachevskij plane. Further deep research in this direction is due mainly to N.V. Efimov. The best known of his results is the following theorem.

Theorem 1.1.1 (Efimov (1964)). *In E^3 there is no C^2-smoothly immersed complete surface with negative Gaussian curvature uniformly separated from zero.*

Completeness is understood here in the intrinsic sense. In other words, the theorem asserts that a C^2-smooth complete Riemannian metric with Gaussian curvature $K(x) \leqslant -c^2$, where $c > 0$, does not admit a C^2-smooth isometric immersion in E^3.

The difficult proof of this theorem was presented in detail by the author and restated by Klotz-Milnor (Klotz-Milnor (1972)) with some improvements in the presentation. We cannot describe all the steps of the proof here, so we just outline below the direction in which it is developed. See also Part II of the present book.

Apart from the fact that the Gaussian curvature is separated from zero, an obstacle to the immersibility in E^3 of a Riemannian metric of negative curvature may be the slow change in the Gaussian curvature from point to point. In this direction Efimov obtained a number of results (see Efimov (1964), Efimov (1966), Efimov (1968)). Let us give one of them.

Theorem 1.1.2 (Efimov (1968)). *If the Gaussian curvature K of a complete surface F, C^2-smoothly immersed in E^3, is everywhere negative, then*

$$\sup_{x \in F} \frac{|\text{grad } K|}{(-K)^{3/2}} = \infty.$$

We can present the statement of this theorem more clearly if we introduce the *radius of Gaussian curvature* $\varkappa = (-K)^{-1/2}$. Then Theorem 1.1.2 is equivalent to the assertion that a complete two-dimensional Riemannian manifold of negative curvature cannot be immersed in E^3 (in the class of C^2-smooth surfaces) if $|\text{grad } \varkappa| \leqslant \text{const} < \infty$.

We say that the change in a function f, defined on a metric space X with metric ρ, admits a *linear estimate* if there are constants c_1, c_2 such that for any $x_1, x_2 \in X$ we have $|f(x_1) - f(x_2)| \leqslant c_1 \rho(x_1, x_2) + c_2$. The next result of Efimov takes in both Theorem 1.1.1 and Theorem 1.1.2.

Theorem 1.1.3 (Efimov (1968)). *If the Gaussian curvature K of an intrinsically complete surface F, C^2-smoothly immersed in E^3, is everywhere negative, then the change in the function $\varkappa = (-K)^{-1/2}$ does not admit a linear estimate.*[1]

The following example of Rozendorn shows that the conditions of Theorem 1.1.3 are in some sense close to being necessary. Namely, when $m > 1$ the metric $ds^2 = du^2 + (1 + u^2)^{m/2} dv^2$, $-\infty < u < \infty$, $-\infty < v < \infty$, cannot be immersed in E^3 (by Theorem 1.1.3). However, the metric obtained when $m = 1$ admits the following immersion:

$$x = \ln(u + \sqrt{1 + u^2}), \qquad y = \sqrt{1 + u^2} \cos v, \qquad z = \sqrt{1 + u^2} \sin v.$$

---

[1] A deep generalization of this theorem has now been proved by Perel'man (1990a), (1990b).

The proofs of Theorems 1.1.1–1.1.3 follow the same plan and are based on the study of the spherical mapping of a surface F in a neighbourhood of a boundary point of the spherical image. If the Gaussian curvature of a C^2-smooth surface F in E^3 is of constant sign (negative in our case), then the spherical mapping of it is an immersion, and so it determines a surface Σ C^1-smoothly immersed in E^3 and lying entirely on the unit sphere S^2 (of course, not necessarily as a single sheet). It is obvious that the surface Σ has an intrinsic metric of constant curvature 1. This metric is necessarily not complete, that is, it is different from its minimal metric completion $\bar{\Sigma}$. The difference $\bar{\Sigma} \setminus \Sigma = \partial\Sigma$ is called the *boundary* of Σ. A Riemannian manifold is said to be (metrically) *convex* if any two points of it can be joined by a geodesic whose length is equal to the distance between its ends. For example, every complete (connected without boundary) Riemannian manifold is convex. It is fairly easy to show that a convex incomplete two-dimensional Riemannian manifold of constant curvature 1 is isometric to a convex domain on the sphere (such a domain always lies in a hemisphere). It turns out (this is the basic and most difficult step in the proof) that under the conditions of Efimov's theorems the surface Σ is certainly metrically convex, and so its area does not exceed 2π. On the other hand, we can calculate that under these conditions the absolute value of the total curvature of the surface F, that is, the area of the surface Σ, is infinite (in the case of Theorem 1.1.1 this is obvious). The resulting contradiction proves Theorems 1.1.1–1.1.3.

To prove that Σ is metrically convex it is sufficient to verify that at no point p of the boundary $\partial\Sigma$ does the surface have "concave support"; Fig. 4 shows what we have in mind. At this stage we can limit ourselves to a small neighbourhood of p. This neighbourhood, like the domain on F corresponding to it under the spherical mapping, can be projected onto the tangent plane to Σ at p. Then the main assertion about the absence of concave supports for Σ can be restated as follows.

Basic Lemma. *Suppose that a simply-connected domain D of the (x, y)-plane contains the set $\{(x, y) | 0 < x^2 + y^2 \leqslant r, y^2 \geqslant cx\}$, where r and c are positive constants but $(0, 0) \notin D$. We assume that a C^1-smooth immersion f of D into the*

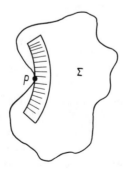

Fig. 4

(p, q)-*plane satisfies the following conditions*: 1) *the Jacobian*

$$\Delta = \frac{D(p, q)}{D(x, y)} \geqslant \frac{1}{a^2(x, y)},$$

where the function $a(x, y)$ *has a change with a linear estimate*; 2) $\dfrac{\partial p}{\partial y} = \dfrac{\partial q}{\partial x}.$
Then the metric induced in D by the immersion f cannot be complete.

1.2. Sufficient Criteria for Immersibility in E^3. Hilbert's theorem and its generalizations obtained by Efimov are assertions "in the large". On the other hand, as Levi proved (Levi (1908)), a Riemannian metric of negative curvature locally always admits an immersion in E^3. The fact that significant (in their dimensions) domains of a two-dimensional Riemannian manifold of negative curvature are actually immersible in E^3 was proved comparatively recently by Poznyak (Poznyak (1966)). In particular, he established the following result.

Theorem 1.2.1 (Poznyak (1973)). *Every domain with compact closure in a complete simply-connected $C^{4,\alpha}$-smooth Riemannian manifold of negative curvature admits a $C^{3,\alpha}$-isometric immersion in E^3.*

Subsequently Poznyak and Shikin also distinguished classes of unbounded domains in two-dimensional Riemannian manifolds of negative curvature that admit regular isometric immersions in E^3. For example, any ε-neighbourhood of a complete geodesic on the Lobachevskij plane can be immersed in E^3; also any "ideal" (that is, with vertices on the absolute) polygon of the Lobachevskij plane can be immersed in E^3 (however, a Lobachevskij half-plane cannot be immersed in E^3; see Efimov (1975)). For comparison we observe that a universal covering of a pseudosphere does not contain any ε-neighbourhood of a complete geodesic. For more complete information about isometric immersions of different domains of Riemannian 2-manifolds of negative curvature in E^3 we refer the reader to the surveys Poznyak (1973), Poznyak and Shikin (1974), and also to Part II of the present book.

1.3. Conjecture About a Saddle Immersion in E^n. Despite the fundamental results of Efimov about non-immersibility and significant progress in the theory of isometric immersions, at present it is not obvious that there are any general properties that distinguish metrics immersible in the large in E^3 among all two-dimensional complete Riemannian metrics of negative curvature, and we can hardly hope for this in the near future. The general arguments expressed in Chapter 1 also give no basis for such a hope. Hence one of the most pressing questions (Conjecture A of Ch. 1) seems to us to be that of the immersibility of every complete simply-connected two-dimensional Riemannian manifold of negative curvature in some Euclidean space E^n by a saddle surface. In this connection we explain how this question can be reduced to the problem of the solubility of a certain system of differential equations[1].

Let F be a simply-connected two-dimensional surface in E^n with intrinsic metric of non-positive curvature. We shall assume for simplicity that a unified system of coordinates (u^1, u^2) has been introduced on F, and that the surface F is specified by a vector-function $r(u^1, u^2)$. As usual, $r_i = \dfrac{\partial r}{\partial u^i}$, $r_{ij} = \dfrac{\partial^2 r}{\partial u^i \, \partial u^j}$. Along F we fix an orthonormal basis of normals $e_1, \ldots, e_{n-2}$. It is well known that the coefficients of the first and second fundamental forms $g_{ij} = r_i r_j$, $b_{ij}^\alpha = r_{ij} e_\alpha$; $i, j = 1, 2$; $\alpha = 1, \ldots, n-2$, and the so-called torsion coefficients $\mu_{\alpha\beta k} = \dfrac{\partial e_\alpha}{\partial u^k} e_\beta$ satisfy the system (1)–(3) consisting of the Gauss equation

$$(\det g_{ij})K = \sum_{\alpha=1}^{n-2} (b_{11}^\alpha b_{22}^\alpha - (b_{12}^\alpha)^2), \tag{1}$$

the $2(n - 2)$ Peterson-Codazzi equations

$$\frac{\partial}{\partial u^1} b_{12}^\alpha - \frac{\partial}{\partial u^2} b_{11}^\alpha = \sum_{k=1}^{2} (\Gamma_{11}^k b_{k2}^\alpha - \Gamma_{12}^k b_{k1}^\alpha) + \sum_{\beta=1}^{n-2} (b_{11}^\beta \mu_{\alpha\beta2} - b_{12}^\beta \mu_{\alpha\beta1}),$$

$$\frac{\partial}{\partial u^2} b_{12}^\alpha - \frac{\partial}{\partial u^1} b_{22}^\alpha = \sum_{k=1}^{2} (\Gamma_{22}^k b_{k1}^\alpha - \Gamma_{12}^k b_{k2}^\alpha) + \sum_{\beta=1}^{n-2} (b_{22}^\beta \mu_{\alpha\beta1} - b_{12}^\beta \mu_{\alpha\beta2}), \tag{2}$$

and the $\frac{1}{2}(n - 2)(n - 3)$ Ricci equations

$$\frac{\partial \mu_{\alpha\beta1}}{\partial u^2} - \frac{\partial \mu_{\alpha\beta2}}{\partial u^1} = \sum_{\gamma=1}^{n-2} (\mu_{\alpha\gamma1} \mu_{\beta\gamma2} - \mu_{\alpha\gamma2} \mu_{\beta\gamma1}) + \sum_{k,p=1}^{2} g^{kp}(b_{1p}^\alpha b_{2k}^\beta - b_{2p}^\alpha b_{1k}^\beta). \tag{3}$$

Assuming that the g_{ij} and the Christoffel symbols Γ_{ij}^k are known, we regard (1)–(3) as a system of equations for the functions b_{ij}^α and $\mu_{\alpha\beta k}$. Then equation (1) is algebraic, and (2) and (3) are quasilinear first-order equations. So long as the b_{ij}^α and $\mu_{\alpha\beta k}$ satisfy (1)–(3), according to Bonnet's multidimensional theorem (see Chen (1973), for example) there is a parametrized surface F in E^n and an ortho-normal basis of normals to it such that the functions g_{ij}, b_{ij}^α and $\mu_{\alpha\beta k}$ are respectively the coefficients of the first and second fundamental forms and the torsion coefficients of F with respect to the basis of normals that we have found.

However, we need to find these functions so that the resulting surface F is automatically a saddle surface. For this we specialize the choice of the b_{ij}^α. A saddle surface is characterized by the fact that all its points are hyperbolic. In all there are three affinely distinct types of hyperbolic points (see 2.1 of Ch. 1), and one of them corresponds to the case of general position, and the other two are

---

[1] We must observe that the fact that there is at least one complete regular saddle surface with Gaussian curvature $K \leqslant \text{const} < 0$ in E^4 is not self-evident. An example of this kind (which shows that Efimov's theorem cannot be generalized to saddle surfaces in E^n, $n > 3$) was constructed recently by Perel'man; see Perel'man (1989).

limiting cases of it. In the first basic case the space of osculating paraboloids at a point of the surface is two-dimensional, and for the generators we can take any two hyperbolic paraboloids for which the asymptotic directions regularly alternate. Conversely, a linear combination of two hyperbolic paraboloids with alternating asymptotic directions is a hyperbolic paraboloid. These arguments give a basis for looking for second fundamental forms

$$\sum_{i,j=1}^{2} b_{ij}^{\alpha} u^i u^j = A^{\alpha}((u^1)^2 + c_1 u^1 u^2 - c_2 (u^2)^2) + B^{\alpha} u^1 u^2. \qquad (4)$$

Substituting in (1)–(3) the expressions for the b_{ij}^{α} in terms of A^{α}, B^{α}, c_j obtained from (4), we arrive at a system of equations with unknown functions A^{α}, B^{α}, $\mu_{\alpha\beta k}$, c_j, and to each solution of this system there corresponds in E^n a saddle surface with a given metric.

We observe that the difference between the number of unknowns and the number of equations (which to some extent characterizes the degree of indefiniteness of the system) does not change here with the growth of the dimension $n > 3$ of the Euclidean space, in contrast to the case of arbitrary (not saddle) immersions, where this difference increases without limit.

1.4. The Possibility of Non-Immersibility When the Manifold is Not Simply-Connected. As we mentioned in 2.4 of Ch. 1, complete non-simply-connected two-dimensional Riemannian manifolds of both negative and non-positive curvature may be non-immersible in any E^n as a saddle surface. The simplest example of this kind consists of two copies of an "ideal" (that is, with vertices on the absolute) triangle of the Lobachevskij plane whose sides are glued together in the natural way (by isometry). The non-immersibility of such a manifold follows easily from the properties of tapering surfaces; see 2.2 below. Other examples can be obtained on the basis of inequality (5) of Ch. 4[2]. Among these examples there are no manifolds homeomorphic to an annulus, but there is such an example in the class of polyhedra; see 4.3 of Ch. 4.

§2. On the Extrinsic Geometry of Saddle Surfaces

2.1. The Variety of Saddle Surfaces. The extrinsic geometry of saddle surfaces has not been studied much, certainly less than in the case of convex surfaces. This is explained by a number of circumstances. Firstly, as we have already emphasized, saddle surfaces in contrast to convex surfaces are objects of a finite-dimensional Euclidean space of arbitrary dimension, which is not traditional for classical differential geometry; even if saddle surfaces are situated in E^3, an understanding of the more general properties of their extrinsic geometry must apparently take this into account. Secondly, although the concept of the saddle

---

[2] Our attention was drawn to this by G.Ya. Perel'man.

shape of a surface may be in some sense no less significant than convexity, the extrinsic geometry of saddle surfaces is much more varied, in particular, there are saddle surfaces of very different topological types. Here an analogy with Riemannian metrics suggests itself: while complete metrics of positive and zero curvature exist only for finitely many topologically distinct two-dimensional manifolds, which are easy to calculate, metrics of negative (even constant) curvature exist both on any two-dimensional open manifold and on any closed manifold with Euler characteristic $\chi < 0$.

Similarly, for any finitely-connected orientable non-compact (without boundary) two-dimensional manifold there is a saddle surface homeomorphic to it and embedded in E^3 that is externally complete (that is, the embedding is proper), and this surface can be chosen to be C^∞-smooth and with everywhere negative Gaussian curvature. A method of constructing such surfaces was first given by Hadamard (Hadamard (1898)). The construction can easily be seen from Fig. 5; we first construct a piecewise smooth (with edges) surface that is the boundary of the union of bodies bounded by hyperboloids of one sheet; then the edges are smoothed out. For the technique of such a smoothing, see Bakel'man, Verner and Kantor (1973) and Rozendorn (1966). This construction is not suitable if the surface is homeomorphic to a closed surface with a point deleted; however, in this case we can modify the construction as shown in Fig. 6: the surface shown in Fig. 5 is cut along the curve ab and to the sides of the cut we glue surfaces homeomorphic to a half-plane that wind asymptotically onto the hyperboloids of one sheet. Less clear is the question of the existence of complete unorientable saddle surfaces of different topological types in E^n.

In passing we give an example of a saddle surface in E^4 that does not fit into any 3-plane. This is the intersection of two hyperbolic cylinders: $x_3 = x_1^2 - x_2^2$, $x_4 = x_1 x_2$.

Since in this section we shall be dealing mainly with complete surfaces, we first dwell on questions of the connection between the *intrinsic and extrinsic com-*

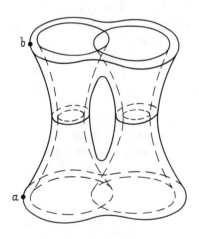

Fig. 5

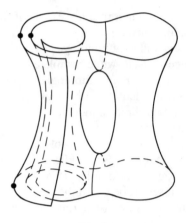

Fig. 6

pleteness of a surface. We recall that for a surface immersed in E^n it makes sense to talk about its intrinsic completeness, that is, about the completeness of the induced metric, and about its extrinsic completeness, that is, about the situation when the immersion is proper (the inverse image of any compactum is a compactum). Clearly, extrinsic completeness always implies intrinsic completeness. For convex surfaces the converse is also true. However, an intrinsically complete saddle surface may not be extrinsically complete. An example is the analytic surface of strictly negative curvature given by the equation

$$z = x \tan y - e^{(\tan y)^2} \sin(\tan y), \qquad -\infty < x < \infty, \qquad -\frac{\pi}{2} < y < \frac{\pi}{2},$$

see Bakel'man, Verner and Kantor (1973). Moreover, in E^3 there are intrinsically complete saddle surfaces that are bounded in space. Such examples were constructed by Rozendorn (Rozendorn (1961)). The main idea of Rozendorn's construction is described in Part II of the present book and explained there by a figure. The surface is constructed from blocks Q_i of the same type, it is not simply-connected, and in the intrinsic sense it has infinitely many "exits to infinity". Transition to the universal covering gives a similar example of an immersed simply-connected surface. We can become acquainted with the details of the construction in Rozendorn (1961), Rozendorn (1966).

We observe that in all examples of this kind constructed up to now there are flat points, which are certainly branch points for the spherical mapping[3]. The assumption that there is no C^2-smooth surface of strictly negative curvature that is intrinsically complete and bounded in E^3 is sometimes called Hadamard's

---

[3] Rozendorn (Rozendorn (1981)), by lowering the smoothness at individual points, succeeded in guaranteeing that the Gaussian curvature of the intrinsic metric is strictly negative; however, this does not alter the basic fact that such points continue to be branch points of the spherical mapping.

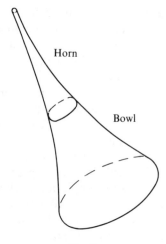

Fig. 7

conjecture. It would be interesting to determine whether it is possible to construct such a surface on the lines of Rozendorn's example mentioned above, using instead of the blocks Q_i similar constructions from Vaigant's example, which we deal with in 2.2 below. Such blocks must then be glued together along non-convex curves.

2.2. Tapering Surfaces. Let us dwell on one class of saddle surfaces for which, in particular, intrinsic completeness implies extrinsic completeness. These are the so-called tapering surfaces. With respect to a number of properties, these surfaces occupy the same place in the theory of saddle surfaces as closed surfaces in the class of all convex surfaces. Tapering surfaces are distinguished by purely intrinsic properties. Before describing them we give some definitions.

A complete Riemannian manifold homeomorphic to a disc with the centre removed is called a *tube*. In the case of metrics of non-positive curvature tubes are naturally divided into "*tapering*" (or *horns*) and "*widening*" (*bowls*), see Fig. 7; a Euclidean half-cylinder belongs to the bowls[4]. The concepts of horn and bowl were introduced by Cohn-Vossen (Cohn-Vossen (1959)). A surface in E^n is called a tube, a horn or a bowl respectively if it is of this kind with respect to its intrinsic geometry. Finally, a horn is said to be *pointed* if there are arbitrarily short loops on it that are not homotopic to zero.

---

[4] A general definition of a horn and a bowl, not connected with the condition that the curvature is non-positive, is as follows. A loop on a tube, homotopic to the boundary of the tube, is called a *belt*. Let a be the greatest lower bound of the lengths of belts on a tube T. If every sequence of belts γ_i for which the lengths $s(\gamma_i) \to a$ is divergent, then the tube is called a *horn*, otherwise a *bowl*. A bowl is said to be proper if it does not contain a horn. In the case of non-positive curvature all bowls are proper. A horn can never contain a bowl.

An intrinsically complete (without boundary) saddle surface in E^n, $n \geqslant 3$, is said to be *tapering* if it admits a partition into a compact surface with boundary and finitely many tapering tubes (horns). It is not difficult to see that this definition has an intrinsic character. We shall be concerned with sufficiently smooth (of class C^2) tapering surfaces.

It is well known that an open Riemannian 2-manifold M satisfies Cohn-Vossen's inequality $\int_M K dS \leqslant 2\pi\chi$ (so long as this integral makes sense). Here K is the Gaussian curvature, dS is the element of area, and χ is the Euler characteristic. For tapering manifolds we have equality, that is, the Gauss-Bonnet formula is true for them in the same form as for closed manifolds. Generally speaking, Cohn-Vossen's inequality becomes an equality not only for tapering complete surfaces (for example, for a cylinder and an elliptic paraboloid). But in the class of complete surfaces of non-positive curvature such that on each convergent tube there are arbitrarily distant points with negative curvature, this equality distinguishes precisely the tapering surfaces; see Verner (1968).

Theorem 2.2.1. *Every C^3-smooth saddle horn T in E^n is unbounded in space.*

This theorem for a non-pointed horn follows from Theorem 1.3.1 of Ch. 4, concerned with general surfaces of bounded extrinsic curvature. In the case of a pointed horn this theorem can be proved in the same way as Verner's theorem (Verner (1970a)) on a pointed horn in E^3. We observe straight away that Theorem 2.2.1 is obviously true for general surfaces of bounded extrinsic curvature in E^n (see 1.3 of Ch. 4); for C^2-smooth surfaces in E^3 this was proved in Burago (1984), and for general surfaces in E^n, but under additional assumptions, which are evidently unconnected with the heart of the matter, in Yu.D. Burago (1968b). However, Verner's proof (Verner (1970a)) is specific for saddle surfaces, while D.Yu. Burago's proof (D.Yu. Burago (1984)) cannot be carried over to surfaces in E^n, $n > 3$.

From Theorem 2.2.1 it follows easily that a saddle horn T has a limiting cone[5] in the form of a ray L; the direction of this ray is called the *direction of the horn*. There is a cylinder with axis parallel to L such that the horn T is entirely contained in this cylinder (Verner (1967–1968)). From what we have said it follows that a tapering surface is unbounded in E^n, and every horn of it exits to infinity in a definite direction. If such a tapering surface has no supporting hyperplane, that is, it does not fit into any half-space of E^n, then there are at least $n + 1$ of these exits to infinity. (If there is a supporting hyperplane, then its intersection with the surface "exits to infinity" non-compactly along certain horns. Examples of such surfaces are tapering saddle surfaces lying in a subspace $E^k \subset E^n$, $2 \leqslant k < n$.) In passing we observe that there are orientable tapering surfaces with any number $N \geqslant n + 1$ of exits to infinity. An example of one of them is the algebraic surface given in E^3 by the equation

---

[5] The *limiting cone* of a surface F is the limiting set (as $n \to \infty$) of the sequence of surfaces F_n obtained from F by means of homotheties with coefficients $1/n$ and common centre.

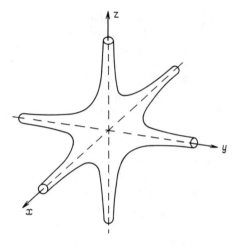

Fig. 8

$$x^2y^2 + y^2z^2 + z^2x^2 = 3a^4.$$

Six of its horns exit along the coordinate axes (Fig. 8), and the Gaussian curvature

$$K = -\frac{6a^4(a^4(x^2 + y^2 + z^2) - 3x^2y^2z^2)}{(a^4(x^2 + y^2 + z^2) + x^2y^2z^2)^2} \leqslant 0$$

vanishes at the eight points where $|x| = |y| = |z| = a$. These points are saddles of the third order; see Bakel'man, Verner and Kantor (1973).

Further investigation of tapering surfaces in E^3 consists in studying the dependence between the presence (and number) of branch points of the spherical mapping of the surface and the behaviour of the surface at infinity.

For simplicity we shall assume that the spherical mapping v of the surface F has only isolated branch points, and outside them v is a local homeomorphism[6]. The spherical mapping of the surface F determines a new surface Φ, which even lies on the sphere S^2, but may cover certain domains of the sphere repeatedly and is ramified like the Riemann surface of a many-valued function of a complex variable. As usual, the surface is endowed with the induced metric, which has constant curvature 1 outside the branch points, since it coincides locally with the metric of the sphere. We complete the surface Φ by its limiting points. The completion $\bar{\Phi}$ of the surface is (in the case of a tapering surface F) a compact

---

[6] The general case reduces to the one we have considered. In fact, let Γ be the topological space whose points are the components of the sets $v^{-1}(x)$, $x \in S^2$, and the topology is induced by the projection $p: M \to \Gamma$, where $y = p(x)$ if $x \in y$. Then v decomposes uniquely as $v = q \circ p$, $M \xrightarrow{q} \Gamma \xrightarrow{p} S^2$. According to Kerekjarto (Kerekjarto (1923)), Γ is also a two-dimensional manifold (since the components $v^{-1}(x)$ do not partition M), and the mapping q has only isolated branch points.

surface with boundary $\partial \Phi$ consisting of finitely many components γ_i correspond-ing to the exits of F to infinity. Each component γ_i from the viewpoint of the intrinsic geometry of Φ is a simple closed geodesic open polygon lying on a great circle of the sphere S^2, possibly with self-overlappings.

The following remark, which is important for what comes later, explains why the boundary $\partial \Phi$ is a geodesic open polygon. With each horn of the surface F we associate a circle O_i on S^2 perpendicular to the direction of the horn. Then the corresponding component γ_i of $\partial \Phi$ lies on O_i and fills either the whole circle O_i or an arc not less than a semicircle. The surface Φ either has no intersections with each polygon P into which $\bigcup O_i$ splits S^2 or it forms a ramified covering over P, so the number of layers is constant over P except at the branch points. Since the area of Φ is finite (it is equal to $2\pi\chi(F)$) the number of branch points is finite.

The open polygon γ_i, considered from the viewpoint of the intrinsic geome-try of Φ, has a definite turn $\tau_i = \tau(\gamma_i)$ from the side of Φ. Verner (Verner (1970b)) proved the following result.

Theorem 2.2.2. *For a tapering surface F in E^3 with l exits to infinity we have*

$$2\chi(F) = \sum_i (1 - s(b_i)) + \frac{1}{2\pi} \sum_{j=1}^{l} \tau_j, \tag{5}$$

where $s(b_i)$ is the multiplicity of the branch point b_i of the spherical mapping of the surface F; the first sum is taken over all such branch points. We always have $\tau_j \leqslant 0$. Equality holds here if and only if for some neighbourhood $U \subset \bar{\Phi}$ of each open polygon γ_i we have $(U \setminus \gamma_i) \cap O_i = \varnothing$.

Suppose that the Gaussian curvature K of a surface F in E^3 is negative in a neighbourhood of a point p, except possibly for the point itself. Then the tangent plane to F at p intersects F close to p in an even number $2m$ of simple arcs with origin at p. The number m is called the *order of saddleness* of F at p. We always have $m \geqslant 2$. If $K(p) < 0$, then obviously $m = 2$. A point with order of saddleness $m > 2$ is a branch point of multiplicity $s = m - 1$ for the spherical mapping.

A tapering surface F, regarded as a topological space, can be compactified by completing each horn by one point (this is called the *point at infinity* of the horn). The completion is a closed manifold. The *order of saddleness* of a horn at a point at infinity a_i is the number $m(a_i) = -(1/2\pi)\tau(\gamma_i)$. Verner (Verner (1970b)) proved that if the spherical mapping of a horn in some neighbourhood of a point at infinity a is univalent, then $m(a)$ is equal to 0 or $+1$. From this and Theorem 2.2.2 it follows that, for example, if a tapering surface F with Euler characteristic $\chi(F) \neq 2$ close to each point at infinity has a one-to-one spherical mapping, then on F there are points with Gaussian curvature $K = 0$ and order of saddleness greater than two.

The simplest tapering surfaces in E^3 are those with $\chi = -2$. Such a surface has four exits to infinity and is homeomorphic to a sphere with four handles. However, in this case the formulae (5) are insufficient to establish whether there are points on F with order of saddleness greater than two, and how many of

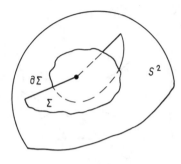

Fig. 9

them there are. The structure of such surfaces has been studied by P. Sh. Rechevskij and S.Z. Shefel'.

It turns out that a tapering surface F in E^3 with $\chi(F) = -2$ has the following structure. Generally speaking, it has four points at which the order of saddleness is equal to three. Correspondingly, the spherical mapping has four branch points. In this general case, close to each point at infinity the spherical mapping of F is one-to-one, and each horn is given explicitly by $z = f(x, y)$. The branch points of the spherical mapping cannot "stick together", so for example instead of two points with order of saddleness three there cannot be one point with order of saddleness four. However, a priori we do not exclude the "limiting" cases when the singular points "exit to infinity", so the spherical mapping has branch points on the boundary; see Fig. 9. It is not known whether all possibilities of this kind can be realized (for example, whether only one singular point exits to infinity). The construction of examples of this kind is very difficult. However, Vaigant constructed an example of an analytic tapering surface F with $\chi(F) = -2$ whose Gaussian curvature is negative everywhere, so all four branch points of the spherical mapping lie on its boundary. The surface in question is given by the equation

$$(z - \sqrt{1 + x^2} + \sqrt{1 + y^2})^2 (8 + \sqrt{1 + x^2} + \sqrt{1 + y^2})^2$$
$$- M^2 [2 - (\sqrt{1 + x^2} - 1)(\sqrt{1 + y^2} - 1)] = 0,$$

where $0 < M < 10^{-4}$.

Figure 10 shows the rough form of this surface; it is essential that its sections perpendicular to the directions of the horns are not convex.

The qualitative description given above is based on the following theorems of Rechevskij and Shefel'. We first observe that if a tapering surface F has Euler characteristic -2, then the circles O_i $(i = 1, 2, 3, 4)$ on the sphere S^2 perpendicular to the directions of the horns split the sphere into eight triangles T_j and six quadrangles Q_k.

Theorem 2.2.3. *The spherical mapping of a tapering surface F in E^3 with $\chi(F) = -2$ satisfies the following condition: in each of the quadrangles Q_k the*

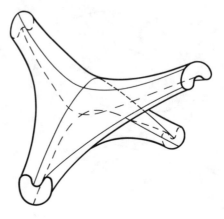

Fig. 10

spherical mapping is bijective; of any two diametrically opposite triangles T_j one is always doubly covered (possibly with a branch point), and the second is empty. The points of intersection $O_{ij} = O_i \cap O_j$ of the circles with $i \neq j$ do not belong to the spherical image of F.

Theorem 2.2.4. *For a tapering surface F in E^3 with $\chi(F) = -2$ the following three conditions are equivalent.*

1. *Of any two diametrically opposite points of the set $S^2 \setminus \bigcup_{i \neq j} O_{ij}$ at least one belongs to the spherical image of F.*

2. *The spherical mapping of F has exactly four branch points (of multiplicity 2), one in each of the triangles.*

3. *The spherical mapping of F is one-to-one close to each point at infinity.*

If conditions 1–3 are satisfied, then every horn can be given by an explicit equation $z = f(x, y)$.

From Theorems 2.2.2 and 2.2.3 it follows that a tapering surface in E^3 cannot have a one-to-one spherical mapping. Hence a complete saddle surface with a one-to-one spherical mapping contains a bowl. In a series of papers (see Verner (1967–1968)) Verner studied the extrinsic geometry of complete regular saddle surfaces with a one-to-one spherical mapping, or *spherically one-to-one saddle surfaces* for short. Before stating the results, let us recall one definition. Suppose that a surface is a horn. The direction of the horn corresponds to a point $p \in S^2$. Let us remove from S^2 the point p' diametrically opposite to p. The degree of p on $S^2 \setminus p'$ with respect to the spherical image of some horn γ is the same for all horns sufficiently far from the boundary of the horn. This degree is called the *turn of the horn F.*

Theorem 2.2.5. *For a spherically one-to-one saddle surface we certainly have $\chi(F) \geqslant -1$. Such surfaces can only be of the following types: a) surfaces homeomorphic to a plane; b) surfaces homeomorphic to a cylinder (these surfaces consist*

Fig. 11

of a horn and a bowl or two bowls); c) surfaces with $\chi(F) = -1$ *consisting of two horns and a bowl.*

However, it is not known whether there is at least one spherically one-to-one saddle surface of the last type.

If a surface consists of a horn and a bowl, then its properties depend strongly on whether the horn has non-zero turn. If the turn of the horn is zero, then the surface certainly has self-intersections; such surfaces exist: an example is the surface given in cylindrical coordinates by the equation

$$\rho = e^{-z} \frac{\sqrt{\cos^2\varphi - e^{2z}\sin^2\varphi}}{\cos^2\varphi + e^{2z}\sin^2\varphi},$$

see Fig. 11. The most completely studied are the surfaces with $\chi = 0$ having a horn with non-zero turn.

Theorem 2.2.6. *If a spherically one-to-one saddle surface F with $\chi(F) = 0$ contains a horn with non-zero turn, then*

1) *in some coordinate system x, y, z (with z-axis parallel to the direction of the horn) the surface F is given by an equation $z = f(x, y)$ over the domain of the (x, y)-plane that is a completion to a closed convex set;*

2) *the surface F has a limiting cone consisting of a ray and a convex horn;*

3) *the closure of the spherical image of F is obtained from S^2 by removing the hemisphere corresponding to the horn and some convex domain (possibly empty) of the interior of the spherical image of the limiting cone of the bowl.*

§3. Non-Regular Saddle Surfaces

3.1. Definitions. Suppose that the surface F is given by a continuous mapping f of a two-dimensional manifold M into E^n, $n \geqslant 2$. A non-empty open set $E \subset M$ with compact closure $\overline{E} \subset \text{int } M$ is called a *crust* if there is a hyperplane P such that E is a component of the set $M \setminus f^{-1}(P)$. In this case we say that the hyperplane P cuts off the crust E. The surface F is called a *saddle surface* if it is

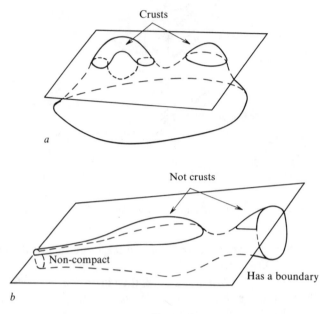

Fig. 12

impossible to cut off a crust from it by any hyperplane. We note that a crust is not assumed to be simply-connected; Fig. 12 explains the visual meaning of this concept. Clearly, a smooth surface in E^3 is a saddle surface if and only if its Gaussian curvature is non-positive.

If in the definition of a crust we replace the words "hyperplane P" by "hyper-sphere S^{n-1}", then we arrive at the definition of a *spherical crust*. A surface F is called an *R-saddle surface*, $0 < R < \infty$, if it is impossible to cut off a spherical crust from it by any hypersphere whose radius is at least R. Finally, a surface is called a *strictly saddle surface* if each point p of it has a neighbourhood in the form of an R-saddle surface for some $R = R(p)$. A smooth saddle surface in E^3 is an R-saddle surface if all its principal curvatures are at least R^{-1} in absolute value, and a strictly saddle surface if the Gaussian curvature of its intrinsic metric is negative everywhere.

If the boundaries of compact simply-connected saddle surfaces lying in E^n form a compact set, then the set of such surfaces itself is compact (in the space of surfaces with Fréchet metric) (S.Z. Shefel' (1967)). Saddle and R-saddle (for fixed R) surfaces form closed sets, but the set of strictly saddle surfaces is not closed.

3.2. Intrinsic Geometry. In the theory of non-regular saddle surfaces we should like to distinguish questions that seem to us to be fundamental for the development of this theory. Although these questions are similar to those that arose in the theory of general convex surfaces, their solution is apparently beset with much greater difficulties. In accordance with the general principles that we discussed in Chapter 1, in the first place two questions arise. Here is the first. Is any saddle surface with rectifiable boundary a manifold of non-positive curvature with respect to its intrinsic geometry? Although in the general case the answer to this question has still not been obtained, the following partial result is true.

Theorem 3.2.1 (S.Z. Shefel' (1964)). *Suppose that a saddle surface F in E^3 bounded by a rectifiable curve can be represented as the graph of a function $z = f(x, y)$. Then F is a manifold of non-positive curvature with respect to its intrinsic geometry.*

It is known that at each point of a surface with finite Lebesgue area there are arbitrarily small neighbourhoods bounded by rectifiable curves (Cesari (1956)). On the other hand, by the isoperimetric inequality for saddle surfaces (S.Z. Shefel' (1963)) a compact saddle surface with rectifiable boundary has finite area. Hence Theorem 3.2.1 can be strengthened as follows.

Theorem 3.2.2 (S.Z. Shefel' (1964)). *If a saddle surface F in E^3 has finite Lebesgue area (or is compact and has rectifiable boundary) and in a neighbourhood of each of its points it can be represented as the graph of a function $z = f(x, y)$, then F is a manifold of non-positive curvature with respect to its intrinsic geometry.*

The proof of Theorem 3.2.2 is based on the possibility of approximating a saddle surface that is the graph of a function by saddle polyhedra, and on a suitable criterion for the curvature to be non-positive due to Reshetnyak (Reshetnyak (1960a)). Since the question of the possibility of approximating a general saddle surface by saddle polyhedra (or sufficiently smooth saddle surfaces) remains open and has independent interest, we explain briefly how to solve it for a surface that is the graph of a function. The essence of the matter is that in this case a polyhedron lying in the layer between the surfaces $P_1: z = f(x, y) - h$ and $P_2: z = f(x, y) + h$ can be transformed into a saddle polyhedron lying in the same layer. It is sufficient first to remove from the polyhedron the crusts E lying over the planes that cut them off for which the flat domain Ω bounded by ∂E does not cut off a crust from P_1, each time replacing the crust by the corresponding domain Ω, and then repeating the procedure, but replacing P_1 by P_2, cutting off "lower" crusts instead of "upper" crusts. In fact this procedure is carried out successively for a countable everywhere dense set of "upper" (and then "lower") planes.

3.3. Problems of Immersibility. The second question is as follows: is every metric of non-positive curvature at least locally the intrinsic metric of some saddle surface in Euclidean space of some finite dimension? For a Riemannian metric this problem, as we explained above, reduces to the question of the solubility of a definite system of differential equations. Here we should just like to draw attention to the existence of an alternative approach in the case of general metrics of non-positive curvature, which consists in solving the analogous problem in the class of polyhedra and then carrying out a limiting process.

It is rather encouraging here that, as Perel'man recently proved (Perel'man (1988b)), every complete simply-connected polyhedral metric of non-positive curvature admits an isometric C^0-embedding in E^3 as a saddle polyhedron (so in particular the vertices of this polyhedron are only the images of the existing vertices of the metric). Although the set of saddle polyhedra in E^n whose metrics approximate a certain metric of non-positive curvature on a closed disc is compact (on condition that the lengths of the bounding curves are uniformly bounded), the convergence of these polyhedra to a limiting surface is not necessarily too "good" in the sense that the areas of the Grassmann images (the analogue of the spherical mapping in E^3) of the approximating polyhedra for a "typical" saddle surface in E^n, $n > 3$, cannot converge to the area of the Grassmann image of the surface itself; for the details see 1.1 and 1.4 of Ch. 4. An additional obstacle in this path is the fact that in contrast to the case of convex surfaces it is not at all clear under which additional conditions convergence of the saddle polyhedra implies convergence of their metrics.

3.4. Problems of Non-Immersibility. If we turn to surfaces in three-dimensional space, then here the question arises first of all of the geometrical conditions that replace the requirement of C^2-smoothness in Efimov's theorem discussed above. Experience accumulated in the theory of convex surfaces says that in the case of geometrically meaningful results smoothness conditions can be successfully replaced by assumptions of "geometrical regularity". The requirement of G-stability can apparently serve as such an assumption here, as in many other cases. The class of affinely stable immersions of metrics of negative curvature is the class of strictly saddle surfaces (see 3.1). This makes the following conjecture plausible. A complete Riemannian[7] metric with Gaussian curvature not exceeding a negative number does not admit immersions in E^3 in the class of strictly saddle surfaces[8].

Similar conjectures were made by Rozendorn in Rozendorn (1966). In them, in contrast to the supposition made here, for the extrinsically geometrical condition that compensates for the absence of C^2-smoothness we put forward the requirement that the order of saddleness of a surface at each point is equal to two, and so the spherical mapping does not have branch points. We observe that this condition does not guarantee that the surface is strictly saddle-shaped. In

---

[7] The question of a possible extension of the class of metrics under consideration does not seem to us so immediate; in this scheme it is a question, for example, of manifolds with specific curvature not exceeding a number $c < 0$. On the other hand, we should possibly limit ourselves initially to C^1-smooth surfaces.

[8] It is possibly advisable to extend the class of strictly saddle surfaces to the class of so-called λ-saddle surfaces, $0 \leqslant \lambda < 1$, introduced by Kozlov (Kozlov (1989)). A surface is called a λ-*saddle surface* at a point x if the null-vector of the normal space at x is contained in an ellipse homothetic to the ellipse of normal curvatures (see 1.1 of Ch. 4) with coefficients of homothety $0 \leqslant \lambda < 1$ and with centre of homothety at the centre of the ellipse.

fact, there is an example of a C^1-smooth surface with the following properties: 1) The surface F is C^∞-smooth outside some point a. 2) The surface F has a C^∞-smooth intrinsic metric of strictly negative curvature. 3) At the point a the surface F has order of saddleness equal to two. Nevertheless, in a neighbourhood of a point the surface F is not a strictly saddle surface, nor C^2-smooth[9].

This example is interesting in another respect. It shows that smoothness of the intrinsic metric of a surface and the fact that the curvature of the intrinsic metric in the class of saddle surfaces is negative do not imply that the surface is smooth (so here again there is no complete analogy with convex surfaces).

On the other hand, from Theorem B of Rozendorn (1966) we have the following result.

Theorem 3.4.1. *A C^1-smooth strictly saddle surface, with C^∞-smooth metric of negative curvature, that is C^∞-smooth everywhere except possibly for isolated points is actually C^∞-smooth everywhere.*

It seems to us that these facts can serve as arguments for using the conjecture stated above.

Generally the question of smoothness of a saddle surface with sufficiently smooth metric, except for theorems on removing isolated singularities, has hardly been studied. Only the following result is known.

Theorem 3.4.2 (Rozendorn (1967)). *Suppose that a simply-connected surface F bounded by a closed curve is C^2-smooth and has a C^k-smooth intrinsic metric of negative curvature. If F is C^k-smooth in some neighbourhood of a bounding curve, then it is C^k-smooth everywhere.*

See also Part II of the present book.

Chapter 4
Surfaces of Bounded Extrinsic Curvature

§ 1. Surfaces of Bounded Positive Extrinsic Curvature

1.1. Extrinsic Curvatures of a Smooth Surface. For surfaces F in E^n, $n \geqslant 4$, the concept of spherical mapping can be generalized in two different ways. For simplicity we assume that the surface is oriented (otherwise it is sufficient to go over to a double covering). Then firstly there is a Grassmann mapping $g: F \to G(n, 2)$ that associates with a point x of the surface the oriented tangent plane to

---

[9] The existence of such a surface follows from Theorems 6, A and B of Rozendorn (1966) and the fact that if $|K| \leqslant c < \infty$ for the surface, and the mean curvature in a neighbourhood of a point a is not bounded, then the absolute values of the radii of curvature are not necessarily bounded above.

F at this point. Let $S_g(E)$ be the area of the Grassmann image of a set E of the surface F with respect to the canonical[1] metric in $G(n, 2)$.

Secondly, let W be the subbundle of unit vectors of the normal bundle of a surface F in E^n. Then in W there are pairs (x, v), where v is the unit normal to F at x. The mapping $\gamma: W \to S^{n-1}$, where $\gamma(x, v) = v$, induces in W an $(n-1)$-dimensional measure $\bar{S}_\gamma$ – the "area" of the surface (W, γ). For a set $E \subset F$ we put $\bar{S}_\gamma(E) = \bar{S}_\gamma(p^{-1}(E))$, where p is the projection of W on F, $p(x, v) = x$. The measure $\bar{S}_\gamma$, usually additionally normalized, is called the *Chern-Lashof curvature* (Chern and Lashof (1957)). We choose the normalization by putting $S_\gamma = (4\pi/\omega_{n-1})\bar{S}_\gamma$, where ω_{n-1} is the $(n-1)$-area of the unit sphere S^{n-1}.

The space W, other than critical points of the mapping γ, splits into subsets W^+, W^-, on which γ preserves or reverses the orientation. The extrinsic positive and negative total curvatures of F are defined by the equalities[2] $\mu^+ = S_\gamma(W^+)$, $\mu^- = S_\gamma(W^-)$. The calculations carried out in Chern and Lashof (1957) show that

$$\mu^\pm(F) = \frac{n-2}{\omega_{n-3}} \int_F \int_{S^{n-3}} \varkappa_v^\pm(x)\, d\sigma_v\, dS_x, \tag{1}$$

where $\varkappa_v(x)$ is the *Lipschitz-Killing curvature* (that is, the product of the principal curvatures of F at x with respect to the normal v), $a^+ = \max(a, 0)$, $a^- = \max(-a, 0)$, and dS_x and $d\sigma_v$ are the elements of area of the surface F and the sphere S^{n-3} respectively.

By Gauss's theorem it follows from (1) that

$$\mu^+ - \mu^- = \omega, \tag{2}$$

where $\omega(E) = \int_E K\,dS$ is the integral curvature of the intrinsic metric of F. Since $\mu^+ + \mu^- = S_\gamma$, it follows from (2) that $S_\gamma \geq \mathrm{var}\,\omega$, and in the case of metrics with curvature of definite sign, equality holds only for convex and saddle surfaces respectively.

Thus, for saddle surfaces in E^n, $n \geq 4$, as for any surfaces in E^3, the Chern-Lashof curvature is completely determined by the intrinsic metric of the surface. If the surface does not lie in E^3 and is not a saddle surface, then a kind of "splitting" of the curvature may take place, for which $\mu^+ - \mu^- = \omega$, but $S_\gamma = \mu^+ + \mu^- > \mathrm{var}\,\omega$. (We recall that it is a question of sufficiently smooth surfaces, say immersions of class C^2.)

As Hoffman and Osserman showed (Hoffman and Osserman (1982)), we always have $S_g \geq S_\gamma \geq \mathrm{var}\,\omega$, and the left inequality may be strict.

To describe the extrinsic geometry of a surface in E^n it is helpful to use also the concept of the *ellipse of normal curvatures*. Consider the geodesics passing

---

[1] That is, invariant with respect to the action of the group $O(n)$ and the normalized metric on the homogeneous space $G(n, 2) = O(n)/SO(n-2) \times O(2)$.

[2] For k-dimensional surfaces in E^n, $0 \leq k < n$, S_γ and S_g are defined similarly. In a significant number of papers (see Willmore (1982), Kuiper (1970)) the dependence of the topology of k-submanifolds of E^n on S_γ has been studied, in particular, immersions with the smallest S_γ allowed by the topology (*tight immersions*). However, when $k > 2$ separate examination of μ^+ and μ^- is apparently less meaningful.

through a fixed point x of a surface F in E^n. Their curvature vectors at x fill an ellipse E lying in a two-dimensional plane of the normal space N of F at x. In other words, the ellipse of normal curvatures E is the image of the unit circle $|X| = 1$ of the tangent space under the mapping $X \mapsto B(X, X)$, where B is the second fundamental form of F with values in N.

If we choose a basis in N so that the first two unit vectors are parallel to the axes of E, and its centre has coordinates $(\alpha, \beta, \gamma, 0, \ldots, 0)$, then (see Hoffmann and Osserman (1982), Aminov (1980))

$$\frac{dS_g}{dS} = [K^2 + 4(\alpha^2 b^2 + \beta^2 a^2)]^{1/2} \geqslant |K|, \tag{3}$$

where a and b are the semi-axes of E, and K and dS are the Gaussian curvature and element of area of F. From (3) it follows that $S_y \geqslant \text{var } \omega$. It is not difficult to verify that saddle surfaces are characterized by the fact that the null-vector of N is contained in the flat domain bounded by E. We also note that the centre of the ellipse is the mean curvature vector of F (see §2 below). For surfaces in E^3, but not only for them, the ellipse of normal curvatures degenerates to an interval.

1.2. Extrinsic Curvatures of a General Surface. To extend the concept of *extrinsic curvatures* to non-regular surfaces, it is useful to characterize μ^+ and μ^- by the same method. We fix a point $O \in E^n$ and identify $f(x)$ with the "radius vector" $Of(x)$. By Sard's theorem the function $\varphi_v(x) = \langle v, f(x) \rangle$, where $\langle \, , \, \rangle$ is the scalar product, has only nondegenerate critical points for almost all $v \in S^{n-1}$. For such v we denote by $m^+(v)$, $m^-(v)$ the number of critical points of φ_v having even (respectively, odd) indices. For two-dimensional surfaces $m^+(v)$ is the number of local maxima and minima of φ_v, and $m^-(v)$ is the number of saddles of index 1. Then

$$\mu^+ = \frac{4\pi}{\omega_{n-1}} \int_{S^{n-1}} m^+(v) \, d\sigma_v, \qquad \mu^- = \frac{4\pi}{\omega_{n-1}} \int_{S^{n-1}} m^-(v) \, d\sigma_v. \tag{4}$$

In the case of a general (not necessarily smooth) surface in E^n we determine μ^+ from the first of the inequalities (4), putting $m^+(v)$ equal to the number of points of the surface at which it has locally a strictly supporting hyperplane with outward normal v. (The possibility $m^+(v) = \infty$ is not excluded.) The hyperplane P, oriented by the choice of its normal v, is called a *strictly supporting hyperplane* to the parametrized surface (M, f) at a point $x \in M$ if $f(x) \in P$ and there is a neighbourhood $U \subset M$ of x such that $f(U) \setminus f(x)$ is not empty and lies in the open half-space with outward normal[3] v. If locally a strictly supporting hyperplane P is moved a little parallel to itself in the direction opposite to its outward

---

[3] Similarly we can generalize the negative extrinsic curvature μ^-. However, technical difficulties arise here, but thanks to the Gauss-Bonnet formula all the basic questions do not require μ^- at all (the situation here is similar in many respects to the foundations of the intrinsic geometry of non-regular surfaces, where it is sufficient to impose conditions only on the positive excesses of triangles; see Aleksandrov and Zalgaller (1962)).

normal v, then the new surface cuts off a "crust" from F (see the definition in 3.1 of Ch. 3). Hence it is easy to conclude that $m^+(v)$ is equal to the largest number of pairwise disjoint crusts that can be cut off by hyperplanes with normal directed towards the crust.

The class of surfaces with $\mu^+ < \infty$ includes all convex and saddle surfaces. It is easy to see that the latter are characterized by the condition $\mu^+(F) = 0$. For C^1-smooth surfaces in E^3 satisfying the condition $\mu^+(F) < \infty$ it is probable that $\mu^+ = \omega^+$, but this has been proved only under certain additional assumptions; see Pogorelov (1956b), S.Z. Shefel' (1974), S.Z. Shefel' (1975). If a surface F in E^3 is not smooth, then the equality $\mu^+ = \omega^+$ may be violated independently of the degree of smoothness of the metric. As an example it is sufficient to consider the surface Q of a convex-concave lens with rim R. A lens of this kind is obtained if we cut a standard sphere $S^2 \subset E^3$ by some plane P not passing through the centre of the sphere, and replace one of the two resulting spherical caps by the surface symmetrical to it about P. On the rim $R = S^2 \cap P$ is concentrated the non-zero portion of μ^+, but the intrinsic curvature $\omega(R) = 0$, since Q is isometric to S^2. If we change the example by gluing together the lens Q from spherical caps of different radii, it is obvious that on R there occurs a kind of "splitting" of the curvature

$$\mu^+(R) - \mu^-(R) = \omega(R) \neq 0.$$

Apparently such a splitting is a general situation. As we mentioned above, for surfaces in E^n, $n > 3$, splitting of the intrinsic curvature into the extrinsic positive and negative curvatures is also typical of C^∞-smooth immersions.

The usual concept of a surface according to Fréchet does not exclude "exotic" cases when for any parametrization $f: M \to E^n$ of the surface F the space Γ of components of the sets $f^{-1}(x)$, $x \in f(M)$, contains components that separate Γ. In these cases the mapping f does not induce the structure of a manifold. With the intention of considering isometric immersions of manifolds of bounded curvature, henceforth as a rule we assume that the surface is a C^0-immersion. We recall that a surface is said to be *metrically connected* if any two points of it can be joined by a rectifiable curve lying on the surface; the intrinsic (induced) metric is defined for such a surface F.

Let Φ denote the class of metrically connected surfaces F that are C^0-immersions and satisfy the condition $\mu^+(F) < \infty$.

The example of a cylinder with a nowhere rectifiable generator shows that a connected surface F with $\mu^+(F) < \infty$ need not be metrically connected. Some sufficient criteria for metrical connectedness are contained in Theorems 1.2.1 and 1.3.2 below. For surfaces of class Φ the topology induced by the metric coincides with the original topology of the manifold M.

The choice of the class Φ of surfaces of *bounded positive extrinsic curvature* was connected with the search for a "successful" class of surfaces corresponding to the class $\mathcal{K}$ of metrics of bounded curvature. In their time different authors (see Aleksandrov (1949), Aleksandrov (1950a), Bakel'man (1956), Pogorelov (1956b), Borisov (1958–1960), Yu.D. Burago (1968a)) considered several classes of surfaces that are manifolds of bounded curvature with respect to their intrinsic geometry, but without a clear statement about the character of the correspondence between classes of metrics and surfaces. If we start from the principle of affine stability, as we did above, then accumulated geometrical observations suggest the conjecture, already mentioned in 4.2 of Ch. 1, about the affine compatibility between the classes $\mathscr{K}$ and Φ. In the scheme for confirming it there is a partial result, which we state now. First of all we observe that for any set E on a surface $F \in \Phi$ it is natural to define the extrinsic positive curvature $\mu^+(E)$ "concentrated on E"; on F there arises the (not normalized) Borel measure μ^+.

Theorem 1.2.1. *Suppose that a compact surface F in E^3 with finite positive extrinsic curvature bounded by finitely many rectifiable curves admits an explicit definition $z = f(x, y)$ and does not contain points x with $\mu^+(x) = 2\pi$. Then F is a manifold of bounded curvature with respect to its intrinsic geometry. For any Borel set $E \subset F$ we have $\omega^+(E) \leqslant \mu^+(E)$, where ω^+ is the positive curvature of the intrinsic metric.*

A certain strengthening of this theorem, under which, in particular, the requirement of explicit definition is imposed only locally, is given in 1.3 below. It is possible that this requirement is quite unnecessary. Difficulties arising in attempts to reject it are probably the same as in the case of saddle surfaces; see 3.2 of Ch. 3. As for the condition $\mu^+(x) < 2\pi$, it can be discarded if we generalize the main result of D.Yu. Burago (1984) to the case of non-regular surfaces.

1.3. Inequalities. An important property of the inequalities given below between the extrinsic and intrinsic characteristics of a surface is that they hold without any smoothness assumptions for all surfaces of the class under consideration, and hence they show that in the class Φ the properties of the metric have an influence on the extrinsic geometry of a surface in E^n for any n.

Theorem 1.3.1 (Yu.D. Burago (1968b)). *Let F be a compact surface in E^n (closed or with boundary). Then for its (Lebesgue) area S, length L of the boundary and radius R of a ball in E^n containing F we have*

$$S < C_n((\mu^+ + |\chi|)R^2 + LR), \tag{5}$$

where χ is the Euler characteristic of F, the constant C_n depends only on n, and if F is C^2-smooth, then it may be replaced by an absolute constant C (see Yu.D. Burago and Zalgaller (1980)).

In particular, a compact surface with finite positive extrinsic curvature and bounded by rectifiable curves has finite Lebesgue area.

Hence the next result follows from Theorems 1.2.1 and 1.3.1.

Theorem 1.3.2 (Yu.D. Burago (1968b)). *Suppose that a surface F lying in E^3 of bounded positive extrinsic curvature has finite Lebesgue area, or is compact and bounded by finitely many rectifiable curves (in particular, is closed). If in addition*

F in some neighbourhood of each of its points admits an explicit definition and does not contain points with $\mu^+(x) = 2\pi$, then F is a manifold of negative curvature with respect to its intrinsic geometry.

Suppose that F is an isometric C^0-immersion of a manifold of bounded curvature, as, for example, in the conditions of Theorems 1.2.1 and 1.3.2. Then the Lebesgue area coincides with the area defined by the intrinsic geometry, see S.Z. Shefel' (1970), and inequality (5) enables us to obtain relations between the extrinsic characteristics μ^+ and R, depending on the parameters S and L of the intrinsic metric.

For C^2-smooth surfaces in E^3, inequality (5) can be somewhat strengthened. However, here it is more important that for such surfaces the extrinsic curvature μ^+ coincides with the positive part of the curvature of the intrinsic metric $\omega^+ = \int K^+ \, dS$.

Theorem 1.3.3 (see Yu.D. Burago and Zalgaller (1980)). *For a C^2-smooth closed surface F in E^3 we have*

$$S \leqslant 2R^2(\omega^+ - \pi\chi). \tag{6}$$

If a C^2-smooth compact surface F in E^3 has a non-empty boundary, then

$$S < \begin{cases} C(\omega^+ R^2 + LR) & \text{when } \chi(F) = 1, \\ C((\omega^+ - 2\pi\chi)R^2 + LR) & \text{when } \chi(F) \leqslant 0, \end{cases} \tag{7}$$

where C is an absolute constant.

Inequalities (6) and (7) show that for any two-dimensional Riemannian manifold M there is an $R_0 = R_0(S, L, \chi, \omega^+) > 0$ such that M does not admit a C^2-smooth isometric immersion f in E^3 under which $f(M)$ is contained in a ball of radius R_0. In particular, this is the basis for some examples of Riemannian metrics that do not admit isometric immersions in E^3; see Gromov and Rokhlin (1970).

From Theorem 1.3.1 it is not difficult to deduce that a horn (tapering tube) F in E^n, $n \geqslant 3$, for which $m^+ < \infty$ and the area $S(F)$ is infinite, is unbounded in E^n. Apparently the requirement $S(F) = \infty$ is caused only by the method of proof, at least, it can be omitted in the case of a C^2-smooth horn in E^3 (see D.Yu. Burago (1984)) and a C^3-smooth saddle horn in E^n; see Theorem 2.2.1 of Ch. 3.

§2. The Role of the Mean Curvature

2.1. The Mean Curvature of a Non-Smooth Surface. Let us recall the definition of the mean curvature of a smooth surface in E^n. Let F be such a surface, specified by a vector-function $r(u^1, u^2)$, and $B(X, Y)$ its second fundamental form at a point p; see 2.1 of Ch. 1. The second fundamental form of a surface F at a point p with respect to the normal v is the form $B^v(X, Y) = \langle B(X, Y), v \rangle$, or in local coordinates

$$B^v(X, Y) = \sum_{i,j=1}^{2} b_{ij}^v X^i Y^j,$$

where X^i, Y^j are the coordinates of the tangent vectors X, Y to F, and $b_{ij}^v = \left\langle \dfrac{\partial^2 r}{\partial u^i \, \partial u^j}, v \right\rangle$. Here $\langle \ , \ \rangle$ is the scalar product.

There is a unique linear transformation A^v such that $B^v(X, Y) = -\langle A^v(X), Y \rangle$. In local coordinates the matrix of the transformation A^v has the form $A_{ij}^v = \sum_{k=1}^{2} b_{ik}^v g^{kj}$. The quantity $H(v) = -\frac{1}{2}$ trace A^v is called the *mean curvature of F with respect to v*. In local coordinates $H(v) = \frac{1}{2} \sum_{i,j=1}^{2} b_{ij}^v g^{ij}$.

The *mean curvature vector* $H(p)$ is the mean value of the vector function $vH(v)$ on S^{n-3}. If $v_1, \ldots, v_{n-2}$ is an orthonormal basis of normals, then $H(p) = \sum_{i=1}^{n-2} v_i H(v_i)$. The number $|H(p)| = (\sum_i H(v_i)^2)^{1/2}$ is called the *mean curvature* of the surface at p.

The *Beltrami-Laplace operator* of the vector function $r(u^1, u^2)$ is understood to be the vector-function $\Delta r = (\Delta x^1, \ldots, \Delta x^n)$, where Δx^i is the Beltrami-Laplace operator of the coordinate function $x^i(u^1, u^2)$ in the intrinsic metric of the surface. If F is specified by the vector-function $r(u^1, u^2)$, then its mean curvature vector $H = \frac{1}{2} \Delta r$.

The last equality enables us to generalize the concept of mean curvature to non-smooth surfaces in the following way. Suppose that some parametrization $r(u^1, u^2)$ of a surface F has first generalized derivatives that are square summable on any compactum, that is, $r \in W_{2,\text{loc}}^1$, and the Laplace operator Δr, understood in the sense of the theory of generalized functions, is completely additive on the ring of Borel sets of a vector-valued set function with finite variation var $\Delta r < \infty$. In this case F is called a surface of *bounded* (or *finite*) *mean curvature*, and $\frac{1}{2} \Delta r(E) = H(E)$ is called its mean curvature vector[4] on the set E.

2.2. Surfaces of Bounded Mean Curvature

Theorem 2.2.1 (S.Z. Shefel' (1970)). *In order that a surface F in E^n, $n \geqslant 3$, with intrinsic metric of bounded curvature, should be a conformally stable immersion of a manifold of bounded curvature, it is necessary and sufficient that F should be a surface of bounded mean curvature.*

(We recall that conformal transformations in E^n, $n \geqslant 3$, are superpositions of finitely many similarities and inversions.)

Theorem 2.2.2 (S.Z. Shefel' (1970)). *Suppose that a surface F is an isometric C^0-immersion in E^n of a Riemannian metric of smoothness $C^{k,\alpha}$, $k \geqslant 2$, $0 < \alpha < 1$. Then the following three assertions are equivalent.*

---

[4] We should observe that if F is a surface of bounded mean curvature, then extrinsically-isothermal parametrizations, that is, those such that $\left(\dfrac{\partial r}{\partial u^1} \right)^2 = \left(\dfrac{\partial r}{\partial u^2} \right)^2$, $\left\langle \dfrac{\partial r}{\partial u^1}, \dfrac{\partial r}{\partial u^2} \right\rangle = 0$ almost everywhere, are "admissible" in the sense of the definition just given. If F has a metric of bounded curvature, then intrinsically-isothermal coordinates are simultaneously extrinsically-isothermal; see S.Z. Shefel' (1970).

1) *The mean curvature of* F (*understood a priori as a generalized vector-function*) *is a vector-function of class* $C^{k-2,\alpha}$.

2) *The image of* F *under every inversion* (*with pole outside* F) *has an intrinsic metric of smoothness* $C^{k,\alpha}$.

3) *The surface* F *has smoothness* $C^{k,\alpha}$.

We recall that a group (or pseudogroup) of transformations of E^n is said to be *geometric* if it contains the group of similarities and is distinct from it. According to G.S. Shefel' (1985), geometric groups and pseudogroups of transformations of smoothness C^∞ are of only three types:

1) The affine group and the group of affine equiareal transformations.

2) The conformal pseudogroup (when $n \geqslant 3$ it is generated by similarities and inversions).

3) The so-called general groups. A general group G is characterized by the fact that for any integer $l \geqslant 2$ there is an element $g \in G$ that has at some point of space any preassigned l-growth of the Taylor series that certainly satisfies the condition that the Jacobian is non-singular, and possibly the condition of being equiareal.

It is not difficult to see that there are general groups other than the group of all diffeomorphisms.

Hence for surfaces in E^n we can consider three types of stability – affine, conformal, and general[5].

According to Theorem 2.2.1 conformally stable immersions of metrics of bounded curvature are surfaces of bounded mean curvature. In 4.2 of Ch. 1 we made the conjecture that affinely stable immersions of metrics of this class are surfaces of class Φ of bounded positive extrinsic curvature. This conjecture has been confirmed in the most important special cases; see Theorem 3.3.2, for example. Thus, for manifolds of bounded curvature there are probably two types of G-stable immersions – conformal and affine. Immersions of these types have, so to speak, the smallest possible "regularity". On the other hand, if a manifold of bounded curvature admits an isometric immersion in E^n that is stable under the whole group of diffeomorphisms D, then such an immersion is maximally "regular".

---

[5] Instead of groups and pseudogroups of transformations of E^n we can consider geometric pseudogroups of *local* transformations. The classification of them differs very little from that obtained by G.S. Shefel' in G.S. Shefel' (1985). Namely, such pseudogroups are closed (in a certain natural topology), and are of only five types: 1) the affine pseudogroup, 2) the conformal pseudogroup, 3) the pseudogroup of local projective transformations, 4) the pseudogroup of transformations that preserve the ratios of volumes, 5) the pseudogroup of all local transformations. This result, obtained by modifications of G.S. Shefel' (1985), was announced by Kreinovich (Kreinovich (1986)).

For pseudogroups of local transformations of *pseudo-Euclidean* space he proved a similar result on classification (an essential lemma was proved by E. Golubeva). One of the important cases of this general classification was considered much earlier by Borisov and Ogievetskij (Borisov and Ogievetskij (1974), Ogievetskij (1973)) in connection with questions of theoretical physics (in fact they proved that the algebra generated by infinitesimal affine and conformal transformations is everywhere dense among local transformations).

It is known (G.S. Shefel' (1985)) that a group G generated by similarities and non-empty sets of inversions and affine transformations other than similarities is general. Hence every diffeomorphism (at any point) can be approximated, together with derivatives up to any fixed order l, by composition of finitely many affine transformations and inversions. Hence it is very likely that an immersion of a manifold of bounded curvature in E^n that is stable under such a group G always has "maximal regularity", that is, it is D-stable. Thus we should expect that for manifolds of bounded curvature that are exactly three types of G-stable immersions in E^n – two independent types with least "regularity" and one with the greatest.

2.3. Mean Curvature as First Variation of the Area. Another approach to the generalization of the concept of mean curvature to non-smooth surfaces is as follows. Let ρ be a smooth vector field with compact support supp ρ in E^n, and F a surface with finite Lebesgue area, where $\partial F \cap \text{supp } \rho = \varnothing$. Consider a variation of F of the form $F_t(x) = F(x) + t\rho(x)$ and the corresponding variation $\delta S(\rho)$ of the area of the surface; this is an additive functional on the set of smooth vector fields of compact support that admits infinite values. If it is bounded, that is, if $|\delta S(\rho)| \leqslant C \max_x |\rho(x)|$, $C < \infty$, then there is a vector-valued measure λ such that $\delta S(\rho) = \int \rho \, d\lambda$. In the case of a smooth surface

$$\delta S(\rho) = -2 \int \langle H, \rho \rangle \, dS, \tag{8}$$

where H is the mean curvature vector.

On this basis we call F a surface of *bounded mean curvature* if the functional δS is bounded, and we call the vector measure $\frac{1}{2}\lambda$ the *mean curvature* of F. We can show that the definition of mean curvature given earlier is equivalent to that just given.

There are several important inequalities that include the mean curvature, in the first place the simple inequality

$$2S \leqslant R(L + Q),$$

where R is the radius of a ball containing F, $Q = \int |H| \, dS$, and L is the length of the boundary of F, and also the isoperimetric inequality

$$S < C(L + Q)^2,$$

where C is an absolute constant (its exact value is known only in special cases). Since the proofs of these inequalities can be obtained on the basis of (8), the inequalities are true in the general class of surfaces of bounded mean curvature; see Allard (1972), Michael and Simon (1973).

In conclusion, let us dwell on the connection between surfaces of bounded mean curvature and the theory of currents and varifolds. As Federer proved (Federer (1961)), to each surface F in E^n that has finite Lebesgue area there naturally corresponds a (unique) *integral current* μ_F. Hence there arises a current-valued function μ that associates with each domain G on F the corresponding current μ_G. It turns out that var μ coincides with the area S of the surface F. It is more convenient to state this in the language of varifolds. Every current T induces a natural varifold V_T. In the given case S is equal to the mass of the varifold V_{μ_F}. Hence the mean curvature of a general surface coincides with the first variation of the mass of the corresponding varifold.

§3. C^1-Smooth Surfaces of Bounded Extrinsic Curvature

3.1. The Role of the Condition of Boundedness of the Extrinsic Curvature. The following classic result of Nash (Nash (1956)) is well known.

Theorem 3.1.1. *Every C^k-smooth $(k \geqslant 3)$ n-dimensional Riemannian manifold admits a C^k-smooth isometric immersion in a Euclidean space E^m of some dimension $m = m(n) < \frac{3}{2}n(n-1) + \frac{11}{2}n + 5$.*

A great deal of attention[6] has been paid to improving the upper bound for the least admissible value of $m(n)$. However, the main content of the theorem is the fundamental fact that the class of Riemannian metrics coincides with the class of intrinsic metrics of sufficiently smooth surfaces in Euclidean spaces.

In Theorem 3.1.1, as in the majority of theorems about the existence of isometric immersions, the class of admissible surfaces is not distinguished by any geometrical property. This leads to the fact that the extrinsic geometry of admissible immersions of a surface "does not correspond" to its intrinsic metric. For example, if a Riemannian manifold M^n admits an isometric immersion in E^m, then it can be isometrically imersed in an arbitrarily small ball of the space E^{m+2}. When $m \geqslant n(n+1)/2 + 3n + 5$ any two free C^∞-isometric immersions of M^n in E^m can be joined by a homotopy of isometric immersions of M^n in E^m; see Gromov and Rokhlin (1970). We can say that here isometric immersions are formed with almost the same arbitrariness as topological ones. Although stricter limitations on the codimension of the immersion enable us to trace some connections between the extrinsic and intrinsic geometries of a surface, the class of Riemannian metrics that can be realized in this way decreases sharply, and as a rule uncontrollably.

For C^1-smooth immersions a similar violation of the connections between the intrinsic and extrinsic geometries occurs even for small codimension. Namely, Nash (Nash (1954)) made the following conjecture and outlined a proof of it, and Kuiper (Kuiper (1955)) completely proved it.

Theorem 3.1.2. *Every n-dimensional compact Riemannian manifold M^n can be C^1-smoothly and isometrically immersed in E^{2n-1} and embedded in E^{2n}.*

In the case $n = 2$ of interest to us, Theorem 3.1.2 guarantees the existence of an isometric C^1-smooth immersion in E^3. Despite the codimension 1, this im-

---

[6] See Gromov and Rokhlin (1970). In the case $n = 2$ the best universal value is $m(2) = 6$; see Gromov (1987). For surfaces homeomorphic to a disc, $m'(2) = 4$; see Poznyak and Shikin (1974).

mersion can be carried out with just as much arbitrariness as in the case of C^∞-smooth isometric immersions with large codimension. For example, if M is a two-dimensional Riemannian manifold and $\varphi: M \to E^3$ is a so-called *short immersion*, that is, a C^1-smooth immersion such that $|d\varphi(X)| < |X|$ for any $X \in TM$, then there is a C^1-smooth isometric immersion of M in E^3 arbitrarily C^0-close to φ. (If M is compact, a short immersion φ can be obtained from an arbitrary C^1-smooth one by a homothety.) In addition, any smooth surface F in E^3 can be continuously bent, in the class of surfaces isometric to it, into a surface lying in an arbitrarily small ball. In particular, a C^1-smooth surface with an analytic intrinsic metric of positive curvature need not be convex.

Apparently these assertions are not intuitively obvious; rather, they even contradict our intuitive ideas. At least in their time they were for many people unexpected and served as the starting point for a number of subtle investigations, the first of which was the book by Pogorelov (Pogorelov (1956b)); see also Borisov (1958–1960).

The assertions stated below show that similar phenomena are impossible if the positive extrinsic curvature μ^+ of the surface is finite, and this is explained by the fact that the finiteness of μ^+ guarantees the affine stability of the corresponding class of surfaces with a metric having curvature of definite sign.

3.2. Normal C^1-Smooth Surfaces. A C^1-smooth surface in E^3 is called a *normal* surface of non-negative curvature if at least one of the following assertions is true for each point x of it: (a) x has a neighbourhood in the form of a convex surface (in particular, this surface may be a flat domain); (b) through x there passes a rectilinear generator along which the tangent plane is stationary[7].

A *normal developable surface* is a normal surface of non-negative curvature that does not have points of strict convexity, that is, it is a saddle surface.

It is well known that a C^2-smooth surface is a normal surface of non-negative curvature (a normal developable surface) if and only if its Gaussian curvature is not negative (equal to zero).

It is not difficult to see that if a point of a normal surface of non-negative curvature does not have a neighbourhood in the form of a convex surface, then the rectilinear generator passing through this point is unique, and no other point of this rectilinear generator has such a neighbourhood; this generator can be extended to the boundary of the surface.

A normal surface of non-negative curvature is a manifold of non-negative curvature with respect to its intrinsic geometry, and a normal developable surface is locally isometric to a plane.

A complete normal surface of non-negative curvature whose intrinsic curvature is not equal to zero is a convex surface, and a complete normal developable surface is a cylinder. Every crust cut off from a normal surface of non-negative curvature by a plane is a convex surface.

---

[7] A normal surface of non-negative curvature in E^n, $n > 3$, can be defined similarly. Such a surface may be essentially n-dimensional, that is, it does not lie in any hyperplane.

3.3. The Main Results. Let us recall that we have denoted by K_0^+, K_0^-, K_0 the classes of two-dimensional manifolds of non-negative, non-positive and zero curvature respectively.

Theorem 3.3.1 (S.Z. Shefel' (1974)). *Suppose that a C^1-smooth surface F in E^3 has finite positive extrinsic curvature μ^+. If the intrinsic metric of F belongs to one of the classes K_0^+, K_0^-, K_0, then F itself is a normal surface of non-negative curvature, a saddle surface, or a normal developable surface, respectively.*

Remark. If under the conditions of Theorem 3.3.1 the intrinsic metric of a surface F has positive specific curvature, then F is a locally convex surface; see Borisov and Shefel' (1971).

The last assertion becomes false if instead of the specific curvature being positive we only assume that the curvature of the intrinsic metric is positive on any open subset of the surface. In fact, the surface F shown in Fig. 1 can be chosen in such a way that the line l that is its axis of symmetry splits F into two convex surfaces whose Gaussian curvatures are positive, but tend to zero as l is approached; see Borisov and Shefel' (1971).

Theorem 3.3.1 follows directly from the next two assertions.

Theorem 3.3.2 (S.Z. Shefel' (1974), S.Z. Shefel' (1975)). *If a C^1-smooth surface F in E^3 has finite positive extrinsic curvature and belongs to one of the classes K_0^+, K_0^-, K_0 with respect to its intrinsic geometry, then the image of F under an affine transformation has intrinsic metric of the same class.*

In other words, the theorem asserts that C^1-smooth surfaces of bounded positive extrinsic curvature are affinely stable immersions for the metrics of the classes listed in the theorem.

Theorem 3.3.3 (S.Z. Shefel' (1974), Shefel' (1975)). *If a C^1-smooth surface F is an affinely stable immersion in E^3 of a metric of one of the classes K_0^+. K_0^-, K_0, then F is a normal surface of non-negative curvature, a saddle surface, or a normal developable surface, respectively.*

3.4. Gauss's Theorem. One of the basic relations between the intrinsic and extrinsic geometries of smooth surfaces in E^3 is Gauss's theorem, which says that the curvature of the intrinsic metric of a surface is equal to the area of its spherical image. Although Gauss's theorem can be generalized to surfaces in E^n, $n > 3$, its role in the theory of such surfaces is incomparably less important than for surfaces in E^3.

As we mentioned above, Gauss's theorem remains true for C^1-smooth normal surfaces of non-negative curvature in E^3, but for C^1-smooth saddle surfaces in E^3 (and, more generally, for C^1-smooth surfaces of bounded positive extrinsic curvature) its truth has not been proved, apparently. Subtle, but not very conclusive, results on the generalization of Gauss's theorem were obtained by Pogorelov in 1955–56.

He studied the class Π of so-called C^1-smooth surfaces of bounded extrinsic curvature. The author defined this class by the following condition: for any finite choice of pairwise disjoint closed sets on a surface $F \in \Pi$, the sum of the areas of their spherical images must not exceed some constant $C(F) < \infty$. For a surface $F \in \Pi$ we necessarily have $\mu^+ < \infty$, and although formally the class Π is narrower than Φ, we know of no examples of C^1-smooth surfaces in $\Phi \backslash \Pi$, for example, C^1-smooth saddle surfaces not belonging to Π.

Points of a C^1-smooth surface admit a natural classification into *non-regular* and *regular*, and the latter into elliptic, hyperbolic, and parabolic; see Efimov (1949), Pogorelov (1956b).

Let E be a Borel subset of F and suppose that the complete inverse image of a point $y \in S^2$ under the spherical mapping restricted to E consists of $n_E(y)$ points, of which $n_E^+(y)$ are elliptic and $n_E^-(y)$ are hyperbolic (the case $n_E(y) = \infty$ is not excluded a priori). It turns out that for a surface of class Π we have $n_E(y) < \infty$ almost everywhere on S^2 and $\int_{S^2} n_E(y)\, dy < \infty$. (Properly speaking, these two properties can be taken as the definition of the class Π.) The set functions $\sigma^+(E) = \int_{S^2} n_E^+(y)\, dy$, $\sigma^-(E) = \int_{S^2} n_E^-(y)\, dy$, $\sigma = \sigma^+ - \sigma^-$, are called, respectively, the *positive*, *negative* and *total extrinsic curvature* of the surface F. Pogorelov (Pogorelov (1956b)) proved that Gauss's theorem, in the form $\sigma(E) = \omega(E)$, is true for closed surfaces, and in the general case it is true for those Borel subsets E of F that do not contain non-regular points. In addition, we always have $\omega^+(E) = \sigma^+(E)$, $\omega^-(E) \geqslant \sigma^-(E)$.

3.5. $C^{1,\alpha}$-Smooth Surfaces. The geometrical properties of $C^{1,\alpha}$-smooth surfaces in E^3, $0 < \alpha < 1$, depend on the value of α. For sufficiently small $\alpha > 0$ such surfaces are similar to C^1-smooth surfaces in their properties. Namely, as Borisov proved (Borisov (1958–1960)), when $0 < \alpha < 1/13$ even an analytic surface is bendable with large arbitrariness in the class of $C^{1,\alpha}$-smooth surfaces. However, as α increases the picture changes. In any case when $\alpha > 1/2$ a surface of class $C^{1,\alpha}$ has a smooth (of class $C^{1,2\alpha-1}$) metric (S.Z. Shefel' (1982)). In addition, for such a surface a parallel displacement along a rectifiable curve can be defined both extrinsically and intrinsically, and both methods lead to the same result; see Borisov (1958–1960), S.Z. Shefel' (1982).

Finally, Borisov succeeded in showing that when $\alpha > 2/3$ the correspondence between the extrinsic form of a surface and the sign of the curvature of its intrinsic metric, which is inherent in C^2-smooth surfaces, is preserved. Using a parallel displacement, he obtained a series of important results in this direction, which were later somewhat strengthened and can now be stated in the form of the next theorem.

Theorem 3.5.1 (S.Z. Shefel' (1975)). *Let F be a $C^{1,\alpha}$-smooth surface in E^3, where $\alpha > 2/3$. If F is a two-dimensional manifold of non-negative, non-positive, or zero curvature with respect to its intrinsic geometry, then F is respectively a normal surface of non-negative curvature, a saddle surface, or a normal developable surface.*

This theorem follows immediately from Theorem 3.3.3 and the next assertion, which also has independent interest.

Theorem 3.5.2 (S.Z. Shefel' (1975)). *If a surface F in E^3 of smoothness class $C^{1,\alpha}$, $\alpha > 2/3$, belongs to one of the classes K_0^+, K_0^-, K_0 with respect to its intrinsic geometry, then its image under an affine transformation has intrinsic metric of the same class.*

§4. Polyhedra

4.1. The Role of Polyhedra in the General Theory. It is well known that in the construction of the theory of convex surfaces an important role was played by the possibility of approximating them by convex polyhedra. The place of polyhedra in the theory of two-dimensional manifolds of bounded curvature is remarkable. Although at present most of the results of these theories can be obtained in principle without going over to polyhedra, enlisting the latter frequently gives simpler proofs and helps to discover their geometrical prerequisites. In addition, the theory of convex polyhedra, independently of its applications to general convex surfaces, is distinguished by great unification, completeness, and beauty.

In the theory of non-regular saddle surfaces (like surfaces of bounded extrinsic curvature) polyhedra are also useful in the study of both the intrinsic geometry of such surfaces and the connections between the extrinsic and intrinsic curvatures. However, many theorems for saddle polyhedra were obtained later than the corresponding results for smooth saddle surfaces, by analogy with them, and there has still been no promotion of the study of general saddle surfaces. Moreover, in a number of questions the theory of non-convex polyhedra, particularly saddle polyhedra, is less developed than the corresponding branches for smooth surfaces. The situation that arises is somewhat similar to the relation between the topology of smooth manifolds and the parallel theory of *PL*-manifolds.

4.2. Polyhedral Metric and Polyhedral Surface. Let us refine our terminology. By a polyhedral metric we mean a triangulable manifold M with a metric ρ specified on it such that every simplex of some triangulation is isometric to a rectilinear simplex of Euclidean space. In the two-dimensional case a polyhedral metric is characterized by the following condition: each point p of it has a neighbourhood isometric to either a neighbourhood of a vertex of some cone in E^3 (in particular, this neighbourhood may be a flat domain) or a flat sector if p is a point of the boundary.

A parametrized polyhedral surface is a mapping of a triangulable manifold M in E^n under which each simplex of some triangulation of M goes into a rectilinear simplex of the same dimension in E^n. Parametrized polyhedral surfaces in E^n can be combined in the usual way into equivalence classes – polyhedral surfaces.

An isometric polyhedral mapping of a polyhedral metric (M, ρ), or equivalently an isometric mapping of it in the form of a polyhedral surface, is a polyhedral surface $f: M \to E^n$ that is simultaneously an isometric mapping. This means that for any curve γ on M its length $s(\gamma)$ in the metric ρ is equal to the length of the curve $f \circ \gamma$ in E^n. In other words, the metric induced by the mapping f coincides with ρ.

4.3. Results and Conjectures. It is natural to expect that the class of polyhedral metrics is affinely connected with the class of polyhedral surfaces in Euclidean spaces. This conjecture has still not been proved completely, but the following results testify to its usefulness.

Theorem 4.3.1 (S.Z. Shefel' (1970), Shefel' (1978)). *Let G be the group of piecewise-affine transformations*[8] *in E^n. Then every G-stable isometric mapping of a polyhedral metric in E^n is a polyhedral surface.*

This theorem was proved in S.Z. Shefel' (1978) directly for k-dimensional surfaces in E^n, $2 \leqslant k \leqslant n$. As for the existence of isometric polyhedral immersions of polyhedral metrics, apparently only the following two results are known.

Theorem 4.3.2 (Zalgaller (1958)). *When $k \leqslant 4$ every k-dimensional polyhedral metric admits an isometric mapping in E^k as a polyhedral surface.*

When $k = 2$ the proof of this theorem is so simple that we give it here. Suppose that a polyhedral metric is specified on a closed manifold M (otherwise it is sufficient to glue together two copies of M along the boundary). Let $p_1, \ldots, p_N$ be all the vertices of the metric, and A_j a "Voronoi domain", that is, the set of points at a smaller distance from p_j than from the other vertices p_i, $i \neq j$. Such a domain A_j is isometric to an open polygon on a cone, and the vertex of the cone corresponds to p_j.

We split each domain A_j into triangles with vertex p_j, adjacent to each other along the sides. The bases of these triangles are the sides of the boundary of A_j. Different triangles can have a common base; this is connected with the fact that some points of the boundary of A_j may be joined to p_j by a non-unique shortest curve. To each triangle T of such a partition there corresponds a triangle T' equal to it with the same base, lying either in a polygon adjacent to A_j or directly in A_j. We cut out each quadrangle $T \cup T'$ from M, "bend" it first along the common base of T and T', and bend the resulting twice covered triangle along the bisector of the angle at the vertex opposite to this base. We place the resulting figures in the plane E^2 so that their vertices p_j coincide, and the sides of the triangles go along one ray; now we can again carry out gluings along all the cuts and obtain an isometric mapping of M in E^2.

Theorem 4.3.3 (Burago and Zalgaller (1960)). *An orientable two-dimensional polyhedral metric admits an isometric C^0-embedding in E^3 as a polyhedral surface.*

---

[8] A *piecewise-affine* (or *simplicial*) transformation is a transformation g of E^n that is affine on every simplex Δ_g of some triangulation of E^n.

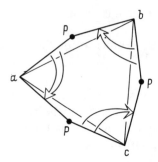

Fig. 13

In both theorems we do not exclude the case when the polyhedral surface that realizes the given metric may be strongly "corrugated", with many "superfluous vertices", that is, vertices at which the curvature of the metric is zero. Moreover, the proof of these theorems, which have a constructive character, leads to "corrugated" polyhedral surfaces. In this connection we note that a polyhedral metric may not admit any isometric immersions in E^n under which the positive extrinsic curvature of the surface is equal to the intrinsic, and obstacles may have a metric (and not only topological) character. An example is the polyhedral metric homeomorphic to S^2 with development shown in Fig. 13, consisting of an equilateral and three "narrow" isosceles triangles; the arrows in the figure mark the sides to be glued together.

In the case of complete metrics with curvature of definite sign the situation changes. Namely, the class of two-dimensional polyhedral metrics of positive curvature (defined on S^2 or E^2) is affinely compatible with the class of complete convex polyhedra. This follows from 3.3 of Ch. 1 and mainly Aleksandrov's theorem on the existence of a convex polyhedron with a given development of positive curvature (Aleksandrov (1950b)).

Apparently the class of complete two-dimensional polyhedral metrics of negative curvature defined on the plane is affinely compatible with the class of complete saddle polyhedral surfaces. This is a special case of Conjecture A (2.5 of Ch. 1), but for polyhedra it has advanced further than in the general case. Namely, Perel'man proved recently (Perel'man (1988a)) that a complete simply-connected polyhedral metric of non-positive curvature admits an isometric C^0-immersion in E^3 as a polyhedral surface. However, the possibility of applying this result to solve the general conjecture by approximation by polyhedra is doubtful. It this work we have investigated the question of isometric immersibility of a surface with complete polyhedral metric of non-positive curvature in E^n as a polyhedron homeomorphic to a cylinder. It turns out that if there is a closed geodesic on such a surface, then the surface admits an isometric *mapping* in E^3, but it may not admit an isometric C^0-*immersion* in any E^n. If there is no closed geodesic, then the metric may not admit an isometric mapping in any E^n. Examples of complete polyhedral metrics of non-positive curvature with Euler characteristic $\chi \leqslant -1$ that do not admit an isometric mapping in any E^n are simpler than in the case $\chi = 0$.

§ 5. Appendix. Smoothness Classes

In the framework of Hölder classes of smoothness $C^{l,\alpha}$ the question of the connection between smoothness classes of Riemannian metrics and surfaces can be assumed to be completely solved in Sabitov and Shefel' (1976), S.Z. Shefel' (1979).

Let $G^{l,\alpha}$ denote the group of $C^{l,\alpha}$-smooth diffeomorphisms[9] of E^n. The definitive result is as follows.

Theorem 5.1.1. *The class of m-dimensional Riemannian manifolds of smoothness* $C^{l,\alpha}$, $m \geqslant 2$, $l \geqslant 2$, $0 < \alpha < 1$, *is* $G^{l,\alpha}$*-compatible with the class of m-dimensional* $C^{l,\alpha}$*-smooth surfaces in Euclidean spaces* E^n, $n \geqslant m$.

In other words, the next three theorems are true, the last of which is a refinement (in relation to the degree of smoothness) of Nash's theorem on isometric immersions.

Theorem 5.1.2 (Sabitov and Shefel' (1976)). *Every* $C^{l,\alpha}$*-smooth m-dimensional surface in* E^n, $2 \leqslant m < n$, *is a Riemannian manifold of smoothness* $C^{l,\alpha}$, $l \geqslant 2$, $0 < \alpha < 1$, *with respect to its intrinsic geometry.*

From the theorem it follows that $C^{l,\alpha}$-smooth surfaces are $G^{l,\alpha}$-stable immersions for the class of $C^{l,\alpha}$-smooth Riemannian manifolds.

We observe that under the conditions of the theorem the usual formulae of differential geometry $g_{ij} = \left\langle \dfrac{\partial r}{\partial u^i}, \dfrac{\partial r}{\partial u^j} \right\rangle$ guarantee only smoothness $C^{l-1,\alpha}$.

The theorem is best possible: for example, the metric of the $C^{l,\alpha}$-smooth surface of revolution $z = x^2 + y^2 - (x^2 + y^2)^{(l+\alpha)/2}$ does not belong to $C^{\alpha+\varepsilon}$ for any $\varepsilon > 0$.

Theorem 5.1.3 (Sabitov and Shefel' (1976)). *When* $l \geqslant 1, 0 < \alpha < 1$, *every* $G^{l,\alpha}$*-stable immersion in* E^n *of an m-dimensional Riemannian manifold of smoothness* $C^{l,\alpha}$ *is a* $C^{l,\alpha}$*-smooth m-dimensional surface,* $2 \leqslant m < n$.

Theorems 5.1.2 and 5.1.3 establish that the class of m-dimensional $C^{l,\alpha}$-smooth surfaces in E^n coincides with the class of $G^{l,\alpha}$-stable immersions of m-dimensional Riemannian metrics of smoothness $C^{l,\alpha}$ in E^n.

Theorem 5.1.4 (Jacobowitz (1972)). *Every* $C^{l,\alpha}$*-smooth Riemannian manifold,* $l \geqslant 2, 0 < \alpha < 1$, *admits a* $C^{l,\alpha}$*-smooth isometric immersion in some* E^n.

---

[9] All the assertions of this section have local character. Hence instead of $G^{l,\alpha}$ we can take the pseudogroup of local diffeomorphisms whose domains of definition contain a neighbourhood of the chosen point. We can also restrict ourselves to diffeomorphisms of smoothness C^∞.

From Theorem 5.1.2 it follows that the assertion of Theorem 5.1.4 is best possible (for Hölder smoothness classes).

Although these results are true for surfaces of any dimension, the case of two-dimensional surfaces has certain peculiarities; in particular, the proofs can be simplified. This refers in the first place to Theorem 5.1.3 and substantially to Theorem 5.1.2 and is connected with the fact that in the two-dimensional case isothermal coordinates always exist and play a fundamental role in our situation (see 4.1 of Ch. 1).

Their advantage is that

1) in isothermal coordinates a Riemannian manifold M has the greatest possible smoothness, that is, if $M \in C^{l,\alpha}$ and in isothermal coordinates $ds^2 = \lambda^2(du^2 + dv^2)$, then $\lambda \in C^{l,\alpha}$;

2) isothermal coordinates are invariants of the group of conformal transformations.

For example, Theorem 5.1.2 for two-dimensional surfaces follows immediately from what we have said: by Gauss's theorem the Gaussian curvature of a surface is equal to the product of its principal curvatures, and so it belongs to $C^{l-2,\alpha}$ if the surface has a parametrization of smoothness $C^{l,\alpha}$. Then in isothermal coordinates the Gaussian curvature K also has smoothness at least $C^{l-2,\alpha}$. But since $\Delta \ln \lambda = -2K\lambda$, where Δ is the Laplace operator, by a property of the solutions of elliptic equations we obtain $\lambda \in C^{l,\alpha}$, and the theorem is proved.

In the general case instead of isothermal coordinates we use the so-called harmonic coordinates; for more details see Sabitov and Shefel' (1976).

Comments on the References

The list of references in this article is far from complete. First of all we mention works that contain a result that is fundamental (in some sense or other), and references are given more often to monographs than to original sources. In the list we also include works required in the course of the exposition. The ideas of Chapter 1 are reflected in S.Z. Shefel' (1985), and the classification of geometric groups in G.S. Shefel' (1984), G.S. Shefel' (1985). The theory of convex surfaces was last summed up in surveys and monographs quite a long time ago; see Cohn-Vossen (1959), Aleksandrov (1948), Aleksandrov (1950b), Bakel'man, Verner and Kantor (1973), Pogorelov (1969). We do not mention surveys on the theory of convex bodies, mixed volumes, and specially convex polyhedra. Of the more recent works we note the investigations of smoothness of convex surfaces (Nikolaev and Shefel' (1982), Nikolaev and Shefel' (1985), Sabitov and Shefel' (1976), S.Z. Shefel' (1979)) and bendings of locally convex surfaces. For answers to particular questions, see Dubrovin (1974), Milka (1974), Usov (1976b), Diskant (1988). The connection with Monge-Ampère equations is developed further in Pogorelov (1975), S.Z. Shefel' (1977). Surveys of the theory of surfaces of negative curvature, Efimov (1948), Efimov (1966), Rozendorn (1966), Poznyak and Shikin (1974), Rozendorn (1989), are devoted almost exclusively to surfaces in E^3; for such surfaces in E^n see Perel'man (1988b). In connection with Chapter 4 we draw attention to Yu.D. Burago (1968b) and Chapter 5 of the monograph Burago and Zalgaller (1988), which are devoted to quantitative connections between extrinsic and intrinsic characteristics of surfaces. For possible immersions in various classes, as well as the classic works Aleksandrov (1948), Nash (1954), Kuiper (1955), Gromov and Rokhlin (1970), Gromov (1987), we mention the works Pogorelov (1956b), Borisov (1958–1960), Poznyak (1973), S.Z. Shefel' (1970).

References*

Ahlfors, L., Sario, L. (1960): Riemann Surfaces. Princeton Univ. Press, Princeton, N.J., Zbl.196,338

Aleksandrov, A.D. (1937–1938): On the theory of mixed volumes of convex bodies. I. Extension of some concepts of the theory of convex bodies. Mat. Sb. 2, 947–972, Zbl.17,426. II. New inequalities between mixed volumes and their applications. ibid. 2, 1205–1238, Zbl.18,276. III. Extension of two theorems of Minkowski on convex polyhedra to arbitrary convex bodies. ibid. 3, 27–46, Zbl.18,424. IV. Mixed discriminants and mixed volumes. ibid. 3, 227–251 (Russian), Zbl.19,328

Aleksandrov, A.D. (1939): Almost everywhere existence of the second differential of a convex function and some properties of convex surfaces connected with it. Uch. Zap. Leningr. Univ. Ser. Mat. No. 37, part 6, 3–35 (Russian)

Aleksandrov, A.D. (1942): Smoothness of a convex surface with bounded Gaussian curvature. Dokl. Akad. Nauk SSSR 36, 195–199 (Russian), Zbl.61,376

Aleksandrov, A.D. (1948): The Intrinsic Geometry of Convex Surfaces. Gostekhizdat, Moscow-Leningrad. German transl.: Die Innere Geometrie der Konvexen Flächen. Akademie-Verlag, Berlin, 1955, Zbl.38,352

Aleksandrov, A.D. (1949): Surfaces representable as a difference of convex functions. Izv. Akad. Nauk KazSSR, Ser. Mat. Mekh. 3, 3–20 (Russian)

Aleksandrov, A.D. (1950a): Surfaces representable as a difference of convex functions. Dokl. Akad. Nauk SSSR 72, 613–616 (Russian), Zbl.39,180

Aleksandrov, A.D. (1950b): Convex Polyhedra. Gostekhizdat, Moscow-Leningrad. German transl.: Konvexe Polyeder. Akademie-Verlag, Berlin, 1958, Zbl.41,509

Aleksandrov, A.D. (1956–1958): Uniqueness theorems for surfaces in the large. I–IV. Vestn. Leningr. Univ. 1956, 11, No. 19, 5–17; 1957, 12, No. 7, 15–44; 1958, 13, No. 7, 14–26; 1958, 13, No. 13, 27–34. Engl. transl.: Am. Math. Soc. Transl., II. Ser. 21, 341–411, Zbl.101,138 and Zbl.101,139

Aleksandrov, A.D., Pogorelov, A.V. (1963): Theory of surfaces and partial differential equations. Proc. 4th All-Union Math. Congr. Leningr. 3–16 (Russian), Zbl.196,411

Aleksandrov, A.D., Zalgaller, V.A. (1962): Two-dimensional manifolds of bounded curvature. Tr. Mat. Inst. Steklova 63, 1–262. Engl. transl.: Proc. Steklov Inst. Math. 76, 183 pp. (1965), Zbl.122,170

Allard, W.K. (1972): On the first variation of a varifold. Ann. Math., II. Ser. 95, 417–491, Zbl.251.49028

Aminov, Yu.A. (1980): On the Grassmann image of a two-dimensional surface in a four-dimensional Euclidean space. Ukr. Geom. Sb. 23, 3–16 (Russian), Zbl.459.53003

Bakel'man, I.Ya. (1956): Differential geometry of smooth nonregular surfaces. Usp. Mat. Nauk 11, No. 2, 67–124 (Russian), Zbl.70,392

Bakel'man, I.Ya., Verner, A.L., Kantor, B.E. (1973): Introduction to Differential Geometry "in the Large". Nauka, Moscow (Russian), Zbl.276.53039

Borisov, A.B., Ogievetskij, V.I. (1974): Theory of dynamical affine and conformal symmetries as the theory of a gravitational field. Teor. Mat. Fiz. 21, 329–342. Engl. transl.: Theor. Math. Phys. 21, 1179–1188 (1974)

Borisov, Yu.F. (1958–1960): Parallel translation on a smooth surface. I–IV. Vestn. Leningr. Univ. 1958, 13, No. 7, 160–171; 1958, 13, No. 19, 45–54; 1959, 14, No. 1, 34–50; 1959, 14, No. 13, 83–92. Corrections, 1960, No. 19, 127–129 (Russian), Zbl.80,151; Zbl.121,171; Zbl.128,163

Burago, D.Yu. (1984): Unboundedness in Euclidean space of a horn with a finite positive part of the curvature. Mat. Zametki 36, 229–237. Engl. transl.: Math. Notes 36, 607–612 (1984), Zbl.565.53034

Burago, Yu.D. (1960): Realization of a two-dimensional metrized manifold by a surface in E^3. Dokl. Akad. Nauk SSSR 135, 1301–1302. Engl. transl.: Sov. Math. Dokl. 1, 1364–1365 (1960), Zbl.100,365

Burago, Yu.D. (1968a): Surfaces of bounded extrinsic curvature. Ukr. Geom. Sb. 5–6, 29–43 (Russian), Zbl.195,516

---

*For the convenience of the reader, references to reviews in Zentralblatt für Mathematik (Zbl.), compiled using the MATH database, and Jahrbuch über die Fortschritte der Mathematik (Jbuch) have, as far as possible, been included in this bibliography.

Burago, Yu.D. (1968b): Isoperimetric inequalties in the theory of surfaces of bounded extrinsic curvature. Zap. Nauchn. Semin. Leningr. Otd. Mat. Inst. Steklova *10*. Engl. transl.: Sem. Math., V.A. Steklov Math. Inst., Leningr. *10* (1968), Zbl.198,550

Burago, Yu.D. (1970): Isometric embedding of a manifold of bounded curvature in a Euclidean space. Uch. Zap. LGPI *395*, 48–86 (Russian)

Burago, Yu.D., Zalgaller, V.A. (1960): Polyhedral embedding of a development Vestn. Leningr. Univ. *15*, No. 7, 66–80 (Russian), Zbl.98,354

Burago, Yu.D., Zalgaller, V.A. (1980): Geometric Inequalities. Nauka, Leningrad, Zbl.436.52009. Engl. transl.: Springer-Verlag, Berlin-New York (1988), Zbl.633.53002

Busemann, H. (1958): Convex Surfaces. Interscience, New York-London, Zbl.196,551

Busemann, H., Feller, W. (1935): Krümmungseigenschaften konvexer Flächen. Acta Math. *66*, 1–47, Zbl.12,274

Cecil, T.E., Ryan, P.J. (1985): Tight and Taut Immersions of Manifolds. Pitman, Boston, London, Zbl.596.53002

Cesari, L. (1956): Surface area. Ann. Math. Stud. *35*, Zbl.73,41

Chen, B. (1973): Geometry of Submanifolds. Marcel Dekker, New York, Zbl.262.53036

Cheng, S.Y., Yau, S.T. (1976): On the regularity of the solution of the *n*-dimensional Minkowski problem. Commun. Pure Appl. Math. *29*, 495–516, Zbl.363.53030

Cheng, S.Y., Yau, S.T. (1977): On the regularity of the Monge-Ampère equation $\det(\partial^2 u / \partial x_i \partial x_j) = F(x, u)$. Commun. Pure Appl. Math. *30*, 41–68, Zbl.347.35019

Chern, S.S., Lashof, R. (1957): On the total curvature of immersed manifolds. Am. J. Math. *79*, 306–313, Zbl.78,139

Cohn-Vossen, S.E. (1927): Zwei Sätze über die Starrheit der Eiflächen. Göttinger Nachrichten, 125–134, Zbl.53,712

Cohn-Vossen, S.E. (1936): Bendability of surfaces in the large. Usp. Mat. Nauk *1*, 33–76 (Russian), Zbl.16,225

Cohn-Vossen, S.E. (1959): Shortest paths and total curvature of a surface. In: Some Questions of Differential Geometry in the Large. Fizmatgiz, Moscow, pp. 174–244 (Russian), Zbl.91,341

Connelly, R. (1978): A counterexample to the rigidity conjecture for polyhedra. Publ. Math. Inst. Hautes Etud. Sci. *47*, 333–338, Zbl.375.53034

Diskant, V.I. (1985): Stability in Aleksandrov's problem for a convex body, one of whose projections is a ball. Ukr. Geom. Sb. *28*, 50–62, Zbl.544.52004. Engl. transl.: J. Sov. Math. *48*, No. 1, 41–49 (1990)

Diskant, V.I. (1988): Refinement of analogues of the generalized isoperimetric inequality. Ukr. Geom. Sb. *31*, 56–59 (Russian), Zbl.672.53053

Dubrovin, A.A. (1974): On the regularity of a convex hypersurface in the neighbourhood of a shortest curve. Ukr. Geom. Sb. *15*, 42–54 (Russian), Zbl.323.53047

Efimov, N.V. (1948): Qualitative questions in the theory of deformations of surfaces. Usp. Mat. Nauk *3*, No. 2, 47–158. Engl. transl.: Am. Math. Soc. Transl. *6*, 274–323, Zbl.30,69

Efimov, N.V. (1949): Qualitative questions in the theory of deformations of surfaces "in the small". Tr. Mat. Inst. Steklova *30*, 1–128 (Russian), Zbl.41,488

Efimov, N.V. (1964): The appearance of singularities on a surface of negative curvature. Mat. Sb., Nov. Ser. *64*, 286–320. Engl. transl.: Am. Math. Soc. Transl., II. Ser. *66*, 154–190, Zbl.126,374

Efimov, N.V. (1966): Surfaces with a slowly changing negative curvature. Usp. Mat. Nauk *21*, No. 5, 3–58. Engl. transl.: Russ. Math. Surv. *21*, No. 5, 1–55 (1966), Zbl.171,199

Efimov, N.V. (1968): Differential criteria for homeomorphy of certain maps with application to the theory of surfaces. Mat. Sb., Nov. Ser. *76*, 499–512. Engl. transl.: Math. USSR, Sb. *5*, 475–488 (1968), Zbl.164,215

Efimov, N.V. (1975): Nonimmersibility of the Lobachevskij half-plane. Vestn.. Mosk. Univ., Ser. I *30*, 83–86. Engl. transl.: Mosc. Univ.. Math. Bull. *30*, 139–142 (1975), Zbl.297.53029

Federer, H. (1961): Currents and area. Trans. Am. Math. Soc. *98*, 204–233, Zbl.187,313

Fenchel, W., Jessen, B. (1938): Mengenfunktionen und konvexe Körper. Danske Vid. Selsk., Mat.-Fyss. Medd. *16*, No. 3, 1–31, Zbl.18,424

Firey, W. (1968): Christoffel's problem for general convex bodies. Mathematika *15*, 7–21, Zbl.162,543

Fomenko, V.T. (1964): Bendings and unique determination of surfaces of positive curvature with boundary. Mat. Sb., Nov. Ser. *63*, 409–425 (Russian) Zbl.163,430

Fomenko, V.T. (1965): Bending of surfaces that preserves congruence points. Mat. Sb., Nov. Ser. *66*, 127–141 (Russian), Zbl.192,272

Gromov, M.L. (1987): Partial Differential Relations. Ergeb. Math. Grenzgeb. (3) *9*. Springer-Verlag, Berlin Heidelberg New York, Zbl.651.53001

Gromov, M.L., Rokhlin, V.A. (1970): Embeddings and immersions of Riemannian manifolds. Usp. Mat. Nauk *25*, No. 5, 3–62. Engl. transl.: Russ. Math. Surv. *25*, No. 5, 1–57 (1970), Zbl.202,210

Hadamard, J. (1898): Les surfaces à courboures opposées et leurs lignes géodésiques. J. Math. Pures Appl. *4*, 27–73

Heinz, E. (1959): Über die Differentialungleichung $0 < \alpha \leqslant rt - s^2 \leqslant \beta < \infty$. Math. Z. *72*, 107–126, Zbl.98,72

Heinz, E. (1962): On Weyl's embedding problem. J. Math. Mech. *11*, 421–454, Zbl.119,166

Herglotz, G. (1943): Über die Starrheit der Eiflächen. Abh. Math. Semin. Hansische Univ. *15*, 127–129, Zbl.28,94

Hoffman, D., Osserman, R. (1982): The area of the generalized Gaussian image and the stability of minimal surfaces in S^n and $\mathbb{R}^n$. Math. Ann. *260*, 437–452, Zbl.471.53037

Isanov, T.G. (1979a): The extension of infinitesimal bendings of a surface of positive curvature. Sib. Mat. Zh. *20*, 1261–1268. Engl. transl.: Sib. Math. J. *20*, 894–899 (1979), Zbl.429.53040

Isanov, T.G. (1979b): The extension of infinitesimal bendings of surfaces of class $C^{m,\lambda}$. Sib. Mat. Zh. *20*, 1306–1307. Engl. transl.: Sib. Math. J. *20*, 929–930 (1979), Zbl.426.53003

Jacobowitz, H. (1972): Implicit function theorems and isometric embeddings. Ann. Math., II. Ser. *95*, 191–225, Zbl.214,129

Kagan, V.F. (1947–1948): Foundations of the Theory of Surfaces. I, II. Gostekhizdat, Moscow-Leningrad (Russian), Zbl.41,487

Kerekjarto, B. (1923): Vorlesungen über Topologie. Berlin, Jbuch 49, 396

Klimentov, S.B. (1982): The structure of the set of solutions of the basic equations of the theory of surfaces. Ukr. Geom. Sb. *25*, 69–82 (Russian), Zbl.509.53021

Klimentov, S.B. (1984): Extension of infinitesimal high-order bendings of a simply-connected surface of positive curvature. Mat. Zametki *36*, 393–403. Engl. transl.: Math. Notes *36*, 695–700 (1984), Zbl.581.53002

Klimentov, S.B. (1986): On a way of constructing solutions of boundary-value problems of the theory of bendings of surfaces of positive curvature. Ukr. Geom. Sb. *29*, 56–82 (Russian), Zbl.615.35014

Klimentov, S.B. (1987): On the extension of infinitesimal higher-order bendings of a simply-connected surface of positive curvature under boundary conditions. Ukr. Geom. Sb. *30*, 41–49 (Russian), Zbl.631.53049

Klotz-Milnor, T. (1972): Efimov's theorem about complete immersed surfaces of negative curvature. Adv. Math. *8*, 474–542, Zbl.236.53055

Kozlov, S.E. (1989): Estimates of the area of spherical images of two-dimensional surfaces in Riemannian manifolds. Mat. Zametki *46*, No. 3, 120–122, Zbl.687.53018

Kreinovich, V.Ya. (1986): Query No. 369. Notices Amer. Math. Soc. *33*, 945

Kuiper, N. (1955): On C^1-isometric imbedding. Nederl. Akad. Wet. Proc., Ser. A *58* (= Indagationes Math. *17*), 545–556, 683–689, Zbl.67,396

Kuiper, N. (1970): Minimal total absolute curvature for immersions. Invent. Math. *10*, 209–238, Zbl.195,511

Levi, E. (1908): Sur l'application des équations integrales au problème de Riemann. Nachr. König. Gesell. Wiss. Göttingen Mat. 249–252

Lewy, H. (1935–1937): A priori limitations for solutions of Monge-Ampère equations. I, II. Trans. Am. Math. Soc. *37*, 417–434, Zbl.11,350; *41*, 365–374, Zbl.17,211

Lewy, H. (1936): On the non-vanishing of the Jacobian in certain one-to-one mappings. Bull. Am. Math. Soc. *42*, 689–692, Zbl.15,159

Lewy, H. (1938): On the existence of a closed convex surface realising a given Riemannian metric. Proc. Natl. Acad. Sci. USA *24*, 104–106, Zbl.18,88

Liberman, I.M. (1941): Geodesic curves on convex surfaces. Dokl. Akad. Nauk SSSR *32*, 310–313 (Russian), Zbl.61,376

Lin, C.S. (1985): The local isometric embedding in $\mathbb{R}^3$ of two-dimensional Riemannian manifolds with nonnegative curvature. J. Differ. Geom. *21*, 213–230, Zbl.584.53002

Michael, J.H., Simon, L.M. (1973): Sobolev and mean-value inequalities on generalized submanifolds in $\mathbb{R}^n$. Commun. Pure Appl. Math. *26*, 361–379, Zbl.252.53006

Milka, A.D. (1974a): A theorem on a smooth point of a shortest curve. Ukr. Geom. Sb. *15*, 62–70 (Russian), Zbl.321.53045

Milka, A.D. (1974b): An estimate of the curvature of a set adjacent to a shortest curve. Ukr. Geom. Sb. *15*, 70–80 (Russian), Zbl.343.53037

Milka, A.D. (1975): A shortest curve with a nonrectifiable spherical image. Ukr. Geom. Sb. *16*, 35–52 (Russian), Zbl.321.53046

Milka, A.D. (1977): A shortest curve, all of whose points are singular. Ukr. Geom. Sb. *20*, 95–98 (Russian), Zbl.429.53041

Milka, A.D. (1980): The unique determination of general closed convex surfaces in Lobachevskij space. Ukr. Geom. Sb. *23*, 99–107 (Russian), Zbl.459.51012

Minkowski, H. (1903): Volumen und Oberfläche. Math. Ann. *57*, Jbuch 34, 649

Nash, J. (1954): C^1-isometric imbeddings. Ann. Math., II. Ser. *60*, 383–396, Zbl.58,377

Nash, J. (1956): The imbedding problem for Riemannian manifolds. Ann. Math., II. Ser. *63*, 20–63, Zbl.70,386

Nikolaev, I.G., Shefel', S.Z. (1982): Smoothness of convex surfaces on the basis of differential properties of quasiconformal maps. Dokl. Akad. Nauk SSSR *267*, 296–300. Engl. transl.: Sov. Math. Dokl. *26*, 599–602 (1982), Zbl.527.53036

Nikolaev, I.G., Shefel', S.Z. (1985): Smoothness of convex surfaces and generalized solutions of the Monge-Ampère equation on the basis of differential properties of quasiconformal maps. Sib. Mat. Zh. *26*, 77–89. Engl. transl.: Sib. Math. J. *26*, 841–851 (1985), Zbl.585.53053

Nirenberg, L. (1953): The Weyl and Minkowski problems in differential geometry in the large. Commun. Pure Appl. Math. *6*, 337–394, Zbl.51,124

Ogievetskij, V.I. (1973): Infinite-dimensional algebra of general covariance group as the closure of finite-dimensional algebras of conformal and linear groups. Lett. Nuovo Cimento (2) *8*, 988–990

Olovyanishnikov, S.P. (1946): On the bending of infinite convex surfaces. Mat. Sb., Nov. Ser. *18*, 429–440 (Russian), Zbl.61,376

Perel'man, G.Ya. (1988a): Polyhedral saddle surfaces. Ukr. Geom. Sb. *31*, 100–108 (Russian), Zbl.681.53032

Perel'man, G.Ya. (1988b): Metric obstructions to the existence of certain saddle surfaces. Preprint P-1-88, LOMI, Leningrad (Russian)

Perel'man, G.Ya. (1989): An example of a complete saddle surface in $\mathbb{R}^4$ with Gaussian curvature different from zero. Ukr. Geom. Sb. *32*, 99–102 (Russian), Zbl.714.53035

Perel'man, G.Ya. (1990a): A new statement of a theorem of N.V. Efimov. Lectures at the All-Union conference of young scientists on differential geometry, dedicated to the 80th anniversary of the birth of N.V. Efimov, Rostov-on-Don, 1990, p. 89. (Russian)

Perel'man, G.Ya. (1990b): Saddle surfaces in Euclidean spaces. Dissertation, Leningrad State Univ., Leningrad, 1990 (Russian)

Pogorelov, A.V. (1949a): Unique determination of convex surfaces. Tr. Mat. Inst. Steklova *29*, 3–99 (Russian), Zbl.41,508

Pogorelov, A.V. (1949b): Regularity of convex surfaces with a regular metric. Dokl. Akad. Nauk SSSR *66*, 1051–1053 (Russian), Zbl.33,214

Pogorelov, A.V. (1949c): Convex surfaces with a regular metric. Dokl. Akad. Nauk SSSR *67*, 791–794. Engl. transl.: Am. Math. Soc. Transl., I. Ser. *6*, 424–429, Zbl.33,214

Pogorelov, A.V. (1951): Bending of Convex Surfaces. Nauka, Moscow-Leningrad. German transl.: Die Verbiegung Konvexer Flächen. Akademie-Verlag, Berlin, 1957, Zbl.45,425

Pogorelov, A.V. (1952a): Unique Determination of General Convex Surfaces. Izdat. Akad. Nauk SSSR, Kiev. German transl.: Die Eindeutige Bestimmung Allgemeiner Konvexer Flächen, Akademie-Verlag, Berlin, 1956, Zbl.72,175

Pogorelov, A.V. (1952b): Regularity of a convex surface with given Gaussian curvature. Mat. Sb. Nov. Ser. *31*, 88–103, Zbl.48,405

Pogorelov, A.V. (1953): Stability of isolated ridge points on a convex surface under bending. Usp. Mat. Nauk 8, No. 3, 131–134 (Russian), Zbl.51,384

Pogorelov, A.V. (1954): Unique determination of infinite convex surfaces. Dokl. Akad. Nauk *94*, 21–23 (Russian), Zbl.55,154

Pogorelov, A.V. (1956a): Nonbendability of general infinite convex surfaces with total curvature 2π. Dokl. Akad. Nauk SSSR *106*, 19–20 (Russian), Zbl.70,168

Pogorelov, A.V. (1956b): Surfaces of Bounded Extrinsic Curvature. Izdat. Gosud. Univ., Kharkov (Russian), Zbl.74,176

Pogorelov, A.V. (1969): Extrinsic Geometry of Convex Surfaces. Nauka, Moscow. Engl. transl.: Am. Math. Soc., Providence, RI, 1973, Zbl.311.53067

Pogorelov, A.V. (1971): An example of a two-dimensional Riemannian metric that does not locally admit a realization in E^3. Dokl. Akad. Nauk SSSR *198*, 42–43. Engl. transl.: Sov. Math. Dokl. *12*, 729–730 (1971), Zbl.232.53013

Pogorelov, A.V. (1975): The Multidimensional Minkowski Problem. Nauka, Moscow. Engl. transl.: J. Wiley & Sons, New York etc., 1978, Zbl.387.53023

Poznyak, E.G. (1966): Regular realization in the large of two-dimensional metrics of negative curvature. Ukr. Geom. Sb. *3*, 78–92 (Russian), Zbl.205,514

Poznyak, E.G. (1973): Isometric immersions of two-dimensional Riemannian metrics in Euclidean space. Usp. Mat. Nauk *28*, No. 4, 47–76. Engl. transl.: Russ. Math. Surv. *28*, No. 4, 47–77 (1973), Zbl.283.53001

Poznyak, E.G., Shikin, E.V. (1974): Surfaces of negative curvature. Itogi Nauki Tekh., Ser. Algebra, Topologiya, Geometriya *12*, 171–208. Engl. transl.: J. Sov. Math. *5*, 865–887 (1976), Zbl.318.53050

Poznyak, E.G., Shikin, E.V. (1980): Isometric immersions of domains of the Lobachevskij plane in Euclidean spaces. Tr. Tbilis. Mat. Inst. Razmadze *64*, 82–93 (Russian), Zbl.497.53006

Reshetnyak, Yu.G. (1956): A generalization of convex surfaces. Mat. Sb., Nov. Ser. *40*, 381–398 (Russian), Zbl.72,176

Reshetnyak, Yu.G. (1959): Investigation of manifolds of bounded curvature by means of isothermal coordinates. Izv. Sib. Otd. Akad. Nauk SSSR *10*, 15–28 (Russian), Zbl.115,164

Reshetnyak, Yu.G. (1960a): On the theory of spaces of curvature not greater than K. Mat. Sb., Nov. Ser. *52*, 789–798 (Russian), Zbl.101,402

Reshetnyak, Yu.G. (1960b): Isothermal coordinates in a manifold of bounded curvature. I, II. Sib. Mat. Zh. *1*, 88–116, 248–276 (Russian), Zbl.108,338

Reshetnyak, Yu.G. (1962): A special map of a cone into a manifold of bounded curvature. Sib. Mat. Zh. *3*, 256–272 (Russian), Zbl.124,153

Reshetnyak, Yu.G. (1967): Isothermal coordinates on surfaces of bounded integral mean curvature. Dokl. Akad. Nauk SSSR *174*, 1024–1025. Engl. transl.: Sov. Math. Dokl. *8*, 715–717 (1967), Zbl.155,303

Rozendorn, E.R. (1961): Construction of a bounded complete surface of nonpositive curvature. Usp. Mat. Nauk *16*, No. 2, 149–156 (Russian), Zbl.103,154

Rozendorn, E.R. (1966): Weakly irregular surfaces of negative curvature. Usp. Mat. Nauk *21*, No. 5, 59–116. Engl. transl.: Russ. Math. Surv. *21*, No. 5, 57–112 (1966), Zbl.173,232

Rozendorn, E.R. (1967): The influence of the intrinsic metric on the regularity of a surface of negative curvature. Mat. Sb., Nov. Ser. *73*, 236–254. Engl. transl.: Math. USSR, Sb. *2*, 207–223 (1967), Zbl.154,212

Rozendorn, E.R. (1981): Bounded complete weakly nonregular surfaces with negative curvature bounded away from zero. Mat. Sb., Nov. Ser. *116*, 558–567. Engl. transl.: Math. USSR, Sb. *44*, 501–509 (1983), Zbl.477.53003

Rozendorn, E.R. (1989): Surfaces of negative curvature. Itogi Nauki Tekh., Ser. Sovrem. Probl. Mat., Fundam. Napravleniya *48*, 98–195. Engl. transl. in: Encycl. Math. Sc. *48*, Springer-Verlag, Heidelberg, 87–178, 1992 (Part II of this volume)

Rozenson, N.A. (1940–1943): Riemannian spaces of class I. Izv. Akad. Nauk SSSR Ser. Mat. *4*, 181–192; *5*, 325–352; *7*, 253–284 (Russian), Zbl.24,282; Zbl.60,383

Sabitov, I.Kh. (1976): Regularity of convex surfaces with a metric that is regular in Hölder classes. Sib. Mat. Zh. *17*, 907–915. Engl. transl.: Sib. Math. J. *17*, 681–687 (1977), Zbl.356.53017

Sabitov, I.Kh. (1989): Local theory of bendings of surfaces. Itogi Nauki Tekh., Ser. Sovrem. Probl. Mat., Fundam. Napravleniya *48*, 196–270. Engl. transl. in: Encycl. Math. Sc. *48*, Springer-Verlag. Heidelberg, 179–250, 1992 (Part III of this volume)

Sabitov, I.Kh., Shefel', S.Z. (1976): Connections between the orders of smoothness of a surface and its metric. Sib. Mat. Zh. *17*, 916–925. Engl. transl.: Sib. Math. J. *17*, 687–694 (1977), Zbl.358.53015

Shefel', G.S. (1984): Geometric properties of transformation groups of Euclidean space. Dokl. Akad. Nauk SSSR *277*, 803–806. Engl. transl.: Sov. Math. Dokl. *30*, 178–181 (1984), Zbl.597.20033

Shefel', G.S. (1985): Transformation groups of Euclidean space. Sib. Mat. Zh. *26*, No. 3, 197–215. Engl. transl.: Sib. Math. J. *26*, 464–478 (1985), Zbl.574.53006

Shefel', S.Z. (1963): Research into the geometry of saddle surfaces. Preprint, Inst. Mat. Sib. Otd. Akad. Nauk SSSR (Russian)

Shefel', S.Z. (1964): The intrinsic geometry of saddle surfaces. Sib. Mat. Zh. *5*, 1382–1396 (Russian), Zbl.142,189

Shefel', S.Z. (1967): Compactness conditions for a family of saddle surfaces. Sib. Mat. Zh. *8*, 705–714. Engl. transl.: Sib. Math. J. *8*, 528–535 (1967), Zbl.161,418

Shefel', S.Z. (1969): Two classes of k-dimensional surfaces in n-dimensional space. Sib. Mat. Zh. *10*, 459–466. Engl. transl.: Sib. Math. J. *10*, 328–333 (1969), Zbl.174,531

Shefel', S.Z. (1970): Completely regular isometric immersions in Euclidean space. Sib. Mat. Zh. *11*, 442–460. Engl. transl.: Sib. Math. J. *11*, 337–350 (1970), Zbl.201,242

Shefel', S.Z. (1974): C^1-smooth isometric immersions. Sib. Mat. Zh. *15*, 1372–1393. Engl. transl.: Sib. Math. J. *15*, 972–987 (1974), Zbl.301.53033

Shefel', S.Z. (1975): C^1-smooth surfaces of bounded positive extrinsic curvature. Sib. Mat. Zh. *16*, 1122–1123. Engl. transl. Sib. Math. J. *16*, 863–864 (1976), Zbl.323.53043

Shefel', S.Z. (1977): Smoothness of the solution of the Minkowski problem. Sib. Mat. Zh. *18*, 472–475. Engl. transl. Sib. Math. J. *18*, 338–340 (1977), Zbl.357.52004

Shefel', S.Z. (1978): Surfaces in Euclidean space. In: Mathematical Analysis and Mixed Questions of Mathematics. Nauka, Novosibirsk, 297–318 (Russian)

Shefel', S.Z. (1979): Conformal correspondence of metrics and smoothness of isometric immersions. Sib. Mat. Zh. *20*, 397–401. Engl. transl.: Sib. Math. J. *20*, 284–287 (1979), Zbl.414.53004

Shefel', S.Z. (1982): Smoothness of a conformal map of Riemannian spaces. Sib. Mat. Zh. *23*, 153–159. Engl. transl.: Sib. Math. J. *23*, 119–124 (1982), Zbl.494.53020

Shefel', S.Z. (1985): Geometric properties of immersed manifolds. Sib. Mat. Zh. *26*, No. 1, 170–188. Engl. transl.: Sib. Math. J. *26*, 133–147 (1985), Zbl.567.53005

Shor, L.A. (1967): Isometric nondeformable convex surfaces. Mat. Zametki *1*, 209–216. Engl. transl.: Math. Notes *1*, 140–144 (1968), Zbl.163,441

Usov, V.V. (1976a): The length of the spherical image of a geodesic on a convex surface. Sib. Mat. Zh. *17*, 233–236. Engl. transl.: Sib. Math. J. *17*, 185–188 (1976), Zbl.332.53036

Usov, V.V. (1976b): Spatial rotation of curves on convex surfaces. Sib. Mat. Zh. *17*, 1427–1430. Engl. transl.: Sib. Math. J. *17*, 1043–1045 (1976), Zbl.404.53050

Usov, V.V. (1977): The indicator of a shortest curve on a convex surface. Sib. Mat. Zh. *18*, 899–907. Engl. transl.: Sib. Math. J. *18*, 637–644 (1977), Zbl.378.53034

Vekua, I.N. (1959): Generalized Analytic Functions. Fizmatgiz, Moscow. Engl. transl.: Pergamon Press, Oxford etc. (1962), Zbl.92,297

Vekua, I.N. (1982): Some General Methods of Constructing Various Versions of Shell Theory. Nauka, Moscow, Zbl.598.73100. Engl. transl.: Pitman, Boston etc. (1985)

Verner, A.L. (1968): Cohn-Vossen's theorem on the integral curvature of complete surfaces. Sib. Mat. Zh. *9*, 199–203. Engl. transl.: Sib. Math. J. *9*, 150–153 (1968), Zbl.159,230

Verner, A.L. (1967–1968): The extrinsic geometry of the simplest complete surfaces of nonpositive curvature. I, II. Mat. Sb., Nov. Ser *74*, 218–240; *75*, 112–139. Corrections: *77*, 136. Engl. transl. Math. USSR, Sb. *3*, 205–224; *4*, 99–123, Zbl.164,216; Zbl.167,194

Verner, A'.L. (1970a): Unboundedness of a hyperbolic horn in Euclidean space. Sib. Mat. Zh. *11*, 20–29. Engl. transl.: Sib. Math. J. *11*, 15–21 (1970), Zbl.212,263

Verner, A.L. (1970b): Constricting saddle surfaces. Sib. Mat. Zh. *11*, 750–769. Engl. transl.: Sib. Math. J. *11*, 567–581 (1970), Zbl.219.53051

Volkov, Yu.A. (1963): Stability of the solution of the Minkowski problem. Vestn. Leningr. Univ., Ser. Mat. Mekh. Astronom. *18*, No. 1, 33–43 (Russian), Zbl.158,197

Volkov, Yu.A. (1968): Estimate of the deformation of a convex surface depending on a change of its intrinsic metric. Ukr. Geom. Sb. 5–6, 44–69 (Russian), Zbl.207,208

Weyl, H. (1916): Über die Bestimmung einer geschlossen konvexen Fläche durch ihr Linielement. Zürich Naturf. Ges. *61*, 40–72, Jbuch 46, 1115

Willmore, T.J. (1982): Total Curvature in Riemannian Geometry. Ellis Horwood, Chichester; Halstead Press, New York, Zbl.501.53038

Zalgaller, V.A. (1950): Curves with bounded variation of rotation on convex surfaces. Mat. Sb., Nov. Ser. *26*, 205–214 (Russian), Zbl.39,181

Zalgaller, V.A. (1958): Isometric embedding of polyhedra. Dokl. Akad. Nauk SSSR *123*, 599–601 (Russian), Zbl.94,360

II. Surfaces of Negative Curvature

E.R. Rozendorn

Translated from the Russian
by E. Primrose

Contents

Preface

This article is devoted to surfaces of negative Gaussian curvature $K < 0$ in three-dimensional Euclidean space E^3 and related problems. These surfaces constitute part of the class of *saddle surfaces* in E^N. Hence the article serves as an extension of the third chapter of Part I of this book, written by Yu.D. Burago and S.Z. Shefel'. At the same time, this article is meant to be read independently, and so together with the references to Alekseevskij, Vinogradov and Lychagin (1988), Alekseevskij, Vinberg and Solodovnikov (1988), Burago and Shefel' (1989), and Sabitov (1989b), we repeat certain facts in the text that are already reflected in these surveys. However, these repetitions are comparatively small.

We pay most attention to surfaces of negative curvature in E^3, because among other saddle surfaces (along with *tapering* surfaces, which are also mentioned in Part I) they are the ones that have been most studied. We shall be dealing with questions that are connected in one way or another with properties of surfaces in the large. The transition from the study of merely local properties to the wider and deeper study of connections between local properties of geometric objects and their global structure is characteristic of 20th century geometry, especially the recent decades.

The main object of further consideration will be classes of surfaces that are distinguished by inequalities, and so, when surfaces are represented by points of a function space, whole regions of it are filled. In this connection it is useful to recall that in the 19th century the attention of geometers was concentrated to a very large extent on the study of classes of surfaces specified by equalities, so from the viewpoint of the function space they form submanifolds of it of positive codimension. The transition from the study of such special objects to objects in general position (in particular, to surfaces of classes that fill regions of function spaces) is also characteristic of 20th century mathematics.

At the same time, one class of surfaces determined by an equality plays a special role below – these are *surfaces of constant negative curvature*. This is connected not only with the peculiarities of the historical development of the branch of geometry discussed here, but also with the fact that for surfaces of constant curvature many properties appear simpler than in the general case, and often far simpler, which enables us to use these surfaces in research as a model object. Nevertheless, the theory of surfaces of negative curvature, even in three-dimensional Euclidean space, is far from complete.

In the transition from E^3 to surfaces in E^N in § 5 below (in contrast to Part I and to § 7 of Ch. 8 of the survey Alekseevskij, Vinogradov and Lychagin (1988)), the main attention will be concentrated on questions to do with immersions in E^N of manifolds of negative curvature and connections known at present between their dimension and N, independently of whether there is an immersion of the saddle surface or not.

§1. Hilbert's Theorem

1.1. Statement of the Problem. Historically the study of surfaces of negative curvature in E^3 was closely connected with the problem of interpretation of non-Euclidean geometry.

At the end of the 1830s Minding (Minding (1839), Minding (1840)) investigated certain properties of surfaces having Gaussian curvature

$$K = \text{const} < 0. \tag{1.1}$$

He discovered *helical surfaces* of constant curvature (1.1) (Minding (1839)). Dini investigated them later (Dini 1965)). Minding (Minding (1839)) obtained, and integrated in quadratures, a differential equation for the meridian of a surface of revolution having curvature (1.1). He showed that these surfaces of revolution can be naturally divided into three types. Two of them are shown in Fig. 1. Now they are often called the "*Minding bobbin*" (Fig. 1a) and the "*Minding top*" (Fig. 1b). In the third type there is (up to similarity) only one surface – the *pseudosphere* (see Fig. 2), studied in more detail by Beltrami (Beltrami (1872)). In addition, under the condition (1.1) Minding (Minding (1840)) found a series of relations between the sides and angles of triangles formed by geodesic curves, drawing attention to the analogy between them and the formulae of spherical trigonometry. The fact that the formulae that he found are equivalent to trigonometric relations on the Lobachevskij plane went unnoticed then, apparently, in view of the unfortunate chain of circumstances and the general lack of preparedness of the overwhelming majority of mathematicians of the time to perceive such an idea. Historians assume that Minding was not interested in the problem associated with non-Euclidean geometry, and that these works of his only accidentally concerned Lobachevskij (in this connection see Norden (1956), and also Galchenkova, Lumiste, Ozhigova and Pogrebysskij (1970)).

Later, in 1868, that is, more than a quarter of a century after the death of Lobachevskij, Beltrami (Beltrami (1868a)), using the results of Minding, showed that on surfaces in three-dimensional Euclidean space under condition (1.1) Lobachevskij planimetry holds locally if angles are understood in the usual sense, and for rectilinear segments we take arcs of geodesic curves. From Beltrami (1868a) it is clear, among other things, that Beltrami had already seen the difference between the local and global statements of the problem, and understood that the question of interpretation of Lobachevskij planimetry in the large, that is, for the whole Lobachevskij plane, was then unsolved. It is well known that the paper Beltrami (1868a) played an important role in the mastering and popularization of the ideas of non-Euclidean geometry; it is perhaps less well known that it served as an important preparatory step in the further research of Beltrami himself: in Beltrami (1868b) he constructed an analytic model of n-dimensional Lobachevskij space.

Returning to surfaces in E^3, we draw attention to the fact that the pseudosphere, the surface obtained by rotating about the asymptote the so-called "*curve*

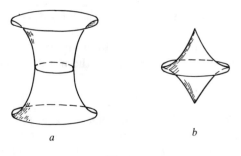

Fig. 1

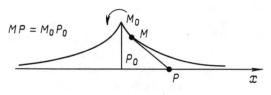

Fig. 2

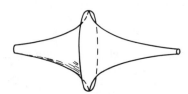

Fig. 3

of pursuit", the *tractrix* (a curve with constant length of subtangent; see Fig. 3) has a singular curve, namely a circular edge. It is traced out by the cusp of the tractrix. On the edge of the pseudosphere its smoothness is violated and its mean curvature becomes infinite. On each of the two smooth parts of the pseudosphere separated by its edge, in the form of a universal covering there is a *horocycle*, cut out from the Lobachevskij plane L^2.

Not only on the pseudosphere, but also on other surfaces of constant curvature the whole of L^2 cannot be moved even as a covering – the singular curves and points move.

All the same, can the whole of L^2 be realized in E^3 as a surface without singularities?

Hilbert (Hilbert (1903)) regarded this question as fundamental; he investigated it and gave a negative answer (Hilbert (1901)).

Hilbert's Theorem. *In E^3 there is no complete analytic surface of constant negative Gaussian curvature.*

We recall that a surface is said to be *complete* if it is a complete metric space in its *intrinsic metric*.

Hilbert noted that in the proof he used not analyticity, but sufficiently high C^r-smoothness of the surface in question. He did not determine precisely which r is required.

1.2. Plan of the Proof of Hilbert's Theorem. Let us suppose that L^2 is realized in the large in E^3 as a surface $\mathscr{F}$ and investigate this surface. From (1.1) we can deduce that the net of its *asymptotic lines* is a *Chebyshev net*. This means that in each net quadrangle $\mathscr{D}$ (generally speaking, curvilinear) the lengths of opposite sides are equal, as in a parallelogram in the Euclidean plane. Also from (1.1) we have the *Hazzidakis formula* (Hazzidakis (1879))

$$[\omega]_{\mathscr{D}} = |K|\sigma(\mathscr{D}), \tag{1.2}$$

where ω is the angle between the *asymptotic directions*, $\sigma(\mathscr{D})$ is the area of $\mathscr{D}$, and the symbol $[\ldots]_{\mathscr{D}}$ denotes the alternating sum

$$[\omega]_{\mathscr{D}} = \sum_{j=0}^{3} (-1)^j \omega(X_j)$$

of values of the angle ω at the vertices X_j of the quadrangle $\mathscr{D}$, numbered in the order of going round its contour; see Fig. 4.

Next, without loss of generality we may assume that $K = -1$: under the condition (1.1) this can always be achieved by changing the scale in E^3. Then, if we introduce the so-called *asymptotic coordinates*, that is, those in which the coordinate lines are the asymptotic lines of the surface, and take for u and v the natural parameters on two intersecting asymptotic lines, then u and v will serve as natural parameters on neighbouring asymptotic lines also, because of the Chebyshev property of the net, and the line element takes the form

$$ds^2 = du^2 + 2\cos\omega\, du\, dv + dv^2, \tag{1.3}$$

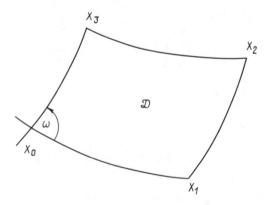

Fig. 4

but so far we can assert this only locally – in a neighbourhood of each specific point on $\mathscr{F}$.

We also need to prove (in this connection see Alekseevskij, Vinogradov and Lychagin (1988), Ch. 8, § 6) that in fact the whole surface $\mathscr{F}$ is covered by one chart with asymptotic coordinates (u, v), that is, the net of asymptotic lines on $\mathscr{F}$ in the large is homeomorphic to a Cartesian net on E^2.

For this we can proceed as in Klotz-Milnor (1972): together with (1.3) by means of these coordinates (still local!) we introduce on $\mathscr{F}$ an auxiliary metric

$$\Lambda = du^2 + dv^2 \geqslant \tfrac{1}{2} ds^2, \tag{1.4}$$

which is obviously *locally Euclidean*. Then it turns out that the manifold $\mathscr{F}$ with the metric Λ will be isometric in the large to the plane E^2, on which (u, v) are the usual Cartesian coordinates[1]. The reader can find the technical details needed for an accurate account of this part of the proof in "Application 1" of Klotz-Milnor (1972). It is also easy to deduce that on $\mathscr{F}$ there must exist net quadrangles $\mathscr{D}$ of arbitrarily large area $\sigma(\mathscr{D})$. But here we have arrived at a contradiction, since from (1.2) we have the uniform upper bound $\sigma(\mathscr{D}) \leqslant 4\pi |K|^{-1}$.

Further investigation shows (Hartman and Wintner (1951), Klotz-Milnor (1972)) that on the plan presented the proof goes through with smaller smoothness of the surface ($\mathscr{F} \in C^2$), which ensures the existence on $\mathscr{F}$ of the Gaussian curvature in its usual interpretation, as the product of the principal curvatures.

Thus we have the following result.

Theorem A. *In E^3 there is no C^2-smooth surface isometric in the large to the Lobachevskij plane L^2.*

Hilbert's theorem, stated in 1.1, is somewhat more general. Hilbert did not require in advance that the unknown surface $\mathscr{F}$ should be homeomorphic to a plane, but only assumed that it is complete.

Hilbert's result can be obtained as a consequence of Theorem A; for this we need to use the fact that $\mathscr{F}$ cannot be compact (see 2.3.1 below) and go over from $\mathscr{F}$ to its *universal covering* $\tilde{\mathscr{F}}$. For the definition of universal covering see Novikov (1986) or Seifert and Threlfall (1934), for example (a reader who is not very familiar with the topological material should first look at §§ 6–7 of Part 3 of the book Novikov and Fomenko (1987)). Efimov mentioned that Hilbert's paper (Hilbert (1901)) was one of the very first mathematical papers in which the concept of a universal covering was actually used, although a precise formulation of it had still not crystallized by then, and this was the cause of the initial difficulties in understanding Hilbert's theorem.

Slightly changing the plan of the proof presented above, we can manage without the auxiliary assertion that the *asymptotic net* on $\mathscr{F}$ is homeomorphic

---

[1] Poznyak mentioned in a lecture (Seminar on geometry in the large (1986)) that the majorization (1.4) leads to a contradiction with well-known properties of the Lobachevskij plane, in particular, with the rate of growth of the area of a disc in L^2 as its radius increases. This gives another way of proving the impossibility of an isometric immersions of L^2 in E^3.

in the large to a Cartesian net on E^2. In fact, it is sufficient to demonstrate the existence of a new quadrangle $\mathscr{D}$ that is so large that we arrive at a contradiction: we obtain (as above) contradictory upper and lower bounds either for its area $\sigma(\mathscr{D})$ or for the range of variation of the angle ω. The upper bound of the range of variation of ω is estimated on the basis of the necessary condition

$$0 < \omega < \pi, \tag{1.5}$$

and the lower bound, which exceeds π when $\mathscr{D}$ is sufficiently large, is obtained from equation (1.6) discussed below.

Here we mention that it is not only impossible to realize the whole of L^2 as a surface of constant curvature in E^3, but even the Lobachevskij half-plane (see[2] Efimov (1975) in Part I), and for the proof, according to Vorob'eva (1976), it is also sufficient to assume only C^2-smoothness of the unknown surface.

In consequence of the well-known results of Nash and Kuiper (see Kuiper (1955) and Nash (1954) in Part I), Theorem A cannot be extended to the class of C^1-smooth surfaces.

1.3. Connection with the Equations of Mathematical Physics. If condition (1.1) is satisfied and $\mathscr{F} \in C^4$, then the system of three *Peterson-Codazzi* and *Gauss equations* can be reduced (see Blaschke (1930)) to a form in which one pair of equations actually implies that the asymptotic lines form a Chebyshev net, and the third equation (under the normalization $K = -1$) in asymptotic coordinates takes the form

$$\omega''_{uv} = \sin \omega, \tag{1.6}$$

where u, v are the natural parameters on the asymptotic lines. Equation (1.6) was used by Hilbert in Hilbert (1903); it is the two-dimensional special case of the "sine-Gordon" equation, now widely known in mathematical physics (see Barone, Esposito, Magee and Scott (1971), Barone and Paternò (1982), for example):

$$\Box \omega = \sin \omega,$$

where $\Box$ is the *d'Alembertian*.

More general than (1.6) is the equation $\omega''_{uv} = f(\omega)$. It is also of interest in theoretical physics (see Kosevich (1972)) and with a specific choice of f it has been studied in Galeeva and Sokolov (1984a), for example.

Poznyak proved (Poznyak (1979)) that every solution $\omega(u, v)$ of (1.6) (except $\omega(u, v) = n\pi$, where n is an integer) generates in E^3 a surface with Gaussian curvature $K = -1$ on which $\omega(u, v)$ is the angle between the asymptotic lines, u and v are the natural parameters on them, and the surface has singular points where $\sin \omega = 0$.

Theorem A shows that there is no solution $\omega(u, v)$ of (1.6) that is defined on the whole (u, v)-plane and satisfies (1.5). Equation (1.6) is used in the theory of

---

[2] A reference of this kind refers to Part I of this book, that is, the part by Burago and Shefel'.

superconductivity (see Barone and Paternò (1982)), and also in other problems of mathematical physics (in this connection see Kosevich (1972), Enz (1964), Lamb (1971)). Popov (Popov (1989)) discussed the physical meaning of those regimes for which $\sin \omega = 0$ for problems of so-called self-induced transparency; on the surface singular curves correspond to them; see also Poznyak and Popov (1991).

1.4. Generalizations. In connection with Hilbert's theorem the following questions arise first of all:

1) Does the p-dimensional Lobachevskij space L^p admit a realization as a surface in E^N?

2) If so, under what relations between p and N?

A positive answer to the first of these questions follows from the results of Nash (see Nash (1956) in Part I); in this connection see Aminov (1982), Poznyak and Sokolov (1977), and Blanuša (1955), and also Gromov and Rokhlin (1970) in Part I, Gromov (1987) in Part I, and Alekseevskij, Vinogradov and Lychagin (1988), Ch. 8, § 7.

For a discussion of the results known at present on the second question see § 5 below.

3) Is the constancy of the curvature essential in Hilbert's theorem?

As far as we know, this question was posed by Hilbert himself. Cohn-Vossen conjectured (see Cohn-Vossen (1936) in Part I) that in Hilbert's theorem (1.1) can be replaced by the inequality

$$K \leqslant \text{const} < 0. \qquad (1.7)$$

It is natural to call this conjecture the *Hilbert-Cohn-Vossen problem*. On transition to E^N, $N > 3$, its statement admits various generalizations; some of them are discussed in § 5 below.

For E^3 the Hilbert-Cohn-Vossen problem was solved by Efimov (see Efimov (1963)), and he then obtained a more general result (Efimov (1968)); see Theorem B below. To formulate it we need an auxiliary concept.

Let $\mathfrak{M}$ be a metric space, and $\rho(x, y)$ the distance in it.

Definition (Efimov (1968)). A function $f(x, y)$, defined on $\mathfrak{M}$, has a *variation with linear estimate* if there are numbers $C_1 \geqslant 0$ and $C_2 \geqslant 0$ such that

$$|f(x) - f(y)| \leqslant C_1 \rho(x, y) + C_2. \qquad (1.8)$$

Remark. Any function f, defined and bounded on $\mathfrak{M}$, has a variation with linear estimate on it, since it satisfies (1.8) with $C_1 = 0$ and some $C_2 > 0$. If $C_2 = 0$ and $C_1 > 0$, then (1.8) turns into a Lipschitz condition.

Suppose, as above, that K is the Gaussian curvature. We put

$$R = |K|^{-1/2}. \qquad (1.9)$$

Sometimes (1.9) is called the *radius of Gaussian curvature*.

Theorem B (Efimov (1968)). *In E^3 there is no complete C^2-smooth surface $\mathscr{F}$ on which $K < 0$ and the radius of Gaussian curvature R has a variation with linear estimate.*

Corollary 1. *On any complete C^2-smooth surface $\mathscr{F}$ in E^3 we have*

$$\sup K \geqslant 0. \tag{1.10}$$

Thus, Theorem A is a special case of Theorem B.

Corollary 2. *If $\mathscr{F} \in C^2$, it is complete in E^3, it has $K < 0$ and $K \in C^1$, then on $\mathscr{F}$*

$$\sup|\operatorname{grad} R| = +\infty. \tag{1.11}$$

Here the gradient is understood in the sense of the intrinsic metric of the surface:

$$|\operatorname{grad} R| = \max_x \left|\frac{\partial R}{\partial s}\right|, \qquad x \in \mathscr{F},$$

$\dfrac{\partial}{\partial s}$ is the derivative with respect to arc length.

To derive Corollary 2 from Theorem B it is sufficient to observe that if the quantity on the right-hand side of (1.11) is finite, then R satisfies a Lipschitz condition on $\mathscr{F}$.

In the theory of surfaces of negative curvature, Theorem B takes a central place after Hilbert's theorem. The plan of proof, presented in a general way in Part I (§ 1 of Ch. 3) is discussed in more detail in 3.3 below.

Theorem B is exact from the viewpoint of the regularity class of the surfaces in question. Namely, if instead of C^2 we require regularity $C^{1,1}$, that is, we suppose that the first derivatives of the radius vector are continuous and non-collinear on the whole surface, and in each compact part of it they satisfy a Lipschitz condition, then we can construct a counterexample – a tapering saddle surface having four horns (Fig. 5a). Visually this surface is like a deformed tetrahedron whose faces are curved, whose edges are smoothed out, and whose vertices are at infinity. C^2-smoothness on it is violated only at four separate points – the centres of the faces of the tetrahedron. The curvature K, understood

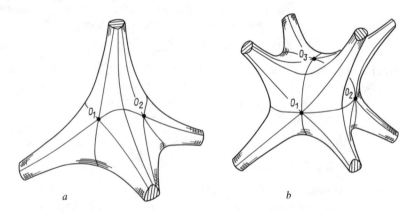

Fig. 5

in a suitable generalized sense (see 2.1.6), is defined on the whole surface, including the points where C^2-smoothness is violated, and satisfies an inequality of the form (1.7). The surface has a high degree of arbitrariness in its structure; see Rozendorn (1962). By ensuring all the listed properties, we can achieve somewhat higher smoothness. Namely, we can arrange that the normal curvature (that is, the curvature of normal sections) exists at each point of the surface in each direction and it is bounded in each compact part of the surface. For this it is sufficient to take a cube instead of a tetrahedron as the original figure, increasing to eight the number of horns going off to infinity, and to six the number of points where C^2-smoothness is violated (Fig. 5b). The construction of such surfaces and other examples connected with them are discussed in more detail in 2.1.8, 2.3.3 and 5.5 below.

To conclude this section we note that on the three-dimensional pseudo-Euclidean space E_1^3 with metric form $ds^2 = dt^2 - dx^2 - dy^2$ the Lobachevskij plane L^2 is realized in the large as an analytic surface $t^2 - x^2 - y^2 = 1$, $t > 0$, which in the "superimposed" Euclidean space E^3, that is, in the same affine space but with metric $ds_0^2 = dt^2 + dx^2 + dy^2$, is one half of an ordinary hyperboloid of revolution of two sheets; in this connection see Alekseevskij, Vinberg and Solodovnikov (1988), Ch. 2, and also Dubrovin, Novikov and Fomenko (1979), and Novikov and Fomenko (1987).

Various geometrical problems, still hardly studied, are connected with questions concerning immersions of metrics of negative curvature in pseudo-Euclidean spaces; for a first acquaintance with them the reader can refer to the article Efimov (1984), and for a more detailed acquaintance Galeeva and Sokolov (1984a) and the survey Sokolov (1980).

§2. Surfaces of Negative Curvature in E^3. Examples. Intrinsic and Extrinsic Curvature. Hadamard's Problem

2.1. Examples of Surfaces of Negative Curvature in E^3, and Their Extrinsic and Intrinsic Geometry. The curvature K being negative implies that the principal curvatures k_1 and k_2 of the surface in question are non-zero and of different signs, so the *osculating paraboloid* is hyperbolic, and the surface itself in a small neighbourhood of an arbitrary point of it looks like an "ordinary saddle": two ascents and two descents in relation to the tangent plane (Fig. 6). The rectilinear generators of the osculating paraboloid specify the *asymptotic directions* of the surface. When $K < 0$ there are two of them at each point. Correspondingly there are two families of asymptotic lines. They are distinguished by the sign of the *geodesic torsion*, whose square is equal to $|K|$ by the *Beltrami-Enneper theorem*; here we have in mind the torsion of *bands of the surface* formed along the asymptotic lines by tangent planes to the surface. It is the same as the torsion of the *asymptotic lines* when the curvature of the latter is non-zero. This curvature in turn is the same (up to the sign) as the *geodesic curvature* of the asymptotic

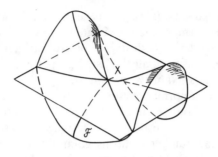

Fig. 6

lines. We shall return to the question of the role of the geodesic curvature of the asymptotic lines in §4.

The second-order surfaces of negative curvature $K < 0$ are the hyperbolic paraboloid and the hyperboloid of one sheet. Only on these surfaces does the net of asymptotic lines consist entirely of two families of rectilinear generators; see Finikov (1952), for example.

2.1.1. Before discussing examples of surfaces of negative curvature, let us fix the terminology that we shall use. In the mathematical literature the term "surface" is needed in different senses. Visually we associate with this term a thin film, the surface of a body or part of it. But here is another example – a front propagated in the space of a wave; in the mathematical description of a physical process a wave front is naturally regarded as a surface. However, if the source of the wave perturbation is not a point, but spread out, or if there are obstacles from which the wave is reflected, then the wave front may have self-intersections and self-overlappings.

We need to give a formal definition so as to include these situations, which are important in applications.

Suppose we are given a differentiable manifold $\mathfrak{M}$ of dimension $p \geqslant 2$ and a map of it $\mathscr{F}: \mathfrak{M} \to E^N$, $N > p$. Let $x_1, \ldots, x_N$ be rectangular Cartesian coordinates in E^N, and $u^1, \ldots, u^p$ local coordinates on $\mathfrak{M}$. We require that the functions $x_j(u^1, \ldots, u^p), j = 1, \ldots, N$, that locally specify the map $\mathscr{F}$ satisfy the two conditions

$$x_j(u^1, \ldots, u^p) \in C^r, \qquad r \geqslant 1 \tag{2.1}$$

and

$$\operatorname{rank}\left(\frac{\partial x_j}{\partial u^i}\right) = p. \tag{2.2}$$

If $\mathscr{F}$ satisfies (2.1) and (2.2), we say that a *p-dimensional C^r-smooth surface $\mathscr{F}$* is specified in E^N. The manifold $\mathfrak{M}$ is specified up to diffeomorphisms, which must be regarded here as C^r-smooth. If $s > r$, then any C^s-smooth surface is C^r-smooth. Preserving this definition for classes of surfaces, in the case of an individual surface we make the concept of its smoothness more precise.

Namely, suppose that the surface is C^m-smooth, in symbols $\mathscr{F} \in C^m$. Do we have $\mathscr{F} \in C^{m+1}$, $\mathscr{F} \in C^{m+2}$, ... ? The highest value of $r \geqslant m$ that we can achieve determines the C^r-smoothness of an individual surface. Here (both for individual surfaces and for classes of them) r may be a natural number or ∞. Finally, a surface $\mathscr{F}$ is assumed to be *analytic* (briefly $\mathscr{F} \in C^A$) if $\mathfrak{M}$ is an analytic manifold and the functions (1.2) are analytic when (2.2) is satisfied.

Similarly we can define the concepts of $C^{r,\alpha}$-*smooth surface* (α is the Hölder index, $0 < \alpha \leqslant 1$) and the class of $C^{r,\alpha}$-smooth surfaces. In the local formulation of the question they are discussed in more detail in Part III of the present book. In considering a surface in the large we need to distinguish uniform $C^{r,\alpha}$-smoothness, when the Hölder constant is the same for the whole surface, and "ordinary" $C^{r,\alpha}$-smoothness, when this constant is different for different compact parts of the surface.

Here, when we speak of compactness we are actually touching on the question of a topology on the surface. We first need to clarify what we mean by a point on it.

A *point X on a surface* $\mathscr{F}$ is a pair $X = \{X', X''\}$, where $X' \in \mathfrak{M}$, $X'' = \mathscr{F}(X') \in E^N$, understanding by this that in the definition of a surface $\mathfrak{M}$ is specified up to a diffeomorphism. The space E^N is said to be the *ambient space* for the surface $\mathscr{F}$.

We can approach the definition of a topology on $\mathscr{F}$ in two ways: we can introduce it either by starting from the topology on $\mathfrak{M}$ and the map $\mathscr{F} : \mathfrak{M} \to E^N$, or by means of the metric of the ambient E^N. For C^r-smooth surfaces with $r \geqslant 1$ these two approaches lead to the same result. In particular, $\mathscr{F}$ is compact when $\mathfrak{M}$ is compact.

We often consider *non-smooth surfaces*, supposing that $r = 0$. Then we assume that $\mathfrak{M}$ is a topological manifold specified up to a homeomorphism, and replace condition (2.2) by the requirement that the map $\mathscr{F}$ is locally homeomorphic; however, the question of a topology on $\mathscr{F}$ is then more complicated (see Part I, Ch. 4).

Similarly we define a p-dimensional C^r-smooth (or analytic) surface in other spaces, for example in an N-dimensional sphere S^N. When $p = N - 1$ we require the term *hypersurface*. Sometimes we allow $N = p$ and then we talk of an N-dimensional surface in N-dimensional space, or equivalently of a "*many-sheeted domain*"; below we shall see that this is convenient.

If in the construction we assume that $\mathfrak{M}$ is a manifold with boundary, then we have a *surface with boundary*.

Example. A pseudosphere with its edge forms a non-smooth surface homeomorphic to an open circular annulus or a plane with a point removed. The edge splits it into two analytic sufaces. Each of them with the edge of the pseudosphere adjointed to it is an analytic surface with boundary.

A compact surface without boundary is called a *closed surface* (a closed hypersurface when $N = p + 1$).

The set $\hat{\mathscr{F}} = \mathscr{F}(\mathfrak{M}) \in E^N$ is sometimes called the *support of the surface* $\mathscr{F}$. This terminology is not the same as that in function theory and functional analysis, where, as we know, the support of a function with compact support is the closure of the set on which it is non-zero, so it is a subset of the domain of definition of the function, but here it is a subset of the range of values of the function $\mathscr{F}$. However, in the theory of surfaces this "non-standard" terminology is quite convenient from the intuitive viewpoint: thus, for example, an infinite-sheeted covering of half a pseudosphere, which, as we have already mentioned in § 1, realizes a horocycle cut out from L^2, is a surface with self-intersections, homeomorphic (and diffeomorphic) to a plane, and the half pseudosphere itself is the support of this surface.

In the non-smooth case we need the map $\mathscr{F}$ to be locally homeomorphic, and in the smooth case this is ensured by condition (2.2). Hence in local questions, and also in those cases when it is known that the map $\mathscr{F}: \mathfrak{M} \to E^N$ is a homeomorphism in the large (for example, for closed convex surfaces; see Part I), we can identify points of the support $\hat{\mathscr{F}}$ with points of the surface, which is usually done; then we arrive at a definition of a surface as a subset of E^N, see 2.1 of Ch. 2 in Alekseevskij, Vinogradov and Lychagin (1988). However, this is only a special case of the general definition and in Chapter 8 of Alekseevskij, Vinogradov and Lychagin (1988) this is tacitly assumed. The reader should bear in mind the relation presented here between the approaches to the definition of a surface "in the small" and "in the large".

In § 1, when we discussed Beltrami's interpretation for Lobachevskij planimetry, we spoke of the "realization in E^3" of the Lobachevskij plane L^2 or part of it. This is a special case of the *problem of immersion* of a Riemannian metric in Euclidean space, since the existence in the ambient space of a rule for measuring distances leads to the fact that a surface on which points can be joined by rectifiable curves is itself a metric space: the distance $\rho(X, Y)$ between points $X, Y \in \mathscr{F}$ is defined as the greatest lower bound of the lengths of paths joining X and Y on $\mathscr{F}$. This is the *intrinsic metric* of the surface. For a C^r-smooth surface $\mathscr{F}$ with $r \geqslant 1$ it is Riemannian:

$$ds^2 = \sum_{i=1}^{N} dx_i^2 = \sum_{i,j=1}^{N} g_{ij}\, du^i\, du^j. \qquad (2.3)$$

From the last equality it is obvious that

$$\sum_{m=1}^{N} \frac{\partial x_m}{\partial u^i} \frac{\partial x_m}{\partial u^j} = g_{ij}. \qquad (2.4)$$

Two surfaces that are isomorphic as metric spaces are said to be *isometric*. The totality of properties preserved by isometries of the surface constitute its *intrinsic geometry*.

If the map $\mathscr{F}$ that specifies the surface depends continuously on the parameter t, we say that there is a *deformation* $\{\mathscr{F}_t\}$ of the surface $\mathscr{F} = \mathscr{F}_{t_0}$; here it is implied that the number t_0 belongs to the interval of the number axis over which t ranges. If all the surfaces of the family $\{\mathscr{F}_t\}$ are isometric, we call the deformation $\{\mathscr{F}_t\}$ a *bending*.

Suppose that the p-dimensional manifold $\mathfrak{M}$ is Riemannian. Then there arises the question of whether there is a surface $\mathscr{F}$ isometric to $\mathfrak{M}$ in the space E^N for a given N or more generally for some $N > p$. If the answer is yes, we say that there is an immersion (more precisely, an isometric immersion) of the given Riemannian manifold $\mathfrak{M}$ in E^N. If, moreover, the map

$$\mathscr{F}: \mathfrak{M} \to \hat{\mathscr{F}} \subset E^N \tag{2.5}$$

is a homeomorphism in the large, we say that it is an *isometric embedding*. Assuming that the metric on $\mathfrak{M}$ is given in the form (2.3) in each local chart, we arrive at an analytic statement of the problem in the form of a system of equations (2.4) – a non-linear system of partial differential equations on which no boundary conditions are imposed, and it is required to find a solution of it defined on the whole of $\mathfrak{M}$ and satisfying the additional condition (2.2). To obtain an isometric embedding, we need to ensure that the map (2.5) is homeomorphic for the unknown solution; in the C^r-smooth case, $r \geqslant 1$, condition (2.2) is necessary for this, but not sufficient.

From the results of Nash (see Nash (1956) in Part I), later developed by other authors (in this connection see Gromov and Rokhlin (1970) in Part I and Ch. 8 of Alekseevskij, Vinogradov and Lychagin (1988)) there follows the remarkable fact that any Riemannian geometry can be realized on p-dimensional surfaces of N-dimensional Euclidean space E^N for sufficiently large codimension $N - p$.

However, if we pose the question of the greatest possible lowering of the codimension $N - p$, then even in the local formulation the problem is far from completely solved, and in the global formulation presented above we are only at the first steps of the development. "The problem of immersing a Riemannian metric in Euclidean space", said A.D. Aleksandrov in one of his public lectures (Moscow State University, May 1970), "is a tangle of non-linear problems".

2.1.2. Turning to surfaces in E^3, let us recall that the planes of symmetry of the osculating paraboloid, intersecting with the tangent plane to the surface, give the directions of the lines of curvature. Hence when $K < 0$ the lines of curvature bisect the angles between the asymptotic directions.

Among the simplest surfaces in E^3 from the intuitive viewpoint are undoubtedly surfaces of revolution. Let us recall that on them the lines of curvature are the meridians and the parallels (Fig. 7), and the inequality $K < 0$ is satisfied where the meridian is convex on the side of the axis of rotation. Using the theorems of Euler and Meusnier, well known in the theory of surfaces, it is not difficult, when we have a specific surface of revolution, to find its asymptotic lines in those regions where $K < 0$. It is intuitively obvious that each surface of revolution $\mathscr{F}$ with axis l_0 has a specific line $\mathscr{L}$ in space that is an asymptotic line for $\mathscr{F}$: one family of asymptotic lines is obtained from $\mathscr{L}$ by rotation about l_0, and the other family is obtained from the first by mirror symmetry. In Fig. 7, and also in Figs. 8–10, we show only those parts of the lines that are on the visible side of the surface. One family of asymptotic lines is shown by solid lines, and the other by dotted lines. The line $\mathscr{L}$ is distinguished in Fig. 7. If the

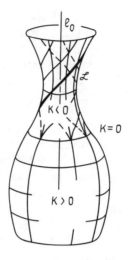

Fig. 7

meridian of a surface of revolution is C^2-smooth and has a point of inflexion (as in Fig. 7), then the point of inflexion traces out a parabolic parallel. On it we have $K = 0$, and it splits the surface into regions in which the curvature K has different signs. In this situation the asymptotic lines go out to the parabolic parallel, touching the meridians (see Fig. 7).[2]

On a hyperboloid of revolution of one sheet, as we have already mentioned, its rectilinear generators are asymptotic lines. Here we draw attention to the fact that the net consisting of them is only homeomorphic to a Cartesian net in the small. An extended net quadrangle can be opened up so that one of its vertices goes off to infinity, and the sides that should have occurred in it do not intersect; see Fig. 8, in which the notation is compatible with Fig. 4, and the vertex X_3 is absent: it has gone off to infinity. Under the conditions of Hilbert's theorem (§ 1) opened up quadrangles are excluded, thanks to the Chebyshev property of the net. In § 4 below we shall see that they are impossible on a simply-connected surface and when $K \approx \text{const} = 0$. But it is only important to understand clearly in what sense the approximate equality is admitted.

On a pseudosphere the asymptotic lines go off to infinity, winding on the horn and intersecting one another infinitely often; see Fig. 9a. It is far from obvious that when we go through the edge of the pseudosphere an asymptotic line does not lose its smoothness, and the whole of it is an analytic space curve; see Gribkov (1977).

It is not difficult to construct a surface of revolution of negative curvature in the form of a non-pointed horn on which the asymptotic lines, as they go off to

---

[2] In the common case there are other possibilities for the structure of the asymptotic net of a surface in a neighbourhood of a point of the parabolic curve; see Arnol'd (1990), section 11.

Fig. 8

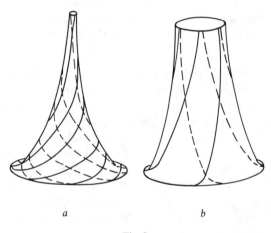

a   b

Fig. 9

infinity, approximate to the meridians as asymptotes (Fig. 9b). Then the situation is like that in Fig. 7, except that the cusps formed by pairs of asymptotic lines meeting on the parabolic parallel are at infinity here.

Among surfaces of revolution we should mention toroidal surfaces with a strictly convex meridian, to which the well-known circular torus belongs. If the meridian does not intersect the axis of rotation, then such a toroidal surface does not have self-intersections; its two bounding parallels, the "upper" and the "lower", are parabolic curves and split the surface into two parts, the "external", where $K > 0$, and the "internal". Depending on the form of its meridian, the asymptotic lines either go out to the parabolic parallel, touching it (Fig. 10), in

Fig. 10

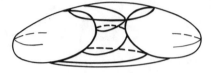

Fig. 11

the same way as an asymptotic line of half a pseudosphere touches the edge, or they wind onto the parabolic parallel as a limit cycle (Fig. 11).

Looking back at the contents of this subsection, we draw attention to a geometrical fact that it is useful to bear in mind later: if we do not count parabolic parallels, then on a tube of revolution there cannot be *closed asymptotic lines*.

2.1.3. Asymptotic lines and lines of curvature are objects of the *extrinsic geometry* of a surface: under a bending of the surface their position on it changes, generally speaking, with the variation of the *second fundamental form*. At the same time, the asymptotic lines play an important role in various questions, for example in the study of unique determination and rigidity of surfaces. Thus, when $K < 0$ a complete C^3-smooth surface on which the asymptotic net is homeomorphic to a Cartesian net in the large is *rigid* and *uniquely determined* by its metric if two of its complete (that is, infinitely extended) intersecting asymptotic lines are fixed in space (see Kantor (1976), Kantor (1978a), and Kantor (1981)). Surfaces of this kind are, for example, the hyperbolic paraboloid and the right helicoid.

Next, suppose we are given a C^3-smooth surface $\mathcal{F}$ of curvature $K < 0$ and we mark on it an asymptotic net quadrangle $\mathcal{D}$. We shall regard $\mathcal{D}$ together with its boundary $\partial\mathcal{D}$ as an independent surface with boundary. Let O denote one of its vertices, and L_1 and L_2 two of its sides (arcs of asymptotic lines on $\mathcal{F}$), starting from O. Suppose that O and the tangent plane to $\mathcal{F}$ at O are fixed in space. Then it follows from Rozendorn (1987) that for rigidity of the surface $\mathcal{D}$ it is necessary and sufficient that the projection of the *bending field* (see Part I, 4.1 of Ch. 2) on L_j in the direction of the asymptotic lines intersecting L_j, $j = 1, 2$, should vanish. A similar problem has been considered in a number of other works, for example in Minagawa and Rado (1952), Khineva (1977), and Mikhailovskij (1988). Quite an extensive literature has been devoted to the study of infinitesimal bendings of surfaces of revolution when $K < 0$ with various fixing condi-

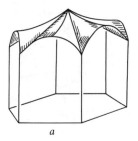 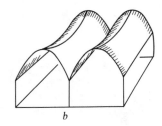

Fig. 12

tions (in particular, Mikhailovskij (1962a), Mikhailovskij (1962b), Mikhailovskij (1962c), Mikhailovskij (1962d)).

As for rigidity and unique determination of non-compact surfaces of negative curvature without additional fixing conditions, these questions have been little studied (see Galeeva and Sokolov (1984b), Seminar on geometry in the large (1986), Ten (1980)). Galeeva proved in 1979 that a smooth half of a pseudosphere that is not fixed but completed by an edge is rigid, and later she prove that it is uniquely determined. Infinitesimal bendings of a complete hyperboloid of one sheet (under various conditions at infinity) were studied in Galeeva and Sokolov (1984b).

Closely connected with the study of rigidity of surfaces are problems of the mechanics of thin elastic *shells* (for more details on this see Goldenveizer (1976), Pogorelov (1967), and Vekua (1959) in Part I), and constructions that use surfaces of negative curvature have become more and more widespread. Thus, Figure 12 shows schematically two types of roofs on supports, built in the form of piecewise smooth surfaces whose smooth parts have $K < 0$. Some problems for thin elastic shells, built in the form of surfaces of negative curvature, were considered in Klabukova (1983).

The best known uses of helicoidal surfaces in engineering are as propellor screws, turbine blades, and so on. But in these cases one is using so-called "thick shells" – their thickness is not small in comparison with the radii of principal curvature and the linear dimensions of the construction. The calculation of them is connected to a lesser extent with the geometrical theory of surfaces. However, thin shells are used that contain parts of negative curvature: the planks of some parts of ships (Fig. 13a) and the wings of certain types of aircraft (Fig. 13b). In various countries experimental models of modern airships have been developed, and also aircraft that combine the features of an airship and an aeroplane. Such "hybrid" aircraft as airships not only have important applications here on Earth, but in the future in learning about Venus and its atmosphere, which is much denser than ours; there, thanks to the greater lifting force for the same cubic content, the advantages of an airship will be more pronounced. Figure 13c shows schematically the form of one of these hypothetical aircraft of "hybrid" (aeroplane-airship) type – a wing-shaped construction with dropped hulls; a significant part of the volume of the wing and the dropped hulls is filled with a

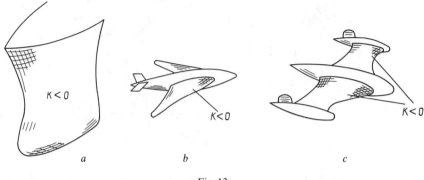

Fig. 13

light gas, which constantly creates a real lifting force. In motion there is also the aerodynamic lifting force of the wing. Such a construction is a thin shell on which a significant part of its area is taken up by parts of negative curvature. In passing we note that problems of rigidity of surfaces with curvature of varying sign in the analytic formulation lead to equations of mixed type. In this connection see Bakievich (1960), and also Ivanova-Karatoprakdieva (1983), Ivanova-Karatoprakdieva (1984a), Ivanova-Karatoprakdieva (1984b), Ivanova-Karatoprakdieva (1985), and Ivanova-Karatoprakdieva (1988) in Part III.

One advantage of convex shells, widely used in practice, is that under load they can play the role of supporting elements of the construction. As for the parts that have negative curvature, in engineering practice they are usually supported by additional struts. For optimization of the construction and the position of the struts on these parts of the shell it would be very useful to have a theory of surfaces of negative curvature if it were sufficiently far advanced.

We should bear in mind that the problem of the connection between the mechanics of thin shells and the geometry of surfaces in the cases of shells of negative curvature and curvature of varying sign is still largely awaiting development. It is known, for example (see Vekua (1959) in Part I), that a strictly convex shell can be calculated according to the moment-free theory if it is geometrically rigid. More precisely, for strictly convex but gently sloping shells, fixed at the boundary, along with the concepts "rigid – non-rigid" we should need to have concepts in geometry that more delicately and quantitively distinguish the amount of non-rigidity of the surface and that adequately reflect the corresponding properties of mechanical shells. Some approaches in this direction were made in Gol'denveizer (1979) and Gol'denveizer, Lidskij and Tovstik (1979). In addition, Gol'denveizer drew attention to the following fact: although an ordinary circular torus is geometrically rigid, a shell in the form of a circular torus can be loaded with a system of external forces so that its equilibrium position cannot be moment-free. This means that the connection between mechanical and geometrical rigidity of shells in the general case is more complicated than for strictly convex shells, and needs further investigation.

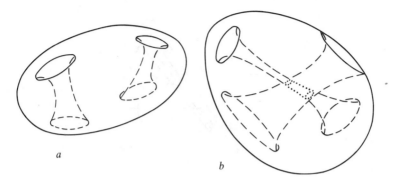

a

b

Fig. 14

Toroidal surfaces of revolution with convex meridian, discussed in the previous subsection, occur in the class of "*T-surfaces*", distinguished by Aleksandrov; see Aleksandrov (1938). Referring the reader to Aleksandrov (1938) for the exact definition, we explain the intuitive crux of the matter. Suppose we are given, in E^3 a closed convex surface on which there is an even number of flat domains. Let us cut them out. We join the resulting apertures in pairs by tubes of negative curvature, situated inside the original convex body. We obtain a "*T*-surface", Fig. 14a. We allow the possibility that the tubes of the "*T*-surface" intersect one another (Fig. 14b). In the analytic case Aleksandrov proved the unique determination of the "*T*-surface".

Nirenberg studied the non-analytic case; see Nirenberg (1963). He needed to impose a series of additional conditions, among them the following: it is required that each tube of the "*T*-surface" contains at most two closed asymptotic lines. Of course, this is satisfied it the tubes themselves are surfaces of revolution (see 2.1.2 above). Some sufficient conditions for the absence of closed asymptotic lines on a surface of negative curvature homeomorphic to a circular annulus can be found in Kantor (1980), but in the general case the question of the possible existence of closed asymptotic lines on the tubes of a "*T*-surface" remains unsolved at present.

It would be interesting to investigate the unique determination, non-bendability and rigidity of tapering surfaces of strictly negative and non-positive curvature. No results in this direction are known to the author. By analogy with "*T*-surfaces" we can expect that in this connection for tapering surfaces questions of the structure of their asymptotic net will also arise. The example of Vaigant (see part I, p. 55 and Fig. 10) visually demonstrates the non-triviality of this problem area.

2.1.4. Let us continue the discussion of examples. All *ruled* and all *minimal* surfaces in E^3 have $K \leqslant 0$. In passing we mention, without going into details, that ruled surfaces, and also surfaces of both constant and variable negative curvature, play an important role in the theory of congruences (Finikov (1950)).

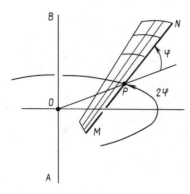

Fig. 15

When studying surfaces with $K < 0$ we need to bear in mind that among them
there are unorientable ones. Thus, suppose we are given a circle, and through its
centre 0 we draw the perpendicular AB to its plane; let P be a moving point on
the circle and suppose that a line MN is drawn through P in the plane PAB. We
shall assume that the line MN in the plane PAB is uniformly rotated about P
and the plane PAB is rotated about AB with twice the angular velocity. When
P completes a full circuit of the circle, the line MN returns to its original position,
but changing its direction to the opposite. We thus obtain an unorientable
surface homeomorphic to an infinitely wide Möbius band. The kinematic construction and a piece of this surface are shown in Fig. 15; the surface is complete;
simple calculation shows that its curvature K is negative everywhere.

The well-known *right helicoid* (Fig. 16) is a complete surface of curvature
$K < 0$ that is simultaneously ruled and minimal. One family of asymptotic lines
on it consists of its rectilinear generators. They are the common binormals of the
spirals situated on it, which in turn form the second family of asymptotic lines.
In general, if a ruled surface is formed by the binormals to a space curve with
non-zero curvature and torsion, then $K < 0$ at all points of it. At first glance such
a surface may appear quite complicated, but topologically it can only be one of
the following three types: a plane, a plane with a point deleted, and an infinitely
wide Möbius band; thus we assume that the curve itself is homeomorphic to a
line or a circle, and if it has a self-intersection in space, this means that the
positions in space of different points of the curve coincide; correspondingly we
understand a surface according to the definitions presented in 2.1.1. We draw
attention to the fact that among ruled surfaces formed by binormals there are
those on which there are *closed asymptotic lines*. This is easy to understand
visually. In fact, let us take a circular torus with external radius R formed by
rotating a circle of radius r, and make a uniform winding on it, in the same way
as a spiral winds on a circular cylinder (Fig. 17). Let h be the pitch of this
winding, measured along the greatest external parallel of the torus. First of all
we need $2\pi R = kh$, where k is a natural number. Then after a simple circuit of

Fig. 16

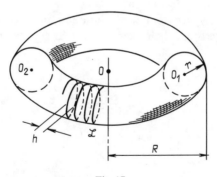

Fig. 17

the torus the winding (let us call it $\mathscr{L}$) closes and is homeomorphic to a circle. Next, for a suitable choice of parameters R, r, h, where $0 < h \ll r \ll R = kh/2\pi$, the winding $\mathscr{L}$ constructed on the torus will have non-zero curvature and torsion everywhere. Then the surface formed by its binormals has curvature $K < 0$; $\mathscr{L}$ itself is an asymptotic line on it.

The construction of a surface of negative curvature homeomorphic to a circular annulus, which may not be ruled but has a closed asymptotic line, is given in Kovaleva (1968).

In connection with the investigation of properties of "T-surfaces" mentioned above, the following more general questions about closed asymptotic lines arise:

If there is one, can there be another homotopic to it?

Can there be a family of homotopic closed asymptotic lines?

Can they completely cover a piece of the surface?

Can there be closed asymptotic lines on a horn, pointed or not, and can they exist on a tube bounded on two sides by parabolic lines?

In fact such tubes, but not completely arbitrary ones, are part of "T-surfaces".

As far as the author knows, none of these questions has been investigated so far.

In E^3 there are surfaces of negative curvature, including complete ones, with a more complicated topological structure than in the examples considered above. Apparently the first to draw attention to this was Hadamard; see Hadamard (1898) in Part I. The geometrical constructions that he presented are described in Part I, p. 49; see Fig. 5 there. The idea of Hadamard that was furthest developed in the work of other authors was that we should first construct a piecewise smooth surface having negative curvature in its regular parts. Then we carry out a smoothing – we construct a surface of the necessary smoothness close to the original and having $K < 0$ everywhere.

We draw attention to the fact that the possibility or impossibility of carrying out such a smoothing is by no means always obvious. For example, in Hadamard's construction why not take as the original piecewise smooth surface not one formed from two hyperboloids but one formed from a hyperboloid and a pseudosphere, as shown in Fig. 18?

In fact, we take a hyperboloid of revolution of one sheet and a pseudosphere situated in such a way that their axes intersect at right angles, and the circular

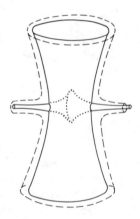

Fig. 18

edge of the pseudosphere is entirely inside the throat of the hyperboloid. As in Hadamard's construction, from each of these two surfaces we remove those parts of it that lie inside the other. It is intuitively quite obvious what "inside" means here. There remains a piecewise smooth ·surface that has two pointed *horns*, from the pseudosphere, and two *bowls*, from the hyperboloid. We now try to "smooth" it so that the resulting surface (schematically shown by a dotted line in Fig. 18) coincides with the original surface outside a ball of sufficiently large radius, it has smoothness at least C^2, and the curvature K is negative everywhere. It turns out that this is impossible, even though $K < 0$ on both the hyperboloid and the pseudosphere. The proof of impossibility is based on a calculation of the index of the fields of asymptotic directions on the horns and the bowls of the original piecewise smooth surface.

2.1.5. Let us discuss another example of a surface with boundary to which we shall need to refer below. Bianchi proved (Bianchi (1927)) that in E^3 there is a surface (with boundary) of constant curvature $K = -1$ on which there are two intersecting rectilinear generators. Amsler investigated this surface in more detail in Amsler (1955). In particular, he showed that the angle ω_0 between its rectilinear generators at a point of intersection can be any angle in the interval $0 < \omega_0 < \pi$, and the boundary of the surface consists of four components homeomorphic to a straight line, and each of the components, as it goes off to infinity in E^3, coils on a ray of the rectilinear generator. See Fig. 19, borrowed from Amsler (1955).

Now imagine that two copies of the Bianchi-Amsler surface shown in Fig. 19 are taken, placed one on the other in E^3 and glued along the boundary, and then air is pumped into the space between them. Suppose that both copies of the Bianchi-Amsler surface, glued along the boundary, are deformed, possibly changing their intrinsic metric, but so that in a neighbourhood of the gluing curve the surface becomes C^r-smooth, $r \geqslant 2$. Then in E^3 there is formed a surface with four pointed horns whose limiting rays lie in one plane. From Chapter 3 of Part I it follows that this cannot be a saddle surface: from it we can certainly cut out a *crust* (see Fig. 12 in Part I).

However, it is possible to alter the construction: simultaneously with an "inflation", which moves the glued copies apart, we "break" the rectilinear

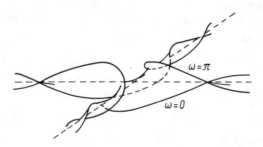

Fig. 19

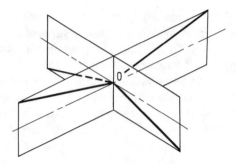

Fig. 20

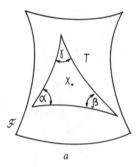

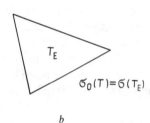

$$\sigma_0(T) = \sigma(T_E)$$

a   b

Fig. 21

generators at a point of intersection so that each of them turns into sides of an obtuse angle close to a straight angle, the planes of these angles are perpendicular to the plane in which the rectilinear generators originally lay, and the interior regions of the obtuse angles lie on opposite sides of this plane (see Fig. 20). Then the limiting rays of the horns will not lie in one plane; hence the obstacle connected with the inevitable cutting out of crusts and which does not make it possible to ensure universal negative curvature is removed. This makes it possible, by considering the specific form of the Bianchi-Amsler surface, to conjecture that in this way we can construct an example of a complete tapering surface of curvature $K < 0$ with four pointed horns, similar to Vaigant's example, see Part I, p. 56; however, this conjecture has still not been verified. We discuss the role of this hypothetical example in 2.3.4 below.

2.1.6. Suppose that on a surface $\mathscr{F}$ we are given a triangle T formed by arcs of shortest curves, where $\sigma(T)$ is its area and α, β, γ are its internal angles (Fig. 21a). Then the quantity

$$\omega(T) = \alpha + \beta + \gamma - \pi \tag{2.6}$$

is called the *excess of the triangle T*. From the Gauss-Bonnet formula it follows that the ratio $\omega(T)/\sigma(T)$ tends to the Gaussian curvature $K = K(X)$ of the surface $\mathscr{F}$ at a point X when the triangle T contracts to X. The usual proof uses C^3-smoothness of the surface (see Blaschke (1930)), and the standard approximation arguments enable us to extend the proof to C^2-smooth surfaces; see Efimov (1949) in Part I.

Let $\sigma_0(T)$ be the area of the triangle T_E on E^2 that has sides of the same length as T, Fig. 21b. We note that if $\mathscr{F} \in C^2$, then $\sigma(T)/\sigma_0(T) \to 1$ as $T \to X$, and the limit of the ratio $\omega(T)/\sigma_0(T)$ (which depends only on the intrinsic metric of the surface) can exist with smaller smoothness than C^2. According to Aleksandrov (see Aleksandrov (1948) in Part I), the *intrinsic Gaussian curvature* K_{int} of the surface $\mathscr{F}$ at the point X is

$$K_{\mathrm{int}} = \lim_{T \to X} \frac{\omega(T)}{\sigma_0(T)}, \qquad (2.7)$$

if this limit exists. Aleksandrov proved (see Aleksandrov (1948) in Part I) that if the limit (2.7) exists for all points of some domain $G \subset \mathscr{F}$, then K_{int} is continuous in G. In the multidimensional case the quantity (2.7) found for a specific two-dimensional direction is a generalization of the concept of sectional curvature.

2.1.7. Example. Let us take a hyperbolic paraboloid, which at the point of intersection O of its planes of symmetry has an angle π/m ($m \geqslant 3$) between the rectilinear generators. At O we draw the tangent plane P to the paraboloid. Of the four parts into which it splits the paraboloid we take the one that is inside the angle π/m, and denote it by V_m. We pave the plane P around O by figures congruent to V_m, taking $2m$ of them, and placing them in turn on different sides of the plane P. We obtain a $C^{1,1}$-smooth piecewise analytic surface whose C^2-smoothness is violated at O and on all the rays l_j ($j = 1, \ldots, 2m$) that separate neighbouring figures V_m from one another. We can show that on the surface obtained by this construction K_{int} exists at all its points; at O and on the rays l_j the curvature K_{int} has the same numerical value as the Gaussian curvature K had at these points on the paraboloid, so $K_{\mathrm{int}} < 0$. For $m = 3$ the surface is shown in Fig. 22. Its asymptotic lines, intersecting the rays l_j, form angular points. The projection of the asymptotic net on the plane P is shown in Fig. 23a (also for $m = 3$). We note that when m is even every normal section at O is either a straight line or a parabola, so at O the curvature of normal sections exists in all directions. We shall use this remark in a later subsection.

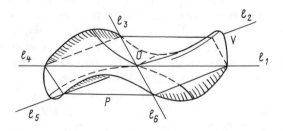

Fig. 22

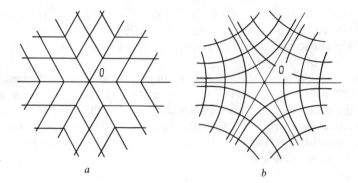

a   b

Fig. 23

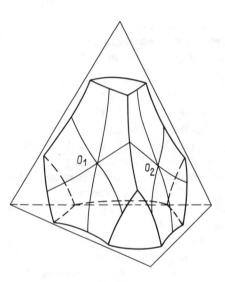

Fig. 24

2.1.8. We can now discuss in more detail the construction of the surface shown in Fig. 5a. First of all, for $m = 3$ we need to take the piecewise analytic $C^{1,1}$-smooth surface whose construction was described in the previous subsection, and subject it to an additional smoothing in a neighbourhood of the open rays l_j. We thus succeed in achieving smoothness C^∞ outside O, preserving the previous value $K_{\text{int}}(O)$ and the inequality $K_{\text{int}} < 0$. The angular points of the asymptotic lines on the rays l_j are also smoothed and the asymptotic net acquires the structure shown schematically in Fig. 23b. We then take a regular tetrahedron, take the plane of a face of it for P, and proceed like this for each of its faces. For a suitable position of the "segments" V_m, after cutting out surplus pieces we obtain a non-smooth saddle surface with boundary, homeomorphic to a sphere with four holes, shown in Fig. 24. For this surface with boundary its edges (situated in neighbourhoods of the edges of the tetrahedron) are smoothed. Then to each component of the boundary we add on a pointed horn, and the transition to it is also smoothed. The details of the construction are presented in Rozendorn (1962) and Rozendorn (1966) in Part I. In the construction we succeed in ensuring not only the existence of K_{int} and the fact that an inequality of type (1.7) is satisfied, but also the differentiability of K_{int} on the whole surface.

To obtain the surface shown in Fig. 5b (which we discussed in 1.4) we need to take $m = 4$, and instead of the tetrahedron to use a cube and carry out similar constructions.

2.1.9. Aleksandrov proved (see Aleksandrov (1948) in Part I) that if the quantity (2.7) exists on a two-dimensional manifold with intrinsic metric, then the metric is Riemannian and in *polar-geodesic* coordinates (ρ, φ), $\rho \geqslant 0$, it takes the form

$$ds^2 = d\rho^2 + B(\rho, \varphi)^2 d\varphi^2, \tag{2.8}$$

where the function $B(\rho, \varphi)$ is twice differentiable with respect to ρ, it is connected with the intrinsic curvature K_{int} by the differential equation

$$B''_{\rho\rho} + (K_{int})B = 0 \tag{2.9}$$

and when $\rho = 0$ it satisfies the initial conditions

$$B(0, \varphi) = 0, \qquad B'_\rho(0, \varphi) = 1. \tag{2.10}$$

In particular, for the Lobachevskij plane L^2 the intrinsic curvature exists and is constant, since excesses of triangles in L^2 are proportional to their areas, and $\sigma(T)/\sigma_0(T) \to 1$ when the triangle T contracts to a point. According to (2.8), (2.9) and (2.10), the metric on L^2 in polar coordinates has the form

$$ds^2 = d\rho^2 + \text{sh}^2\left(\frac{\rho}{R}\right) d\varphi^2, \tag{2.11}$$

where R is the "radius of curvature" and $K_{int} = -R^{-2} = \text{const} < 0$.

From (2.11) it is obvious that on L^2 the length of a circle and the area of a disc increase exponentially as the radius ρ increases. From (2.8), (2.9) and (2.10) it is also obvious that on two-dimensional manifolds whose intrinsic curvature in the sense of Aleksandrov exists and satisfies the inequality

$$K_{int} \leqslant \text{const} < 0, \tag{2.12}$$

the length of a geodesic circle and the area of a geodesic disc have no less than exponential growth as ρ increases. In an attempt to immerse such manifolds isometrically in E^3 (under the condition that they are complete and the immersion is C^2-smooth) the space E^3 turns out to be too tight for them according to Theorem B (see 1.4).

In the more general case, when $K \leqslant 0$, the intrinsic geometry of complete surfaces (and also of non-compact two-dimensional Riemannian manifolds) was studied by Eberlein (Eberlein (1979)).

2.2. Some Remarks on a C^1-Isometric Embedding of L^2 in E^3 According to Kuiper

2.2.1. The picture changes significantly if instead of C^2-smoothness we restrict ourselves to the assumption of just C^1-smoothness. Namely, according to Kuiper (1955), to obtain an embedding of a given metric ds^2 it is sufficient to construct first for it a so-called *strictly short embedding*, that is, a surface $\mathscr{F}_1$ whose metric ds_1^2 satisfies the requirement that the differential quadratic form $ds^2 - ds_1^2$ is positive definite. As applied to the specific problem of embedding L^2 in E^3 it is sufficient to take for $\mathscr{F}_1$ the Euclidean plane E^2 with polar coordinates (r, ψ) and put $r = \frac{1}{2}\rho$, $\psi = \varphi$. Then under the parametrization $x_1 = r \cos \psi$, $x_2 = r \sin \psi$, $x_3 = 0$ the plane E^2 is a strictly short embedding in E^3 for the metric (2.11). Another, also strictly short, embedding in E^3 for the same metric (2.11) can be taken, for example, as the open disc $r < 1$ on E^2.

Next, according to Kuiper (1955), we carry out corrugations, as a result of which there arises a sequence of surfaces with metrics $ds_n^2 \to ds^2$. In the specific example of constructing an embedding for L^2 we can imagine that the plane $E^2 = \mathscr{F}_1$ is covered by waves, from which shallower waves go out in different directions, from them still shallower waves, and so on, as happens on the surface of the sea in the presence of ripples and wind. If for the original short embedding for L^2 we use not the whole plane E^2, but the disc $r < 1$ cut out from it, then these waves must be "denser" and "steeper" close to its boundary, the circle $r = 1$.

The fact that under corrugations the metric of the original short embedding $\mathscr{F}_1$ is lengthened is obvious. Much less obvious is the fact that it can be "tightened" so much that in each compact part in all directions the difference $ds^2 - ds_n^2$ tends to zero. If an immersible metric is C^∞-smooth (as in the example of L^2 under consideration), then all surfaces of the sequence $\{\mathscr{F}_n\}$ can also be made C^∞-smooth.

Not at all obvious and very unexpected for geometers in its time was the fact that in this way we can ensure the existence of a C^1-smooth limiting surface on which the ambient space induces a given metric ds^2.

2.2.2. Let us recall (see Vinberg and Shvartsman (1988) or Novikov and Fomenko (1987), for example) that by factorizing L^2 we can obtain compact manifolds with a metric of constant negative curvature. The simplest of them is obtained from an equilateral octagon in L^2 having angles of $45°$ by gluing the sides as shown schematically in Fig. 25. Topologically this is a sphere with two handles, or a pretzel, as we sometimes say; see Fig. 26.

A pretzel with a metric of constant negative curvature, like other compact orientable two-dimensional Riemannian manifolds, admits a C^1-smooth isometric embedding in E^3 by Kuiper's method. We draw attention to the fact that in this way in E^3 we obtain a closed surface whose intrinsic curvature K_{int} is defined at all its points, and $K_{int} = \text{const} < 0$. From the visual point of view this surface must apparently be represented as a pretzel, like that shown in Fig. 26, but covered by shallow waves.

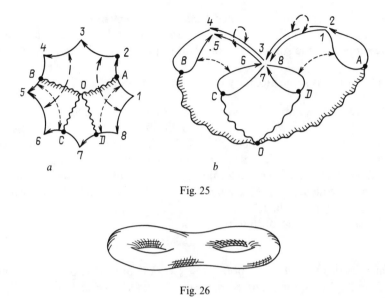

Fig. 25

Fig. 26

Kuiper's constructions show that Theorems A and B, formulated in §1, both lose their force in the class of regularity C^1. In this connection it is of interest to study classes of surfaces intermediate between C^2 and C^1. We shall return to this question frequently later.

2.3. Hadamard's Conjecture

2.3.1. Theorem. *In E^3 there are no C^2-smooth closed surfaces with a metric of non-positive curvature.*

Corollary. *Let $\mathscr{F}$ be a C^1-smooth Kuiper embedding (or immersion) in E^3 of a compact (without boundary) two-dimensional Riemannian manifold with metric ds^2 of non-positive curvature $K \leqslant 0$. Suppose also that $\mathscr{F}$ is obtained as the limit of short embeddings (respectively, immersions) $\{\mathscr{F}_n\}$ that are C^r-smooth surfaces, $r \geqslant 2$. Then all the $\mathscr{F}_n$ have Gaussian curvature of variable sign.*

Thus, on all the $\mathscr{F}_n$ there are points where $K > 0$, although their metrics ds_n^2 converge to a metric ds^2 of non-positive (or even strictly negative) intrinsic curvature.

The proof of the theorem is by contradiction: suppose such a surface $\mathscr{F}$ exists. We include its support $\hat{\mathscr{F}}$ in a sphere of least possible radius. This sphere will have a point A in common with $\hat{\mathscr{F}}$. Parallel to the tangent plane to the sphere at A we draw a sufficiently close cutting plane. It cuts off a crust from $\mathscr{F}$, which is impossible, since $\mathscr{F}$ is a saddle surface.

This argument is actually due to Hadamard, who assumed (see Hadamard (1898) in Part I) that he had proved the following assertion:

In E^3 there are no bounded complete surfaces of negative curvature.

This assertion is now often called *Hadamard's conjecture*.

Below we discuss in detail why the argument about the smallest ball containing the support of the surface is insufficient to prove the validity of Hadamard's conjecture.

We first discuss some auxiliary definitions and examples.

2.3.2. Suppose we are given a surface $\mathscr{F}: \mathfrak{M} \to E^N$, where $\mathfrak{M}$ is a manifold of dimension p, $2 \leqslant p < N$ (understood up to diffeomorphisms), and $X = \{X', X''\}$ is an arbitrary point of $\mathscr{F}$, where $X' \in \mathfrak{M}$, $X'' \in E^N$. Suppose that the surface $\mathscr{F}$ is incomplete in its intrinsic metric. In the standard way we complement $\mathscr{F}$ to a minimal complete metric space $\overline{\mathscr{F}}$, taking as points of $\overline{\mathscr{F}}$ classes of equivalent fundamental sequences on $\mathscr{F}$. Then $\partial\mathscr{F} = \overline{\mathscr{F}} \backslash \mathscr{F}$ is the *metric boundary* for $\mathscr{F}$.

Having a surface $\mathscr{F}$ in E^N, we can consider a map, which we denote by the same letter $\mathscr{F}$, acting according to the rule

$$\mathscr{F} \to E^N, \qquad X = \{X', X''\} \to X'' \in E^N. \tag{2.13}$$

This map associates with each point $X = \{X', X''\}$ of $\mathscr{F}$ that point of E^N where X "is found", that is, the point X'' on the support $(X'' \in \hat{\mathscr{F}} \subset E^N)$.

Because distance on the surface is always at least equal to that in the ambient space, the map (2.13) can be extended by continuity from $\mathscr{F}$ to $\overline{\mathscr{F}}$. As a result in E^N there arises the image $\hat{\partial\mathscr{F}}$ of the metric boundary $\partial\mathscr{F}$. We shall call $\hat{\partial\mathscr{F}}$ the *intrinsic geometric boundary* of $\mathscr{F}$ in E^N.

Example. $x_1 = a\,\mathrm{ch}\,t\cos\varphi$, $x_2 = a\,\mathrm{ch}\,t\sin\varphi$, $x_3 = b\,\mathrm{sh}\,t$; $a > 0, b > 0$; $|t| < t_1$, $-\infty < \varphi < +\infty$. This surface is the universal covering of part of the hyperboloid of revolution of one sheet

$$\frac{x_1^2}{a^2} + \frac{x_2^2}{a^2} - \frac{x_3^2}{b^2} = 1 \tag{2.14}$$

in E^3. Its intrinsic geometric boundary consists of two circles $x_1^2 + x_2^2 = (a\,\mathrm{ch}\,t_1)^2$, $x_3 = \pm B$, $B = b\,\mathrm{sh}\,t_1$, lying in parallel planes. The metric boundary $\partial\mathscr{F}$ is homeomorphic (and isometric) to a pair of straight lines, which under the map (2.13) extended to $\overline{\mathscr{F}}$ wind in E^3 on these parallel circles.

Now consider on $\mathscr{F}$ all possible sequences of points $\{X_n\}$, $X_n = \{X'_n, X''_n\}$, that are not convergent on $\mathscr{F}$ but for which $\{X''_n\}$ are fundamental in E^N. All limiting points of such sequences $\{X''_n\}$ form in E^N a closed set (possibly empty), which we call below the *extrinsic geometric boundary* of $\mathscr{F}$ in E^N. Clearly, the intrinsic geometric boundary is part of the extrinsic geometric boundary. Thus, in the previous example the extrinsic geometric boundary fills a whole piece $|x_3| \leqslant B$ of the hyperboloid (2.14). Of course, the extrinsic geometric and intrinsic geometric boundaries may coincide. Thus, for example, if for $\mathscr{F}$ we take an open disc in E^2, both boundaries then coincide with the circle. If $\mathscr{F} = E^2$, then both boundaries (the intrinsic and the extrinsic) are empty. But the extrinsic geometric boundary may be non-empty when $\mathscr{F}$ is complete (and consequently $\hat{\partial\mathscr{F}} = \varnothing$). Thus, if we alter the previous example and instead of the condition $|t| \leqslant t_1$ we

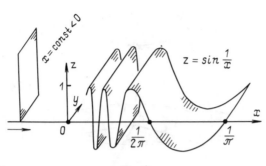

Fig. 27

allow $-\infty < t < +\infty$, that is, we take for $\mathscr{F}$ the universal covering of the whole hyperboloid (2.14), then the extrinsic geometric boundary of this covering fills the whole hyperboloid (2.14). This is a rather obvious example; here for $\mathscr{F}$ its extrinsic geometric boundary coincides with the support $\widehat{\mathscr{F}}$ because the number of sheets of the covering is infinite. However, we can give a rather different but also completely visual example, when the extrinsic geometric boundary of $\mathscr{F}$ is not empty and does not intersect $\widehat{\mathscr{F}}$, and $\mathscr{F}$ is complete. Namely, in E^3 we consider the surface

$$x_3 = \sin \frac{1}{x_1}, \qquad x_1 > 0, \qquad -\infty < x_2 < +\infty. \tag{2.15}$$

The surface (2.15) does not have self-intersections, so we can assume that it coincides with its own support, and its extrinsic geometric boundary consists of a closed strip $|x_3| \leqslant 1$ on the vertical (x_2, x_3)-plane (see Fig. 27).

Turning to a discussion of Hadamard's conjecture, we draw attention to the fact that, for a bounded complete surface, by contracting the ball containing it we can meet its sphere not at a point of the surface but at a point of its extrinsic geometric boundary, and if the ball is contracted further, it may happen that its bounding sphere cuts out not spherical crusts but parts of the surface that are non-compact in the intrinsic metric. We see a similar picture if in Fig. 27 we take in the left half-space $x_1 < 0$ a vertical plane $x_1 = $ const and start to move it to the right. When we get to the position $x_1 = 0$, this plane still does not meet the surface $\mathscr{F}$ itself, but only its extrinsic geometric boundary, and a further (arbitrarily small) shift of this plane to the right leads to a non-compact part of the surface between the plane and the extrinsic geometric boundary. By the way, the surface (2.15) is developable and therefore a saddle surface (like any C^2-smooth developable surface in E^3). Of course, in the case (2.15) $K = 0$, and the plane in the argument above cannot be replaced by a sphere. But before discussing a more complicated example, to which the next subsection is specially devoted, let us draw attention to the fact that if the argument that justifies the impossibility of a closed surface of non-positive curvature $K = k_1 k_2 < 0$ in E^3 were also a justification of Hadamard's conjecture, then in Hadamard's conjecture itself

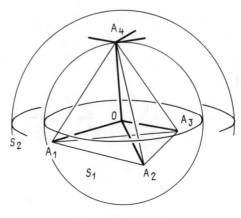

Fig. 28

there would be no need to talk about K being negative: the conjecture would be true for any saddle surfaces in E^3, and this is certainly not so: see 2.3.3.

2.3.3. Example. *A Bounded Complete Saddle Surface in* E^3. Let $\{S_n\}$ be a sequence of concentric spheres in E^3 with common centre O and monotonically increasing radii r_n, where $r_n \to r < +\infty$, so there is a limiting sphere with the same centre O and radius r. In the sphere S_1 we inscribe a regular tetrahedron $A_1 A_2 A_3 A_4$ and draw all the segments OA_j. Then from each point A_j we draw three segments inside the spherical layer between S_1 and S_2. Fig. 28 shows the tetrahedron $A_1 A_2 A_3 A_4$, the segments $OA_j, j = 1, \ldots, 4$, and the three segments going out from one of the vertices of the tetrahedron inside this spherical layer; each of these segments is extended until it meets S_2, and from there three more branch out, and so on.

We shall build up the resulting branched spatial open polygon tier by tier, making sure that at each point of intersection of the polygon with the spheres S_n all four segments make obtuse angles with one another.

In addition, we take care that no self-intersections of the polygon arise. This is not difficult to do: segments can only meet each other in one spherical layer, and if such intersections occur, they need to be removed by a small displacement of the segments before proceeding to the construction of the next tier.

Elementary calculation shows that if $r_{n+1} - r_n = O(1/n^2)$, then the lengths of the segments situated between S_n and S_{n+1} can be made of order $O(1/n)$. It then turns out that all the paths going along the open polygon, more precisely along the infinite graph obtained from this polygon as a result of a countable process of construction from some point of it to the limiting sphere S, have infinite length. In other words, the ambient space E^3 induces a metric on this graph in which it is a complete metric space.

The part of the construction described here is elementary. Next, imagine that each of the segments occurring in the constructed graph is slightly inflated and turns into a tube of negative curvature, and those points where the segments meet in fours are also inflated and become surfaces like that shown in Fig. 5a. Around the central point O and on the first layer, around the points $A_1, \ldots, A_4$ shown in Fig. 28, we can use surfaces as in 2.1.8, and for use as fragments of the construction on tiers further from the centre O we first need to subject them to affine transformations chosen in a suitable way.

As a result we obtain a bounded and at the same time complete saddle surface that is C^1-smooth and even $C^{1,1}$-smooth. In addition, since we have used as the original material for fragments of the construction the surfaces concerned in 2.1.8, here also we can ensure the existence of an intrinsic curvature K_{int} that satisfies an inequality of the form (1.7), and the differentiability of K_{int}. The resulting surface is infinitely connected; all its extrinsic geometric boundary is concentrated on the limiting sphere S, and any concentric sphere of smaller radius cuts off from the surface a non-compact part, although it is impossible to cut off a crust by any plane. Part of the surface thus constructed is shown in Fig. 29. The reader can find details of this construction in Rozendorn (1981) in Part I.

Thus, if in Hadamard's conjecture we replace K by K_{int} and instead of C^2-smoothness of the surface we assume $C^{1,1}$-smoothness, then the conjecture in this form is false.

2.3.4. In connection with what we have said, let us make some more remarks. It is known (see Xavier (1984), for example) that in E^3 there are C^∞-smooth bounded complete surfaces of non-positive Gaussian curvature. Of course,

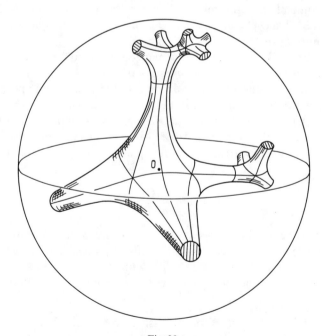

Fig. 29

sup $K = 0$ on them, by Theorem B of 1.4. In addition, inf $K = -\infty$ on them, in view of the next theorem.

Theorem (Baikoussis and Koufogiorgos (1980)). *A complete C^∞-smooth surface with curvature $-\infty < -a^2 \leqslant K \leqslant 0$ in E^3 is unbounded.*

This theorem is one of the most powerful results in the problem area connected with Hadamard's conjecture. As for Hadamard's conjecture itself, at present (1990) there are no counterexamples to it, but it has not been proved. Moreover, arguments were given in Part I that enable us to cast doubt on it. Namely, we propose to study the possibility of a construction similar to that in the previous subsection by using in the details of the construction the surfaces obtained by affine transformations from Vaigant's example. The difficulty of realizing this plan lies in constructing transitional tubes of negative curvature between them. It may happen that it is constructively simpler to use parts of tapering surfaces constructed on the basis of the Bianchi-Amsler example; see 2.1.5 above. If these constructions can finally be carried out, making sure that the curvature is negative on the whole surface, then Hadamard's conjecture will be disproved.

2.4. Surfaces of Negative Intrinsic and Bounded Extrinsic Curvature in E^3

2.4.1. In the theory of surfaces an important role is played by the *Gaussian spherical map*. For a hypersurface $\mathscr{F}$ in E^N it associates with each point of the surface a point of the unit sphere S^{N-1} according to the rule $\Phi: \mathscr{F} \to S^{N-1}$, $Y = \Phi(X) \in S^{N-1}$, if $OY = \mathbf{n}(X)$, where O is the centre of the sphere and $\mathbf{n} = \mathbf{n}(X)$ is the unit normal vector to $\mathscr{F}$ at the point X.

For surfaces in E^3 by means of a map onto the sphere Gauss (see Gauss (1823–1827)) introduced the *total curvature* of a surface as the area $\Omega(F)$ of the image $\Phi(F)$ on S^2, found by taking account of the orientation and multiplicity of the covering, and the curvature of the surface at a point X as the limit of the ratio $\Omega(G)/\sigma(G)$ when the domain $G \subset \mathscr{F}$ is contracted to the point X. Here $\sigma(G)$ denotes the area of the domain G on $\mathscr{F}$. An easy calculation shows than in the C^2-smooth case this limit exists and is equal to the product of the principal curvatures:

$$K = \lim_{G \to X} \frac{\Omega(G)}{\sigma(G)} = k_1 k_2. \tag{2.16}$$

The problems of constructing a surface when we are given information about its spherical image are at the very first stage of study when $K < 0$. Some results in this direction for the simplest situations (Φ one-to-one, $\Phi(\mathscr{F})$ is contained in a hemisphere) were obtained in Fomichëva (1978), Fomichëva (1979a), Fomichëva (1979b).

2.4.2. Definiton. A point X on a C^1-smooth surface $\mathscr{F}$ in E^3 is said to be *Pogorelov regular* if there is a neighbourhood $U \subset \mathscr{F}$ of it such that in $U \backslash X$ the tangent plane is never parallel to the tangent plane at X; a surface is said to be *quasiregular* if each point of it is Pogorelov regular.

Theorem (Efimov (1949) in Part I). *The intersection of a C^1-smooth surface with the tangent plane at a regular point X of it consists of a sufficiently small neighbourhood of it, or of the single point X, or of X and $2m$ simple arcs starting from it. The small neighbourhood can be chosen so that these arcs split it into (curvilinear) sectors situated in turn on opposite sides of the tangent plane.*

The proof based on an application of the implicit function theorem.

The number $m = m(X)$ is called the *saddle order* of $\mathscr{F}$ at X. For regular points of a C^1-smooth curface in E^3 Pogorelov (see Pogorelov (1969) in Part I) suggested the following classification: a point is *elliptic* if $m = 0$, *parabolic* if $m = 1$, *hyperbolic* if $m = 2$, and a *flat point* if $m \geqslant 3$. We need to bear in mind that this classification is not quite the same as the usual classification of points of a C^2-smooth surface according to the type of osculating paraboloid; however, when $K < 0$ in the C^2-smooth case a point is hyperbolic both according to the type of osculating paraboloid and according to the saddle order.

Let G be an arbitrary open domain on $\mathscr{F}$ and $\sigma^*(G)$ the Lebesgue measure of its image $\Phi(G)$. Here $\Phi(\mathscr{F})$ must be understood either as a surface in E^3 whose support is a subset of S^2 or as a two-dimensional surface in the two-dimensional space S^2. This means that σ^* is defined by taking account of the multiplicity of the covering; in contrast to the quantity Ω, here we do not take account of the orientation: if $\Phi(G)$ is measurable and has non-zero Lebesgue measure, then $\sigma^*(G) > 0$. The *extrinsic Gaussian curvature* of $\mathscr{F}$ at a Pogorelov regular point X can be defined by the formula

$$K_{\text{ext}} = \lambda(X) \lim_{G \to X} \frac{\sigma^*(G)}{\sigma(G)}, \tag{2.17}$$

where $\lambda(X) = \operatorname{sgn}(1 - m(X))$, if this limit exists. If, besides regularity, we assume C^2-smoothness, then the quantities (2.17) and (2.16) coincide. For a Pogorelov non-regular point, in the C^2-smooth case we can take $\lambda(X) = 1$. Then the classical *Gauss theorem* (for a C^2-smooth surface) can be written in the form of the equalities

$$K = k_1 k_2 = K_{\text{ext}} = K_{\text{int}}. \tag{2.18}$$

However, K_{ext} may exist under smaller smoothness than C^2.

2.4.3. Using the Gaussian spherical map, Pogorelov distinguished and investigated the class of *surfaces of bounded extrinsic curvature* intermediate between C^2 and C^1.

Suppose we are given in E^3 a C^1-smooth surface $\mathscr{F}$, let $H_1, \ldots, H_r$ be closed sets on $\mathscr{F}$, and $\hat{\sigma}(H_j)$ the Lebesgue measures on the sphere S^2 of the images $\Phi(H_j)$, understood there as point sets (that is, as sets of points on the support $\hat{\Phi}(\mathscr{F})$ of the surface $\Phi(\mathscr{F})$).

Definition. A surface $\mathscr{F}$ has *bounded Pogorelov extrinsic curvature* if $\hat{\sigma}(H_j)$ exist and

$$\bar{\Omega}(\mathscr{F}) = \sup_j \sum_j \hat{\sigma}(H_j) < +\infty \qquad (2.19)$$

over all possible finite systems of pairwise disjoint closed sets $\{H_j\}$ on $\mathscr{F}$.

Referring the reader for the details to §§ 1, 3 of Ch. 4 and § 4 of Ch. 1 of Part I, and also to the original source Pogorelov (1956b) in Part I (or the monograph Pogorelov (1975) in Part I), we present briefly here the facts concerning surfaces with metric of negative curvature and unsolved problems related to them.

Let W be an arbitrary Borel set on a surface $\mathscr{F}$ of bounded extrinsic curvature. Starting from (2.19), we construct $\bar{\Omega}(W)$. Then we introduce the *positive* and *negative* extrinsic curvatures $\Omega^+(W)$ and $\Omega^-(W)$ as the values of $\bar{\Omega}$ on the subsets of elliptic and hyperbolic points in W and the *total extrinsic curvature* $\Omega(W) = \Omega^+(W) - \Omega^-(W)$. It turns out that $\bar{\Omega}(W) = \Omega^+(W) + \Omega^-(W)$, and the subsets of parabolic points, non-regular points and flat points make a zero contribution to $\bar{\Omega}$, and the set of flat points on any surface of bounded extrinsic curvature is no more than countable. For an open set G in the C^2-smooth case $\Omega(G) = \iint_G K d\sigma$ and coincides with the Gauss total curvature (see 2.4.1), $\bar{\Omega}(G) = \iint_G |K| \, d\sigma$ ($d\sigma$ is the element of area on $\mathscr{F}$). For closed and for quasiregular surfaces of bounded extrinsic curvature Pogorelov proved the generalized Gauss theorem

$$\Omega^+(W) = \omega^+(W), \qquad \Omega^-(W) = \omega^-(W), \qquad (2.20)$$

where ω^+ and ω^- are the positive and negative parts of the intrinsic curvature (see Aleksandrov and Zalgaller (1962) in Part I), and he proved the first of the equalities (2.20) for any surface of bounded extrinsic curvature.

Pogorelov also proved that if a surface of bounded extrinsic curvature admits the cutting out of a crust, then the set of elliptic points on it is so "vast" that $\Omega^+ > 0$. From this and (2.20) we have the following corollary.

Corollary. *A Kuiper C^1-smooth isometric embedding in E^3 of a compact (without boundary) two-dimensional Riemannian manifold of non-positive Gaussian (intrinsic) curvature cannot be a surface of bounded extrinsic curvature.*

Next, from (2.20) and the *Gauss–Bonnet formula* for manifolds of bounded extrinsic curvature (see Aleksandrov and Zalgaller (1962) in Part I) it follows that if on a quasiregular surface of bounded extrinsic curvature K_{ext} exists at each point of some open domain G, or K_{int} exists at each point of G, then the second of these quantities exists in G, and the generalized Gauss theorem holds in a local form:

$$K_{ext}(X) = K_{int}(X), \qquad X \in G. \qquad (2.21)$$

In other words, the last equality in (2.18) carries over exactly to the special case considered here. This is how things stand in the example analysed in 2.1.7. The surfaces shown in Fig. 5, whose construction was discussed in 2.1.8, are also quasiregular surfaces of bounded extrinsic curvature, K_{ext} exists on them, and (2.21) is satisfied.

The case when one of the two quantities K_{int} or K_{ext} exists not in a domain but at an isolated point is more complicated. The question of whether the second of these quantities then exists and whether (2.21) is true under the original assumption that the surface has bounded extrinsic curvature has apparently not been investigated.

Turning to the surfaces shown in Fig. 5, we note that they have flat points in the sense of Pogorelov. Namely, a surface with four horns (Fig. 5a) has four of them and $m = 3$ there. The surface in Fig. 5b has six of them (the centres of the faces of the original cube) and $m = 4$ at each of them. These points are important in that the locally homeomorphic property of the Gaussian spherical map is violated there; in this connection see 3.3 and 4.3 below.

In the non-compact case it is natural to define surfaces of *locally bounded extrinsic curvature*, requiring that a condition of the form (2.19) holds for each domain having compact closure on the surface. Then it is clear that for Borel subsets of such domains the first of the equalities (2.20) holds, and under the additional assumption of quasiregularity so does the second. The question of the validity of the first of the equalities (2.20) on surfaces of bounded (or locally bounded) extrinsic curvature without the assumption of quasiregularity (which is apparently unsolved) reduces to the following: can the non-zero negative intrinsic curvature $\omega^- \neq 0$ be concentrated on the set of non-regular points?

As an example of a quasiregular surface of locally bounded extrinsic curvature we can mention the bounded (in E^3) complete saddle surface discussed in 2.3.3.

To conclude this section we draw attention to one more unsolved question: can Theorem A, discussed in § 1, be extended to the class of surfaces of locally bounded extrinsic curvature?

Let us make some comments on this question. As in 2.1.7, using instead of "segments" taken from hyperbolic paraboloids parts of the Bianchi-Amsler surface with $\omega_0 = \pi/m$, we can construct a surface that is an isometric embedding in E^2 of some open disc of L^2. This surface has bounded extrinsic curvature, is quasiregular, and is even $C^{1,1}$-smooth and piecewise analytic, and at O, by an isometry corresponding to the centre of the disc, it has a preassigned saddle order $m \geqslant 3$. It has a net of asymptotic lines that has the structure shown schematically in Fig. 23a. Hence it is obvious that in the given class of surfaces the proof according to the plan presented in § 1 does not go through.

The well-known proof of the more general Theorem B (see 3.3 below) also does not go through here because the locally homeomorphic property of the spherical map is violated.

Using as starting point for further constructions the embedding of a disc of L^2 in E^3 mentioned here, it is tempting to construct a counterexample on the plan of 2.1.8, for example. However, the existing lemmas on "smoothing" of surfaces and joining parts of them, keeping the curvature negative (see Rozendorn (1962) and Rozendorn (1981) in Part I), are insufficient to construct such a counterexample: they do not allow the possibility of ensuring that the intrinsic curvature is constant.

Hence to solve the question of whether Theorem A can be extended to the class of surfaces of locally bounded extrinsic curvature we need new approaches.

§ 3. Surfaces of the Form $z = f(x, y)$; Plan of the Proof of Efimov's Theorem

3.1. Some Results on Surfaces that Project One-to-one on the Plane E^2. The investigation of surfaces of the form

$$z = f(x, y) \tag{3.1}$$

is at the meeting point of geometry and function theory. In this subsection we discuss some results about surfaces of the form (3.1) defined when $-\infty < x < +\infty$, $-\infty < y < +\infty$.

3.1.1. *The Theorem of Bernstein and Adel'son-Vel'skij.* Suppose that a function $f(x, y)$ is defined and continuous on the whole (x, y)-plane, and that

$$\lim_{r \to \infty} \frac{1}{r} |f(x, y)| = 0, \qquad r = \sqrt{x^2 + y^2}. \tag{3.2}$$

If the surface (3.1) is a saddle surface, then it is cylindrical with generators parallel to the plane $z = 0$ (Adel'son-Vel'skij (1945), Bernstein (1960a), Bernstein (1960b)).

The proof relies on the definition of a saddle surface and is based on an analysis of its sections by different planes. We first establish that the surface has rectilinear generators. Then we prove that they are parallel to one another and to the plane $z = 0$ and cover the whole surface.

Corollary 1 (Liouville's theorem for a *harmonic* function). *Suppose that the function $f(x, y)$ is harmonic and bounded on the whole (x, y)-plane. Then it is constant.*

In fact, the graph of a harmonic function is a saddle surface. From the boundedness condition there follows (3.2), so the graph is a cylindrical surface with horizontal rectilinear generators. From this and the harmonic property it follows that the function is linear, and from the boundedness condition it follows that the function is constant.

Corollary 2. *A C^2-smooth surface of the form* (3.1) *of negative curvature $K < 0$, defined for $-\infty < x < +\infty$, $-\infty < y < +\infty$, cannot be situated between planes of the form $z = $ const.*

Remarks. 1) The geometrical fact stated here as Corollary 2 can be applied, in particular, to the investigation of unique determination of complete convex surfaces; see Pogorelov (1952a) in Part I.

2) If we take account of the results of Pogorelov (1956b) in Part I, then in the statement of Corollary 2 we can replace C^2-smoothness and the inequality $K < 0$ by the assumption that the surface is C^1-smooth, but has bounded extrinsic and non-positive intrinsic curvature: $\bar{\Omega} < +\infty$, $\omega^+ = 0$.

3.1.2. We now present some results about the rigidity of complete surfaces of negative curvature in which there is also a connection with function theory, and quite an unexpected one.

A function $W = W(\mathfrak{z}) = U + jV$ of a *hyperbolic complex variable* $\mathfrak{z} = x + jy$, where $j^2 = 1$, is said to be *h-analytic* if in some domain of variation of the argument $\mathfrak{z}$ it is differentiable with respect to $\mathfrak{z}$. For a continuous function $W(\mathfrak{z})$ a criterion for it to be h-analytic is the system of equations

$$U'_x = V'_y, \qquad U'_y = V'_x,$$

the analogue of the Cauchy-Riemann equations, and as in the case of analytic functions it can also be written briefly in the form

$$W'_{\bar{\mathfrak{z}}} = 0. \tag{3.3}$$

It is known (see Lavrent'ev and Shabat (1973)) that an h-analytic function can be represented in the form

$$W = \varphi_1(x + y) + \varphi_2(x - y) + j(\varphi_1(x + y) - \varphi_2(x - y)), \tag{3.4}$$

where φ_1 and φ_2 are arbitrary differentiable functions of one argument.

Definition (Ten (1980)). A complete surface is said to be *B-rigid* if from the fact that its bending field Z is bounded and vanishes at one point it follows that $Z = 0$ identically.

In other words, *B-rigidity* of a surface $\mathscr{F}$ means that it has no other bounded infinitesimal bending fields apart from parallel displacements. Hence for such a surface the condition that the bending fields are bounded, which it is sufficient to impose only at infinity (that is, outside an arbitrary compact set), replaces the condition that it is fixed, which guarantees rigidity.

On a surface $\mathscr{F}$ defined in rectangular Cartesian coordinates by an equation of the form (3.1), let us consider a bending field $Z = \{\xi, \eta, \zeta\}$ written in the same coordinates (x, y, z), and following Pogorelov (see Pogorelov (1969) in Part I) put $\lambda = \xi + p\zeta$, $\mu = \eta + q\zeta$, where

$$p = f'_x(x, y), \qquad q = f'_y(x, y), \tag{3.5}$$

and then introduce the function $W(\mathfrak{z}) = \lambda - j\mu$, where $\mathfrak{z} = x + jy$, $j^2 = 1$. Then for the special case of the hyperbolic paraboloid $f(x, y) = \frac{1}{2}(x^2 - y^2)$ the well-known *differential equation of infinitesimal bendings dr dZ = 0*, where $r = \{x, y, z\}$ is the radius vector of a moving point of the surface, takes the form (3.3).

Thus, h-analytic functions are connected with infinitesimal bendings of a hyperbolic paraboloid and parts of it. For comparison we recall (see Vekua (1959) in Part I) that ordinary analytic functions are connected with infinitesimal bendings of a sphere and parts of it.

Using (3.4), by similar arguments we can establish (see Ten and Fomenko in Seminar on geometry in the large (1986)) the following two facts. One is geometrical: a hyperbolic paraboloid is B-rigid. The other is analytic, the analogue of Liouville's theorem for an h-analytic function: if an h-analytic function $W(\mathfrak{z})$ is defined on the whole $\mathfrak{z}$-plane and $\lim_{\mathfrak{z}\to\infty} W(\mathfrak{z}) = A$, then $W(\mathfrak{z}) = A$. Here we can weaken the requirements and suppose that at infinity either Re $W = U(\mathfrak{z})$ or Im $W = V(\mathfrak{z})$ tends to a constant; this is also sufficient for $W(\mathfrak{z})$ to be constant on the whole $\mathfrak{z}$-plane. Examples show that in the usual formulation (constancy, which follows from boundedness) Liouville's theorem is false for an h-analytic function.

Using additional arguments, mainly concerned with the structure of the asymptotic net of the surface, we can establish a more general geometrical result.

Theorem (Ten (1980)). *Suppose that a C^3-smooth surface $\mathscr{F}$ of the form (3.1), where $-\infty < x, y < +\infty$, has curvature $K < 0$, and outside its compact part it coincides with a hyperbolic paraboloid. Then $\mathscr{F}$ is B-rigid.*

We note that the property of B-rigidity cannot be extended to all complete surfaces of negative curvature. Thus, for example, a right helicoid is not B-rigid.

3.2. A Theorem of Efimov and Heinz on the Extent of a One-to-one Projection onto the Plane of a Surface with Negative Curvature Separated from Zero.

From what follows it will be clear that a surface of the form (3.1) for which K satisfies the inequality (1.7) cannot be defined for $-\infty < x < +\infty$, $-\infty < y < +\infty$. However, the region of the plane that covers the one-to-one projection of a surface with curvature of the form (1.7) may have infinite extent and infinite area; an example is $z = e^x \sin y$. On this surface an inequality of the form (1.7) is satisfied between parallel planes $x = $ const.

Under the conditions (1.7) and (3.1) the surface cannot extend far in all directions. Namely, suppose that

$$K \leqslant -\mu^2, \qquad \mu = \text{const} > 0. \tag{3.6}$$

Then we have the following result.

Theorem (Efimov (1953)). *There is a number $\hat{a} > 0$ such that if a C^2-smooth function $f(x, y)$ is defined on a square with side a, and its graph (3.1) has Gaussian curvature (3.6), then $a \leqslant \hat{a}/\mu$.*

Corollary. *There is a number $\hat{r} > 0$ such that if a C^2-smooth function $f(x, y)$ is defined on a disc of radius r, and its graph (3.1) has Gaussian curvature (3.6), then $r \leqslant \hat{r}/\mu$ $(\hat{r} \leqslant \hat{a}/\sqrt{2})$.*

Let $a_0 = \inf \hat{a}$, $r_0 = \inf \hat{r}$; these are universal positive constants. Heinz proved (Heinz (1955)) that

$$r_0 \leqslant e\sqrt{3}. \tag{3.7}$$

Consequently, $a_0 \leqslant 2e\sqrt{3}$; the original upper bound for a_0 in Efimov (1953) was somewhat rougher. The examples of a hyperbolic paraboloid and a pseudosphere give a lower bound: $r_0 \geqslant \frac{1}{2}$. The precise values of the constants r_0 and a_0 are not known.

The results of Efimov and Heinz can be stated in terms of universal estimates for the dimensions of the domains of regularity of solutions of certain non-linear partial differential equations and inequalities. Namely, taking account of the well-known formula

$$K = (z''_{xx}z''_{yy} - (z''_{xy})^2)(1 + (z'_x)^2 + (z'_y)^2)^{-2} \tag{3.8}$$

for the Gaussian curvature of a surface of the form (3.1), we can assert that the differential inequality

$$\frac{z''_{xx}z''_{yy} - (z''_{xy})^2}{(1 + (z'_x)^2 + (z'_y)^2)^2} \leqslant -\varphi^2(x, y) \tag{3.9}$$

and the Monge-Ampère hyperbolic equation

$$z''_{xx}z''_{yy} - (z''_{xy})^2 + \varphi^2(x, y)(1 + (z'_x)^2 + (z'_y)^2)^2 = 0 \tag{3.10}$$

do not have C^2-smooth solutions that are defined on a disc of radius $r > r_0\mu^{-1}$ or on a square with side $a > a_0\mu^{-1}$ if the (continuous) function φ satisfies the inequality $\varphi(x, y) \geqslant \mu = \text{const} > 0$.

These results can be extended to hyperbolic equations and differential inequalities of a more general form than (3.9) and (3.10); see Azov (1983), Azov (1984), Brys'ev (1985). In addition, Efimov's theorem on the square can be extended (with a change in the estimates) to the case of a rectangular region of the plane (Efimov (1976)), and its analogues hold for hypersurfaces and for vector fields; see Aminov (1968), Aminov (1971).

Let us dwell in a little more detail on an example from geophysics, where we meet inequalities of the form (3.9) and equations of the form (3.10). We consider the following simplified model of the motion of air in the atmosphere. Suppose that air is an ideal gas and moves in the half-space $z > 0$. The vertical speed of motion is assumed to be small, so we neglect it and consider the vector field of the horizontal velocity $\{u, v\}$, written in Cartesian coordinates (x, y). The magnitude of this velocity is $V = \sqrt{u^2 + v^2}$. The field $\{u, v\}$, considered on the isobaric surface $p = \text{const}$, where p is the atmospheric pressure, is assumed to be solenoidal. Then there is a *flow function* $\psi = \psi(x, y; p, t)$,

$$u = -\psi'_y, \qquad v = \psi'_x, \tag{3.11}$$

where t is the time. Below we use the so-called *geopotential* $\Phi = \Phi(x, y; p, t)$, the potential energy of an experimental unit mass in the gravity force field placed on an isobaric surface, found in relation to sea level. Neglecting viscosity and other "small" physical effects, under the assumptions we have made we obtain as a consequence of the system of thermohydrodynamical equations the so-called balance equation for wind and pressure (Bolin (1956)):

$$\psi''_{xx}\psi''_{yy} - (\psi''_{xy})^2 + A(y)\Delta\psi + B(y)\psi'_y = \tfrac{1}{2}\Delta\Phi, \tag{3.12}$$

where Δ is the Laplacian in the variables x, y.

The wind speed occurs in (3.12) through the flow function ψ, and the scalar pressure field through the geopotential Φ. Equation (3.12) is written with a special choice of coordinate axes: the x-axis is directed along a parallel to the East, that is, in the direction of rotation of the Earth, and the y-axis along a meridian to the North, $A(y)$ is the projection of the angular velocity of the Earth's rotation on the vertical direction, and $B(y)$ its derivative with respect to arc length of the meridian. In geophysics the quantities $l = 2A$ and $\beta = 2B$ are called respectively the *Coriolis parameter* and the *Rossby parameter*. Aerological observations show that the assumptions we have made are justified with satisfactory accuracy for the mean troposphere over a sea or a flat country; for the details see Gisina et al (1976). In this model equation (3.12) is satisfied identically in p and t, which play the role of parameters in it. For a given right-hand side it is the *Monge-Ampère equation* for ψ.

Let Ω be the rotor of the vector field $\{u, v\}$ for fixed p and t. In meteorology Ω is called the "relative vorticity of the wind speed". From (3.11) we have $\Omega = \Delta\psi$.

Equation (3.12) enables us to write the expression for the Gaussian curvature $K_{[\psi]}$ of the surface $\psi = \psi(x, y)$, obtained for fixed p and t:

$$K_{[\psi]} = \frac{\frac{1}{2}\Delta\Phi - A\Omega + Bu}{(1 + V^2)^2}. \qquad (3.13)$$

If the quantity on the right-hand side of (3.13) is known, then for ψ we obtain an equation of the form (3.10), and the right-hand side of (3.13) can be negative and then it is hyperbolic. This situation occurs for regions of high atmospheric pressure – anticyclones – particularly often for anticyclones in tropical and subtropical latitudes, where $A(y)$ is small in modulus. The negativity of the right-hand side of (3.13) is then ensured by the sign of the Laplacian of the geopotential $\Delta\Phi$, as shown schematically in Fig. 30, where g is the acceleration of free fall in the gravitational force field.

The right-hand side of (3.13) can be assumed to be known in certain problems of numerical modelling of atmospheric processes. In real meteorological situations u, v and Ω are unknowns together with ψ, but they can be estimated, starting from physical arguments and given aerological observations.

If the right-hand side of (3.13) is bounded above by a negative quantity, we obtain an inequality of the form (3.9), and the results of Efimov and Heinz then

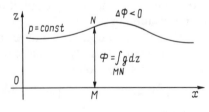

Fig. 30

enable us to obtain upper bounds for the possible extent of an anticyclone. It is interesting that these bounds are roughly only one order greater than the dimensions of anticyclones actually observed in the Earth's atmosphere (and not millions of times greater, as we might think!).

In this connection there is additional interest in finding the precise values of the constants a_0 and r_0.

Concerning the relation (3.13) we note that the situation here is apparently unusual for physicists: in the denominator the dimensionless quantity unity is added to the square of the velocity, that is, to a dimensional physical quantity. However, this is not surprising: the surface $\psi = \psi(x, y)$, constructed in the auxiliary Euclidean space in Cartesian coordinates (x, y, ψ), is not invariant under change of scale of length and time. When these scales are changed the surface $\psi = \psi(x, y)$ undergoes an affine transformation, and the numerical values of $K_{[\psi]}$ on it change, but the sign of $K_{[\psi]}$ is preserved.

We note that the estimate of Heinz relies on his identity

$$
\frac{d}{d\rho}\left(\frac{1}{\rho}\int_0^{2\pi} \tilde{z}_\varphi'(\rho, \varphi)^2\, d\varphi\right)
$$

$$
= \int_0^{2\pi} \tilde{z}_\varphi'(\rho, \varphi)^2\, d\varphi + 2\iint_{x^2+y^2\leqslant\rho^2} (z_{xx}''z_{yy}'' - (z_{xy}'')^2)\, dx\, dy, \quad (3.14)
$$

which has appeared in a different connection in the mathematical literature (see Aminov (1968)) under the name of Bernstein's integral formula. Here $z(x, y)$ is a C^2-smooth function, and $\tilde{z}(\rho, \varphi) = z(\rho\cos\varphi, \rho\sin\varphi)$, where (ρ, φ) are polar coordinates on the (x, y)-plane.

Suppose that the surface $z = z(x, y)$ satisfies (3.6) when $x^2 + y^2 < R^2$. According to Heinz (1955) we can construct an auxiliary function

$$
g(r) = \int_0^r \rho\, d\rho \int_0^{2\pi} (1 + \rho^{-2}\tilde{z}_\varphi'(\rho, \varphi)^2)\, d\varphi. \quad (3.15)
$$

It is immediately obvious that

$$
g(0) = 0; \quad g(r) \geqslant \pi r^2, \quad g'(r) > 0 \quad \text{when} \quad 0 < r < R. \quad (3.16)
$$

Also, by means of (3.14) and (3.6) we can verify that on $(0, R)$ the function $g(r)$ satisfies the differential inequality

$$
g''(r) \geqslant \frac{2\mu}{\pi} r^{-2} g(r)^2. \quad (3.17)
$$

Suitable estimates show that from (3.16) and (3.17) we have the chain of inequalities

$$
\frac{1}{\sqrt{\pi r}} \geqslant g(r)^{-1/2} \geqslant g(r)^{-1/2} - g(\tilde{r})^{-1/2} \geqslant \frac{\mu}{\sqrt{3\pi}} \ln\left(\frac{\tilde{r}}{r}\right) \quad (3.18)
$$

if $0 < r < \tilde{r} < R$. Proceeding to the limit as $\tilde{r} \to R$ in (3.18), we see that

$$\frac{\sqrt{3}}{\mu r} \geqslant \ln\left(\frac{R}{r}\right). \tag{3.19}$$

Suppose that $R \geqslant \sqrt{3}\,\mu^{-1}$. Then in (3.19) we take $r = \sqrt{3}\,\mu^{-1}$ and obtain (3.7). We can summarize the estimates presented here: condition (3.6) for the curvature implies (3.17), which forces the function $g(r)$ to increase so rapidly that it no longer exists when $r \leqslant r_0 \mu^{-1} \leqslant \sqrt{3}\,e\mu^{-1}$.

We also present the geometrical construction used by Efimov in Efimov (1953) to estimate a side of the square, since we shall need to refer to it later. Together with the surface (3.1) we consider a map of the form $E^2 \rightarrow E^2$ defined in rectangular Cartesian coordinates by (3.5). It is called the normal map for the surface (3.1). It is closely connected with the Gaussian spherical map Φ. The fact is that the vector

$$\mathbf{n} = \{-p, -q, 1\}(1 + p^2 + q^2)^{-1/2} \tag{3.20}$$

is the unit normal vector to the surface (3.1) if p and q are as in (3.5). From (3.20) it is obvious that we can construct the normal map geometrically. We first set up a correspondence between a point (x, y) of the plane E^2 and the point $(x, y, z(x, y))$ on the surface given by $z = z(x, y)$. We then carry out a spherical map $\Phi: \mathscr{F} \rightarrow S$, and after this a central projection from the centre O of the sphere S onto its tangent plane E_*^2 at the "South pole" of the sphere. We assume that the Cartesian axes p, q on E_*^2 are respectively parallel to (and in the same direction as) the x and y axes in $E^3 \supset E^2$. The composition of these three maps gives the normal map (3.5). Its Jacobian is

$$\Delta = \frac{\partial(p, q)}{\partial(x, y)} = (1 + p^2 + q^2)^2 K. \tag{3.21}$$

From (3.21) it is obvious that the normal and spherical maps are locally homeomorphic when $K \neq 0$. Suppose that

$$K = -k^2 < 0. \tag{3.22}$$

The asymptotic directions on the surface (3.1) are determined by $d^2 z = 0$, from which it follows that along the asymptotic lines

$$\frac{dq}{dx} = q_x' + q_y'\frac{dy}{dx} = \pm(1 + p^2 + q^2)k, \tag{3.23}$$

where we have the plus sign on one family and the minus sign on the other. The equality (3.23) enables us to construct so-called "*chains*", namely piecewise smooth curves formed on the (x, y)-plane from projections of arcs of asymptotic lines chosen in turn from the first and second families in such a way that under a motion along the chain in one direction q decreases monotonically. On each part of the chain the variation of q in modulus is bounded below by the total variation of x on this part, since q is monotonic and (3.23) and (3.6) hold. From the intuitive viewpoint the existence of such a bound means that on the surface there are paths along which the tangent plane rotates "strongly" in space and "soon" takes up a vertical position, which prevents an extension of the surface while preserving the one-to-one property of its projection on the plane $z = 0$.

To obtain a quantitative estimate of the extent of the surface we need to observe here that we can turn the x and y axes on E^2 and the p and q axes on E_*^2 through the same angle, and then construct new chains. Simultaneous consideration of the chains and level lines of the functions (3.5) leads to an upper bound for the constant a_0; see Efimov (1953).

3.3. Plan of the Proof of Theorem B. We present here the main steps in the proof of this theorem (see p. 95 above), bearing in mind the possible further development of the given problem area. All the details relating to the proof can be found in Efimov (1964), Efimov (1968), and Klotz-Milnor (1972).

3.3.1. Let us begin with an auxiliary construction. Suppose that on the Euclidean plane E^2 or on a two-dimensional sphere S^2 of fixed radius we are given a circle C bounding a disc Q of radius ρ, where on S^2 the disc Q must be smaller than a hemisphere. Suppose that in Q we take an open sector V with central angle γ and a concentric circle C' of radius ρ', $0 < \rho' < \rho$. We denote the intersection of the sector V with the circular annulus between C' and C by $\mathcal{U}$ and put $\Gamma_{\mathcal{U}} = C' \cap V$. Suppose that on S^2 (or on E^2) we are given a domain G with metric boundary ∂G, possibly many-sheeted. In other words, G must be understood either as a two-dimensional surface in E^3 having support $\hat{G} \subset S^2 \subset E^3$ (respectively, $\hat{G} \subset E^2 \subset E^3$) or as a two-dimensional surface in the two-dimensional ambient space S^2 (respectively, E^2).

We say that ∂G admits *concave support* at a point $N \in \partial G$ if G contains a figure $\mathcal{U}$ constructed for certain values of the parameters ρ, ρ', γ and situated in such a way that $N \in \Gamma_{\mathcal{U}}$ and on $\Gamma_{\mathcal{U}}$ there are no other points of ∂G apart from N.

Definition (Klotz-Milnor (1972)). If ∂G does not admit concave support at any of its points, then G is said to be *pseudoconvex*.

Lemma 1 (Efimov (1963), Efimov (1964)). *Suppose that a Riemannian metric of constant positive curvature is specified in G and that G is homeomorphic to E^2 and is pseudoconvex. Then G is a convex domain in this Riemannian metric.*

Corollary. *If under the conditions of Lemma 1 the curvature of the metric specified in G is equal to $+1$, then G is isometric to a convex domain on a sphere S^2 of unit radius and its area*

$$\sigma(G) \leqslant 2\pi. \tag{3.24}$$

Remark. Lemma 1 is used in the proof of Theorem B formulated in 1.4 in connection with the Gaussian spherical map of the surface under investigation. In the literature there are different versions of the proof of Lemma 1. One of these (see Klotz-Milnor (1972)) uses the local isometry $G \to S^2$ and geometrical constructions on S^2 and in $E^3 \supset S^2$. Others (see Efimov (1964), Burago and Zalgaller (1974)) rely on intrinsic geometric constructions on a manifold with a metric of constant positive curvature. Efimov gave preference to an intrinsic geometric version of the proof, assuming that in the course of time someone would find a generalization of Theorem B to the case of a surface in Riemannian space, where there is no Gaussian spherical map in the classical understanding of the term.

3.3.2. From what follows it will be seen that for the proof of Theorem B it is important to obtain a contradiction to the estimate (3.24). Lemmas 2 and 3 given below assist in this. Let $\mathscr{F}$ be a surface of negative curvature $K < 0$ and let $k = \sqrt{-K} = R^{-1}$; see (3.22) and (1.9). Following Efimov (Efimov (1966b), Efimov (1968)), together with the original intrinsic metric ds^2 we shall consider on $\mathscr{F}$ an auxiliary Riemannian metric dl^2, where

$$dl = kds \tag{3.25}$$

and call (3.25) the l-metric, and the lengths of curves in the l-metric their l-lengths (similarly, the areas of figures in the l-metric will be called their l-areas, and so on).

Lemma 2. *Suppose that a metric ds^2 is specified on a plane, it is complete, it has negative curvature $K < 0$, and the radius of curvature R has a variation with linear estimate in it. Then the metric (3.25) constructed on the same plane is also complete.*

The proof is based on the fact that if a curve has infinite length, then along it the integral that expresses its l-length is divergent (under the conditions of Lemma 2).

Remark. The length of an arc of an asymptotic curve in the l-metric is the ordinary length of its spherical image. The area in the l-metric of domain U of $\mathscr{F}$ is equal to the ordinary (unoriented) area of its spherical image $\Phi(U)$.

Lemma 3 (conditional). *If a C^2-smooth surface $\mathscr{F}$ in E^3 is complete, its curvature $K < 0$, and the radius of curvature R has a variation with linear estimate on $\mathscr{F}$, then the area of the universal covering $\tilde{\mathscr{F}}$ of the surface $\mathscr{F}$ in the l-metric is infinite.*

Corollary. *Under the conditions of Lemma 3 the area of the spherical image $\Phi(\tilde{\mathscr{F}})$ of the universal covering $\tilde{\mathscr{F}}$ of $\mathscr{F}$ is infinite:*

$$\sigma(\Phi(\tilde{\mathscr{F}})) = +\infty. \tag{3.26}$$

The proof of Lemma 3 is based on the construction of $\tilde{\mathscr{F}}$ of the unique *polar-geodesic* coordinate system. In it the l-area is expressed by an integral and its divergence is established.

3.3.3. One of the key steps in the proof of Theorem B (and the special case of it when $K \leqslant \text{const} < 0$) is the lemma on maps in §1 of Ch. 3 of Part I of the present book. We shall not state it again here, but instead we single out (as an auxiliary statement for Theorem B) the assertion stated in the present subsection as a "theorem". It refers to surfaces of negative curvature in E^3 and is of independent interest.

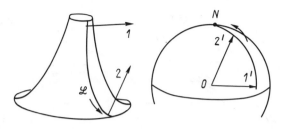

Fig. 31

Suppose, as above, that **n** is the unit normal vector to $\mathscr{F}$, that N is a point of the metric boundary $\partial\Phi$ of the spherical image $\Phi(\mathscr{F})$ of the surface $\mathscr{F}$, and that O is the centre of the Gaussian sphere.

We shall say that a path $\mathscr{L}$ on $\mathscr{F}$ leads to N if $\mathscr{L}$ leaves any compact set on $\mathscr{F}$ that contains its origin, and for a point $X \in \mathscr{L}$ going along $\mathscr{L}$ there is a limit $\lim \mathbf{n}(X) = ON$. Then we have the following result.

Theorem. *Suppose that in E^3 we are given a C^2-smooth surface $\mathscr{F}$ of negative curvature $K < 0$, possibly incomplete, on which the radius of Gaussian curvature R has a variation with linear estimate. If on the metric boundary $\partial\Phi$ of its spherical image $\Phi(\mathscr{F})$ there is a point N where $\partial\Phi$ admits concave support, then on $\mathscr{F}$ there is a path $\mathscr{L}$ leading to N whose length is finite in the metric ds^2 of the surface $\mathscr{F}$.*

Example. The spherical image of a smooth half pseudosphere is a hemisphere with its pole deleted. $\partial\Phi$ consists of two components. One of them – the equator of the hemisphere – corresponds to the exit to infinity along the tapering tube of the pseudosphere. The other consists of one point and therefore admits concave support. This component corresponds to the edge of the pseudosphere. Paths of finite length lead to it (Fig. 31).

3.3.4. We suppose that the boundary $\partial\Phi$ of the spherical image of the surface $\mathscr{F}$ admits concave support $\mathscr{U}_0$ at some point N and give some auxiliary constructions.

We shall assume that N is the North Pole of the Gaussian sphere $S^2 \subset E^3$, O is its centre, and N_* is the South Pole. We denote the tangent plane to S^2 at N_* by E_*^2. We take the diameter NN_* as z-axis, and direct the x and y axes parallel to the axes p and q on E_*^2. We call E^2 the coordinate plane $z = 0$. We direct the p-axis (and the x-axis) parallel to the tangent to the bounding circle of the concave support $\mathscr{U}_0$ at the point N. Let $P: E^3 \to E^2$ be an orthogonal projection, and $P_*: S^2 \to E_*^2$ the central projection from the centre O of the sphere S^2. Without loss of generality we may suppose that $\mathscr{U}_0$ is situated in the ε_0-neighbourhood of N on S^2, where $\varepsilon_0 > 0$ is chosen to be sufficiently small and fixed. Let us put $\mathscr{U}_{0*} = P_*(\mathscr{U}_0)$; see Fig. 32. For sufficiently small $c > 0$ and $r > 0$, $\mathscr{U}_{0*}$ contains the domain

$$p^2 + q^2 < r^2, \qquad q < cp^2. \tag{3.27}$$

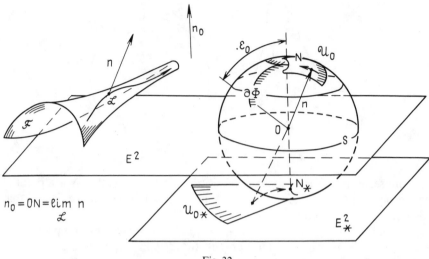

Fig. 32

To the open domain (3.27) on E^2_* we adjoin its boundary, except for the point N_*, and call the resulting figure $\mathcal{U}_*$.

We put

$$\mathcal{U} = P^{-1}_*(\mathcal{U}_*), \qquad \mathcal{G} = \Phi^{-1}(\mathcal{U}), \qquad G = P(\mathcal{G}),$$

and regard G as a two-dimensional surface (with boundary) in E^2. Later we shall use the restrictions of the maps P and P_* to $\mathcal{G}$ and $\mathcal{U}$ respectively, keeping the previous notation P and P_*. The composition

$$P_* \circ \Phi \circ P^{-1} \colon G \to \mathcal{U}_* \tag{3.28}$$

is the normal map constructed for the part $\mathcal{G}$ of the surface $\mathcal{F}$. Let $\varDelta = \dfrac{\partial(p, q)}{\partial(x, y)}$ be its Jacobian. We draw attention to the fact that the ambient space E^2 induces on G an intrinsic metric that is locally Euclidean on $G \setminus \partial G$ and which we can write locally (inside $G \setminus \partial G$) as $ds^2_0 = dx^2 + dy^2$; however, in the large this intrinsic metric (we call it the s_0-metric) is not Euclidean. The fact is that shortest curves constructed in it can go along ∂G, without being rectilinear there. Using the fact that R on $\mathcal{F}$ has a variation with linear estimate, and taking account of (3.21) and (3.22) and the position of $\mathcal{U}$ on S^2, we can specify a function $a = a(\mathcal{M})$ on G such that

I) $0 < a(\mathcal{M}) \leqslant 1$; $a(\mathcal{M})^2 \leqslant |\varDelta(\mathcal{M})|$, where $\mathcal{M}$ is a moving point in G;

II) $1/a(\mathcal{M})$ on G has a variation with linear estimate in the s_0-metric with certain constants C_1, $C_2 \geqslant 0$ (see (1.8)). These constants for $1/a(\mathcal{M})$ differ, in general, from those that occur in the linear estimate of the form (1.8) for R on $\mathcal{F}$.

We denote the map inverse to the normal map (3.28) by ψ:

$$\psi \colon \mathcal{U}_* \to G, \qquad \psi = P \circ \Phi^{-1} \circ P^{-1}_*. \tag{3.29}$$

Let Ψ be its graph in $E^4 = E^2 \times E^2_*$; this is a two-dimensional surface with boundary. It is obvious from the preceding work that in the (x, y, p, q) coordinates it is given by

$$(x, y) = \psi(p, q); \qquad (p, q) \in \mathcal{U}_*.$$

Instead of considering together two plane figures ($G \subset E^2$ and $\mathcal{U}_* \subset E^2_*$) and the map (3.28) we shall consider the graph Ψ of the map (3.29). We shall carry out the auxiliary constructions on the surface Ψ situated in E^4, and follow them by means of projections (either on E^2 or on E^2_*, whichever is convenient in each specific case).

Let $\Pi_*: \Psi \to \mathcal{U}_*$ be the restriction to Ψ of the orthogonal projection $E^4 \to E^2_*$. We put $\Pi = \psi \circ \Pi_*$; $\Pi: \Psi \to G$.

We call the Euclidean metric $ds^2_* = dp^2 + dq^2$ on E^2_* the s_*-metric, and in G together with the s_0-metric we introduce the α-metric by putting $d\alpha = a(\mathcal{M}) \, ds_0$.

Next, by means of the maps Π^{-1}_*, Π^{-1} and $\Pi^{-1} \circ P$ we carry over the metrics ds^2_*, ds^2_0, $d\alpha^2$ and ds^2 introduced above to the surface Ψ. Using them there, we shall need the terms s_0-length, α-area, and so on. We shall also assume that the function $a = a(\mathcal{M})$ is carried over to Ψ by the map Π^{-1}.

3.3.5. For what follows it is important that the α-area of figures on Ψ is majorized by their s_*-area by virtue of the inequality $|\Delta| \geqslant a^2$, and the α-length is majorized by the s_0-length by virtue of the inequality $a \leqslant 1$.

Let $\xi = x \cos \theta + y \sin \theta$, $\eta = p \cos \theta + q \sin \theta$. On Ψ we consider an arbitrary piecewise smooth arc L. We denote by γ_ξ the angle on E^2 between the ξ-axis and the tangent to $\Pi(L)$, and by γ_η the angle on E^2_* between the η-axis and the tangent to $\Pi_*(L)$, and introduce the functionals

$$s_{0\xi}\langle L \rangle = \int_L |\cos \gamma_\xi(\mathcal{M})| \, ds_0 \geqslant 0,$$

$$\alpha_\xi\langle L \rangle = \int_L a(\mathcal{M})| \cos \gamma_\xi(\mathcal{M})| \, ds_0 \geqslant 0$$

and

$$s_{*\eta}\langle L \rangle = \int_L |\cos \gamma_\eta(\mathcal{M})| \, ds_* \geqslant 0; \qquad \mathcal{M} \in L.$$

In particular, when $\theta = 0$ we have $s_{0\xi} = s_{0x}$, and when $\theta = \pi/2$ we have $s_{0\xi} = s_{0y}$ and $s_{*\eta} = s_{*q}$.

Clearly, $s_{0\xi}$, α_ξ and $s_{*\eta}$ are majorized by the s_0-length, the α-length and the s_*-length respectively. We draw attention to the fact that $s_{0x} + s_{0y}$ majorizes the s_0-length, and $\alpha_x + \alpha_y$ majorizes the α-length. In addition, we have the lower bound

$$\alpha_\xi\langle L \rangle \geqslant \alpha_y\langle L \rangle |\sin \theta| - \alpha_x\langle L \rangle \cos \theta \qquad (3.30)$$

when $|\theta| \leqslant \pi/2$.

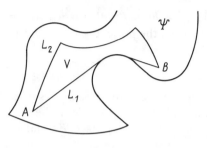

Fig. 33

Lemma 4. *Suppose that points A and B on Ψ are joined by an s_0-shortest arc L_1 and a piecewise smooth arc L_2 without self-intersections; V is the domain on Ψ between L_1 and L_2. Then*

$$\alpha_\xi\langle L_1 \rangle \leqslant (1 + C_2)(\alpha_\xi\langle L_2 \rangle + C_1\sigma_\alpha(V)), \tag{3.31}$$

where the symbol σ_α denotes the α-area.

In the special case shown schematically in Fig. 33, when L_1 and L_2 have no points in common other than their ends A and B, the proof is carried out by estimates usual in mathematical analysis (taking account of the linear estimate for the variation of the function $a(\mathcal{M})$; see Efimov (1968)). The general case of the mutual position of L_1 and L_2 reduces to this special case.

We mention that in the case (1.7) for the proof of Theorem B a simpler "comparative lemma" is sufficient, according to which $s_{0\xi}\langle L_1 \rangle \leqslant s_{0\xi}\langle L_2 \rangle$; see Efimov (1964), Klotz-Milnor (1972).

3.3.6. As in 3.2, we shall construct chains in G, and by means of the map Π^{-1} carry them over to Ψ. We have the following facts:

I) In motion along a chain the coordinate q varies monotonically, so we can take as the positive direction on it the direction in which q decreases, for example.

II) A chain starting from a point where $q \leqslant 0$ goes out to the arc $\Gamma_0 \subset \partial\Psi$ on which $p^2 + q^2 = r^2$.

III) On any part L of a chain we have

$$\alpha_x\langle L \rangle \leqslant s_{*q}\langle L \rangle. \tag{3.32}$$

IV) A similar inequality holds if the x and y axes on E^2 and the p and q axes on E_*^2 are turned through the same angle θ and chains are constructed in the new position of the axes.

3.3.7. Definition. We call a curve L_0 on Ψ homeomorphic to a closed ray a *special ray* if L_0 goes out from each part of Ψ that is compact in the s_0-metric and contains the origin of L_0, and if each part of L_0 between a pair of its points is an s_0-shortest curve.

We can show that a special ray goes out from each point on Ψ and each special ray is piecewise smooth, and its C^1-smoothness can be violated at two points at most, those that occur on $\mathscr{U}_*$ at corner points of its boundary under the projection Π_*. From the intuitive viewpoint, a special ray is a geodesic in the s_0-metric that goes to an improper point.

If the s_0-length of a special ray is infinite, then its α-length is also infinite: from the existence of a linear estimate for the variation of the function $1/a(\mathscr{M})$ it follows in this case that the integral that expresses its α-length is divergent (like it is for l-length in the proof of Lemma 2).

Lemma 5. *Under the conditions of the theorem of 3.3.3 (in the presence of concave support)*

$$\alpha_x\langle L_0\rangle < +\infty \tag{3.33}$$

for any special ray L_0 on Ψ.

We draw attention to the fact that Lemma 5 emphasizes the disparity of the directions x and y that arises because of the existence of concave support.

From Lemma 5 it follows that if the α-length of L_0 is infinite, then its projection $\Pi(L_0)$, and together with it the whole domain G, is so elongated along the y-axis that

$$\alpha_y\langle L_0\rangle = +\infty. \tag{3.34}$$

For the proof of Lemma 5 we need to consider different possible a priori situations separately.

I) The α-length of L_0 is finite. Then we immediately have (3.33), according to 3.3.5.

II) The α-length of L_0 is infinite. This case splits into subcases.

IIA) There is a circular neighbourhood of the point N_* on E_*^2 such that the part of the projection $\Pi_*(L_0)$ that falls into it is situated entirely in the half-plane $q < 0$ (Fig. 34a).

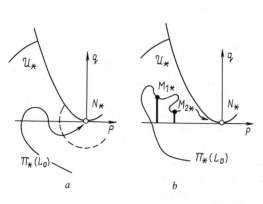
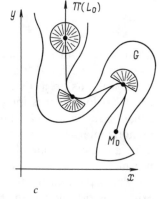

Fig. 34

IIB) Condition IIA is not satisfied, but on L_0 arbitrarily far from its origin in the sense of the α-metric there are points M_1, M_2, ..., M_r, ... such that their projections M_{*r} on E_*^2 ($r = 1, 2, ...$) can be joined to the p-axis by segments for which all the arcs $M_r M_r'$ on Ψ that project into these segments have α-length not exceeding some fixed number $\alpha_1 > 0$ ($q = 0$ at the points M_r'). This case is shown schematically in Fig. 34b, where the segments $\Pi_*(M_r'M_r)$ are distinguished.

IIC) Neither IIA nor IIB is satisfied, although the α-length of a special arc is infinite.

In cases IIA and IIB the proof of Lemma 5 goes through in the same way by means of Lemma 4 and the properties of chains listed in 3.3.6 (of course, we need to construct suitable chains; property IV of 3.3.6 is not used here).

Case IIC means (as additional investigation shows) that the projection of Ψ on E^2 is very "extensive" – in G along $\Pi(L_0)$ we can then mark off an infinite sequence of pairwise non-intersecting open domains of a special kind (like discs or half-discs, see Fig. 34c) such that their α-areas admit a uniform lower bound of positive value. Then the α-area of G (and hence of Ψ) is infinite, which is impossible, since it is majorized by the Euclidean area of the bounded closed domain $\overline{\mathcal{U}}_*$.

Remark. In the special case (1.7) instead of Lemma 5 it is sufficient to obtain the estimate $s_{0x}\langle L_0 \rangle < +\infty$. Technically this is somewhat simpler, but nevertheless it requires consideration of different cases similar to those listed above.

3.3.8. Next, for the proof of the theorem of 3.3.3, there is the prospect, by making the assumption (3.34), of leading to a contradiction of it. In this we are helped by Lemmas 6, 7 and 8 formulated below. To formulate them we need further auxiliary concepts and geometrical constructions, which we now present.

Let M_0 be the origin of a special ray L_0, and M a moving point of it. We put $t = \alpha_y \langle \bigcup M_0 M \rangle$, $\bigcup M_0 M \subset L_0$, and write $M = M_t$. Under the condition (3.34) we have

$$0 \leqslant t < +\infty, \qquad (3.35)$$

when M_t runs through L_0. As usual, we denote the ray (3.35) by $\mathbb{R}_+$.

Following Klotz-Milnor (1972), we say that a simple arc $T \subset \Psi$ is *prehorizontal* if its projection $\Pi(T)$ on E^2 is rectilinear and parallel to the x-axis. For what follows it is important that $\Pi(T)$ cannot contain a ray, that is, the situation shown schematically in Fig. 35 is impossible; if it were to hold, then the prehorizontal arc T (or part of it) would be a special ray on which $s_{0x} = \infty$; then, using a linear estimate of the form (1.8) for $1/a$, we could show that $a_x = \infty$ on it, which is impossible by Lemma 5. Let Y_0 be the set of those values $t \in \mathbb{R}_+$ for which the tangent to the special ray L_0 at the point M_t either does not exist or is orthogonal to the y-axis. Using Sard's theorem we can show that mes $Y_0 = 0$.

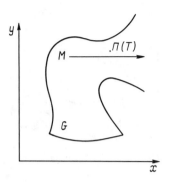

Fig. 35

Suppose that $t \in \mathbb{R}_+ \setminus Y_0$. Through the point M_t we draw a prehorizontal arc. If $M_t \in \partial \Psi$, we draw and extend it on one side – inside Ψ – until it first meets $\partial \Psi$ again. If M_t is an internal point of Ψ, we extend the prehorizontal arc on both sides of M_t until it first meets $\partial \Psi$ again. By what we said above, such meeting points must exist, and we obtain a prehorizontal arc homeomorphic to an interval, which we denote by T_t and call a *cross-section*, following Efimov (1964).

A cross-section splits Ψ into two parts, for one of which its closure in Ψ is non-compact – it contains a sequence of points whose projections on E_*^2 converge to N_*; we denote this part by S_t and consider its α-area $\sigma_\alpha(S_t) = \mathfrak{A}(t)$. It is finite and uniformly bounded with respect to t by virtue of the estimate

$$0 < \mathfrak{A}(t) \leqslant \sigma(\Pi_*(S_t)) \leqslant \sigma((\overline{\mathcal{U}}_*) < \pi r^2. \tag{3.36}$$

We denote the α-length of the cross-section T_t by $z(t)$.

Let Y_1 be the set of those values of t for which $\Pi(T_t)$ is an interval tangent to ∂G drawn at the end-point of $\Pi(T_t)$. By means of Sard's theorem we can establish that mes $Y_1 = 0$.

Suppose that $t \in \mathbb{R}_+ \setminus (Y_0 \cup Y_1)$. Then the derivative $\mathfrak{A}'(t)$ exists, and it is not difficult to find it by geometrical arguments. In fact,

$$d\mathfrak{A} = -\left(\int_{T_t} a^2 \, ds_0 \right)|dy|, \qquad dt = \mathring{a}(t)|dy|$$

and so

$$\mathfrak{A}'(t) = -\mathring{a}(t)^{-1} \int_{T_t} a^2 \, ds_0 = -\mathring{a}(t)^{-1} \int_{T_t} a \, d\alpha, \tag{3.37}$$

where $\mathring{a}(t)$ is the value of a at the point $M_t \in L_0$ through which we have drawn the cross-section T_t. The differential $d\mathfrak{A}$ is shown conventionally in Fig. 36. The part of the figure S_t corresponding to it is hatched.

The boundary $\partial \Psi$ of the surface Ψ in $E^4 = E^2 \times E_*^2$ (simultaneously intrinsic and extrinsic geometrical) consists of three parts, which we denote by Γ_0, Γ_+ and Γ_-:

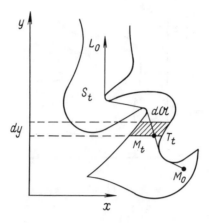

Fig. 36

$$p^2 + q^2 = r^2 \quad \text{on } \Gamma_0, \text{ see above, } 3.3.6(\text{II});$$

$$q = cp^2, \qquad p < 0 \text{ on } \Gamma_-;$$

$$q = cp^2, \qquad p > 0 \text{ on } \Gamma_+.$$

Under the conditions of the theorem of 3.3.3 Lemmas 6–8 are true.

Lemma 6. *There is a monotonically decreasing function $\chi = \chi(\varepsilon) > 0$ ($\varepsilon > 0$) such that if $z(t) \leqslant \varepsilon$, then on the cross-section T_t the function a satisfies the inequality*

$$a(\mathcal{M}) \geqslant \mathring{a}(t)\chi(\varepsilon), \qquad \mathcal{M} \in T_t. \tag{3.38}$$

Lemma 7. *The set $Y_\varepsilon \subset \mathbb{R}_+ \setminus Y_0$ of those values of t for which $z(t) \geqslant \varepsilon$ has finite measure for every $\varepsilon > 0$.*

Lemma 8. *$\Pi_*(T_t) \to N_*$ as $t \to +\infty$, and for all sufficiently large values $t \in \mathbb{R}_+ \setminus Y_0$ one of the ends of the cross-section T_t is on Γ_- and the other is on Γ_+.*

Lemma 6 is proved by estimates usual in mathematical analysis, and the function $\chi(\varepsilon)$ can be written in explicit form, but we do not need it below. The proof of Lemma 7 relies on (3.36), (3.37) and Lemma 6.

Remark. Under the condition (1.7) we can take $t = s_{0y}\langle \bigcup M_0 M \rangle$, and for $z(t)$ and $\mathfrak{R}(t)$ we can take the s_0-length of the cross-section T_t and the s_0-area of S_t; then instead of (3.37) we have the simpler formula $\mathfrak{A}'(t) = -z(t)$.

The proof of Lemma 8 is carried out by contradiction. The assumption that its assertion is false leads to a contradiction with Lemma 5.

3.3.9. In the proof of the theorem of 3.3.3 we need some more elementary facts. Suppose we have drawn the tangents QB^- and QB^+ to the parabola $y = cx^2$ from the point $Q(0, -h)$; see Fig. 37. Together with the arc of the parabola they bound a curvilinear triangle with area

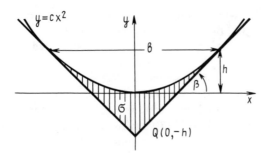

Fig. 37

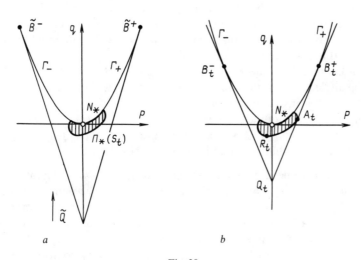

Fig. 38

$$\sigma = \tfrac{1}{3}bh = \tfrac{1}{12}cb^3, \tag{3.39}$$

where $b = |B^- B^+|$ is the length of the chord of the parabola joining the points of contact. Then

$$h = \tfrac{1}{4}cb^2, \tag{3.39'}$$

and for the angle β between the x-axis and the tangent QB^+ we have the estimate

$$\sin \beta > \tan \beta \cos \beta_0 > \tfrac{1}{2}cb \tag{3.40}$$

if $\beta \in (0, \beta_0)$, where $0 < \beta_0 \leqslant \pi/3$.

3.3.10. Let us carry out some auxiliary constructions. Taking $t \in \mathbb{R}_+ \setminus Y_0$ we construct $\varPi_*(S_t)$, and then on the negative q-axis we take a point $\tilde{Q}$ so far down that $\varPi_*(S_t)$ is included in the curvilinear triangle $\tilde{Q}\tilde{B}^+ N_* \tilde{B}^-$, where $\tilde{Q}\tilde{B}^+$ and $\tilde{Q}\tilde{B}^-$ are tangents to the parabola $q = cp^2$, and $\tilde{B}^- N_* \tilde{B}^+$ is an arc of this parabola (Fig. 38a). Then we make $\tilde{Q}$ tend to N_* along the q-axis and denote the upper-most position of this point for which such an inclusion holds by $Q = Q_t$; we denote the points of contact corresponding to it by $B^+ = B_t^+$ and $B^- = B_t^-$ (Fig. 38b). The figure $\Pi_*(S_t)$ in Fig. 38 is hatched. Also let

$\sigma = \sigma(t)$ be the area of the curvilinear triangle bounded by the open polygon $B^- QB^+$ and the arc $B^- N_* B^+$ of the parabola;

$h = h(t)$ be the length of the segment QN_*;

$b = b(t)$ be the length of the chord $B^- B^+$;

$\beta = \beta(t)$ be the acute angle between the p-axis and the tangent QB^+ (Fig. 38b).

Let $R = R_t$ be the point on $\Pi_*(T_t)$ having the smallest value of the coordinate q and $A = A_t$ one of the points of contact of $\Pi_*(T_t)$ with the open polygon $B^- QB^+$. For definiteness this is shown in Fig. 38b on the right-hand tangent QB^+. The notation in Figs. 38–41, which illustrate these and subsequent constructions, is chosen so that we can imagine we see a surface (with boundary) Ψ with figures on it from the side of the two-dimensional plane E_*^2. Points and figures on Ψ and their projections on the plane E_*^2 will be denoted by the same symbols—this does not cause any ambiguity.

Using Lemma 8, we choose $t_0 > 0$ so large that when $t > t_0$:

1) the angle $\beta > 0$ is sufficiently small, for example $\beta < \pi/3$;

2) $\Delta B_t^- QB_t^+$ is inside the disc $p^2 + q^2 < r^2$;

3) one of the ends of the cross-section T_t is on Γ_- and the other on Γ_+.

Then T_t intersects $Q_t N_*$ and so at R_t the coordinate q is negative. This guarantees that the chain starting from R_t reaches the arc Γ_0.

3.3.11. Next we need constructions carried out for two different values $t = t_1 > t_0$ and $t = t_2 > t_1$. For simplicity and greater clarity we proceed as follows. When $t = t_i$ $(i = 1, 2)$ the values of the functions and the points are marked with the index i instead of t_i. The constants C_1 and C_2 in (3.31) are re-denoted by $C_1 = C_I$, $1 + C_2 = C_{II}$. As well as the original coordinates (x, y) on E^2 and (p, q) on E_*^2, we need the coordinates obtained from them by rotation through an angle $\tilde{\beta} = \pm \beta$ (the plus sign if $A_2 \in Q_2 B_2^+$, the minus sign if $A_2 \in Q_2 B_2^-$). Correspondingly, on E^2 we shall have a new abscissa $\xi = \xi(\theta)$ when $\theta = \tilde{\beta}$, and on E_*^2 a new ordinate $\eta = \eta(\theta)$ when $\theta = \tilde{\beta} + \pi/2$, which will be used in estimates of the form (3.30) and (3.31).

Having constructed the cross-section T_1 and T_2, from the points R_2 and A_2 we produce chains, the first in the original position of the coordinate systems, and the second for the coordinates rotated through an angle $\tilde{\beta}$. We extend both of them until they first meet T_1 (Fig. 39). Then the part of the first of these chains from the point R_2 to its meeting with T_1, which we denote by λ, is inside the triangle bounded by the open polygon $B_1^- QB_1^+$ and the p-axis. The part of the second chain from A_2 to its meeting with T_1 is denoted by μ. We extend that tangent $Q_2 B_2^{\pm}$ on which A_2 lies to its intersections at P^- and P^+ with $Q_1 B_1^-$ and $Q_1 B_1^+$. We observe that $\mu \subset \Delta Q_1 P^- P^+$, and the heights of the two triangles concerned here, dropped from the vertex Q_1, do not exceed h_1 (see Figs. 39–41). Taking account of (3.32) and (3.39') we have

$$\left. \begin{array}{l} \alpha_x \langle \lambda \rangle \leqslant s_{*q} \langle \lambda \rangle \\ \alpha_\xi \langle \mu \rangle \leqslant s_{*\eta} \langle \mu \rangle \end{array} \right\} \leqslant h_1 = \tfrac{1}{4} cb_1^2. \tag{3.41}$$

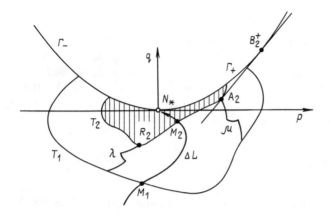

Fig. 39

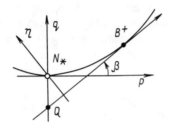

Fig. 40

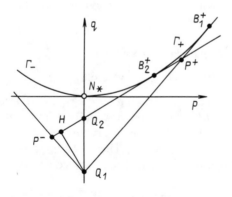

Fig. 41

Suppose that $r < \frac{1}{2}$. We can arrange this in advance in constructing the figure $\mathcal{U}_*$. Then $b(t) < 1$ when $t > t_0$ and from (3.39) we obtain the estimate

$$\sigma(t) < \tfrac{1}{12} c b(t)^2 \qquad \text{when } t > t_0. \tag{3.42}$$

Next, let ΔL be the part $M_1 M_2$ of the special ray L_0. Then by the definition of the quantity t we have

$$\alpha_y \langle \varDelta L \rangle = t_2 - t_1. \tag{3.43}$$

We join the ends M_1 and M_2 of the arc $\varDelta L$ by two different piecewise smooth curves. We form one of them from the chain λ and parts of the cross-sections T_1 and T_2. We form the other from parts of the same cross-sections and the chain μ. Fig. 39 shows both these piecewise smooth curves, and also the arc $\varDelta L$ of the ray L_0. Let us estimate the quantities $\alpha_x \langle \varDelta L \rangle$ and $\alpha_\xi \langle \varDelta L \rangle$, applying Lemma 4. Taking account of (3.41) and (3.42), we find that

$$\left. \begin{array}{l} \alpha_x \langle \varDelta L \rangle \leqslant C_{\mathrm{II}}(z_1 + \alpha_x(\lambda) + z_2 + C_1 \mathfrak{A}_1) \\ \alpha_\xi \langle \varDelta L \rangle \leqslant C_{\mathrm{II}}(z_1 + \alpha_\xi(\mu) + z_2 + C_1 \mathfrak{A}_1) \end{array} \right\} \leqslant C_{\mathrm{II}}(z_1 + z_2 + \tfrac{1}{4}(1 + \tfrac{1}{3}C_1)cb_1^2). \tag{3.44}$$

In addition, using (3.30) and (3.40), we estimate $\alpha_\xi \langle \varDelta L \rangle$ from below:

$$\alpha_\xi \langle \varDelta L \rangle \geqslant \tfrac{1}{2} cb_2 \alpha_y \langle \varDelta L \rangle - \alpha_x \langle \varDelta L \rangle. \tag{3.45}$$

The two-sided estimate for $\alpha_\xi \langle \varDelta L \rangle$ in which $\alpha_y \langle \varDelta L \rangle$ takes part is one of the most important features of the proof of the theorem in 3.3.3. It enables us, having used the form of concave support and having "shaken" the coordinate system (Fig. 40), to look at the special ray L_0 from a different angle and finally to obtain an upper bound for $\alpha_y \langle L_0 \rangle$. However, for this we need some additional arguments.

3.3.12. Below we need the auxiliary function

$$v(t) = \sqrt[3]{\mathfrak{A}(t)} > 0.$$

Using it, we can conveniently write the modulus of $\mathfrak{A}'(t)$ as

$$|\mathfrak{A}'(t)| = 3v(t)^2 |v'(t)|. \tag{3.46}$$

Taking account of (3.36) and (3.39), we have

$$v(t) < C_0 b(t), \qquad \text{where } C_0 = (c/12)^{1/3}. \tag{3.47}$$

Our final goal is to find an upper bound for the right-hand side of (3.44) in terms of a quantity of order $O(b_1^2)$. So far the possibility of large values of $|v'(t)|$ and $z(t)$ appearing has prevented this. Geometrically this is connected with the possibility of long cross-sections T_t appearing. They need to be excluded from consideration. Efimov called this part of the auxiliary constructions "cutting-back of boughs".

Let Y_v denote the set of those values $t \in \mathbb{R}_+$ for which $|v'(t)| > 1$. The function $v(t)$ is positive and monotonically non-decreasing, so mes $Y_v < +\infty$. Let us fix $\varepsilon_1 > 0$; using Lemma 7 and the fact that mes $Y_0 = $ mes $Y_1 = 0$ we construct a closed set $Y \subset \mathbb{R}_+$ *of finite measure* such that

$$Y_0 \cup Y_1 \cup [0, t_0] \cup Y_{\varepsilon_1} \cup Y_v \subset Y.$$

Its complement $W = \mathbb{R}_+ \setminus Y$ is open and

$$\text{mes } W = +\infty, \tag{3.48}$$

because of (3.34) and (3.35). From (3.37), (3.38) and (3.46) we see that

$$z(t) \leqslant \chi(\varepsilon_1)^{-1}|\mathfrak{A}'(t)| \leqslant 3\chi(\varepsilon_1)^{-1}v(t)^2 \tag{3.49}$$

on W. Hence, assuming that $t_1, t_2 \in W$ and taking account of the fact that $v(t)$ is monotonic, from (3.47) and (3.49) we find that

$$z_1 + z_2 \leqslant 6C_0^2\chi(\varepsilon_1)^{-1}b_1^2. \tag{3.50}$$

3.3.13. Now there is one more step—a rather unexpected one.

Lemma 9. *Suppose that on a positive ray of the t-axis we are given an open set W on which there is specified a monotonically decreasing function $g(t) > 0$ such that*

$$(t - t_0)g(t) \leqslant Cg(t_0)^2 \tag{3.51}$$

for any $t_0, t \in W$ and $t > t_0$ for some $C = \text{const} > 0$. Then W has finite measure.

Apparently Lemma 9 was first proved by Efimov, in connection with the solution of a problem of Hilbert and Cohn-Vossen. But it is undoubtedly of independent interest for mathematical analysis, and the proof of it admits an intuitive geometrical interpretation.

Proof of Lemma 9. Let W_j be the constituent intervals of the set W. We shift all the W_j to the left close to each other together with the parts of the graph $y = g(t)$. We denote the set W, its constituent intervals W_j and the function g after this transformation by $\tilde{W}$, $\tilde{W}_j$ and $\tilde{g}$, and let $\overline{W}$ be the closure of $\tilde{W}$ on the ray $t > 0$. At all points of discontinuity and at the ends of the intervals W_j we redefine $\tilde{g}(t)$ by left continuity. After this we complement the graph $y = \tilde{g}(t)$ at points of discontinuity by vertical segments joining the left and right limiting values of this function; see Fig. 42. The curve obtained in this way, sloping downward and to the right relative to the t-axis, we call L. We draw attention to the fact that for the coordinates of points $(t, y) \in L$ an inequality of the form (3.51) also holds. We take an arbitrary $q \in (0, 1)$ and a point $t_1 \in \tilde{W}_1$. We put $y_1 = \tilde{g}(t_1)$ and construct a sequence $\{t_n, y_n\}$ according to the following rules: $t_{n+1} = t_n + \Delta t_n$,

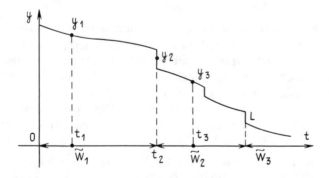

Fig. 42

$n \geqslant 1$; if $t_n \in \overline{W}$, then $(t_n, y_n) \in L$, and $\Delta t_n = Cq^{-1}y_n$; if $\tilde{W}$ is bounded and $t_n \geqslant$ sup $\tilde{W}$ for some $n \geqslant 2$, we put $\Delta t_p = 0$ and $y_p = 0$ for all $p \geqslant n$. A calculation, using (3.51), then shows that $y_n \to 0$ as $n \to \infty$. The domain of definition of the function $\tilde{g}(t) > 0$ cannot extend further to the right than the point that is approached as $y_n \to 0$. Hence mes W is bounded above by the sum of the series consisting of the Δt_n. Using (3.51) again, we discover that it converges: mes $W \leqslant t_1 + \sum_{n=1}^{\infty} \Delta t_n < +\infty$.

Remark. Lemma 9 can be generalized. For example, we can take $Cg(t_0)^{1+\varepsilon}$ with $\varepsilon > 0$ on the right-hand side of (3.51) instead of $Cg(t_0)^2$. Then only the choice of Δt_n changes, otherwise the proof goes though in the same way.

3.3.14. From (3.43)–(3.45) and (3.50) we conclude that the function $b(t)$, the length of the chord $B_t^- B_t^+$ in Fig. 38b, satisfies the conditions of Lemma 9, so mes $W < +\infty$, which contradicts (3.48).

Consequently, (3.34) is impossible. Thus,

$$\alpha_y \langle L_0 \rangle < +\infty. \tag{3.52}$$

From (3.33) and (3.52) it follows that the α-length of the special ray L_0 is finite, and so are its s_0-length and s-length. The fact is that the part $\mathscr{G}$ of the surface $\mathscr{F}$ lies "almost horizontally" in E^3 and $s < s_0\sqrt{1 + \varepsilon_0^2}$ on $\mathscr{G}$, and from the fact that the s_0-length is infinite, as we remarked above, it follows that the α-length is infinite.

Thus, on $\mathscr{F}$ there is a path $\mathscr{L} = P^{-1} \circ \Pi(L_0)$ of finite length that leads to N.

3.3.15. Turning to Theorem B (in 1.4), we now see the following.

a) Either there is a point $N \in \partial \Phi$ where $\partial \Phi$ admits concave support, but then we arrive at a contradiction with the theorem of 3.3.3, by virtue of which the surface is incomplete.

b) Or there is no concave support. Then by the corollary of Lemma 1 the area of the spherical image $\sigma(\Phi) \leqslant 2\pi$. We note in passing that for a hyperbolic paraboloid its spherical image is an open hemisphere, so the equality $\sigma(\Phi) = 2\pi$ for the convex spherical image of a complete surface of negative curvature must be attained. However, under the conditions of Theorem B we again arrive at a contradiction, now between (3.24) and (3.26).[3]

Concluding Remark. Efimov emphasized that in the proof of this theorem we were unable to construct the spherical and normal maps, and all the considerations were carried out on $\mathscr{F}$ by introducing suitable auxiliary metrics. Thus, the surface $\mathscr{F}$ on which we take as the metric form the third fundamental form $d\mathbf{n}^2$, well known in differential geometry, is exactly that Riemannian manifold Φ whose metric boundary $\partial \Phi$ was investigated in the topic of concave supports. Also, the auxiliary surface Ψ can be useful in generalizing Theorem B to classes

---

[3] Using analogous methods Perel'man has established the non-immersibility in E^3 of a more extensive class of complete metrics and even of some incomplete metrics with $K \leqslant 0$; see Perel'man (1990a) in Part I.

of surfaces of smaller smoothness than C^2. Namely, we can hope that then we can "well" approximate the surface Ψ in E^4, possibly non-smooth, by a smoother two-dimensional surface $\tilde{\Psi}$, and then carry out constructions and estimates similar to those presented above on $\tilde{\Psi}$. In connection with the assumptions made here, see 4.7.4 below.

3.4. Sufficient Local Conditions for Plane Maps to be Homeomorphic in the Large. From 3.3 it is obvious that in the proof of Theorem B a large part of the auxiliary material refers not to surfaces, but to maps of the form

$$F: E^2 \rightarrow E^2. \tag{3.53}$$

As above we shall mark the second of these planes by an asterisk.

The map given by formulae of the form

$$Q = F(X), \qquad X = (x, y) \in E^2, \qquad Q = (p, q) \in E_*^2, \tag{3.54}$$

which was considered in 3.3, was a potential map, that is, it had the special form (3.5). However, the approaches presented there can be used to study wider classes of maps. Namely, in this way we obtain the following result.

Theorem (Efimov (1968)). *Suppose that on E^2 we are given a C^1-smooth map* (3.54), *where* $\Delta = \dfrac{\partial(p, q)}{\partial(x, y)} < 0$, *and that there is a function* $a = a(X) > 0$ *such that* $1/a(X)$ *has variation with linear estimate on E^2 and*

$$|\Delta(X)| \geqslant a(X)|J(X)| + a(X)^2, \qquad X \in E^2, \tag{3.55}$$

where $J = q_x' - p_y' = \mathrm{rot}\{p, q\}$. *Then F is a homeomorphism in the large, and $F(E^2)$ is a convex domain on E_*^2.*

The proof in Efimov (1968) is based on geometrical constructions using chains, a special ray and concave support. The arguments presented in Klotz-Milnor (1972) show that the requirement $\Delta < 0$ can be weakened and we can state sufficient conditions for a map of the form (3.54) to be homeomorphic in the large in terms of the eigenvalues.

Special Case. We can guarantee that the map (3.54) is homeomorphic in the large if there are numbers A_1 and A_2 such that

$$\Delta(X) \leqslant A_1 < 0, \qquad |J(X)| \leqslant A_2. \tag{3.56}$$

We draw attention to the fact that conditions (3.56), which are local, guarantee that the map is homeomorphic in the large. Under the conditions (3.56) we can establish a stronger assertion than (3.55); see Efimov (1968): $F(E^2)$ can only be either the whole plane, or a half-plane, or a strip between parallel lines.

Let $D = p_x' + q_y' = \mathrm{div}\{p, q\} = \mathrm{tr}(F')$ and $r = \sqrt{x^2 + y^2}$. Very recently new progress has been attained in this problem area.

Theorem (Aleksandrov (1990)). *Suppose we are given a function* $l: [0, +\infty) \rightarrow (0, +\infty)$ *such that* $\int_\alpha^{+\infty} l(t)\, dt = +\infty$ *for any $\alpha > 0$ and a C^1-smooth map satisfying* *the conditions*

$$\left. \begin{aligned} |\Delta(X)| &\geqslant l(r)|J(X)| + l(r)^2 \\ |D(X)| &\leqslant l(r), \qquad X \in E^2. \end{aligned} \right\} \tag{3.57}$$

Then F is a homeomorphism, and $F(E^2)$ is the whole plane.

The proof in Aleksandrov (1990) is carried out by analytic methods: it relies on an analogue of John's theorem (John (1968)) on homeomorphism of Banach spaces and results of Pourciau (1988). There is another way of reasoning which uses Miller's method for proving the existence of a global inverse function (Miller (1984)) and a theorem of Wintner on the extendability of solutions of a system of ordinary differential equations (Wintner (1945)) side by side with John (1968). In connection with the questions considered in 3.4, see also Geisberg (1970) and Kantor (1978a).

§4. Surfaces with Slowly Varying Curvature. Immersion of Metrics of Negative Curvature in E^3. The Influence of the Metric on the Regularity of a Surface

4.1. Analytic Apparatus

4.1.1. The main analytic means of investigating and solving the problems discussed in this section is the system of Gauss-Peterson-Codazzi equations, written in special ways that are convenient when $K < 0$; however, we shall also discuss here situations in which the a priori smoothness of the surface is less than C^4, and then we shall have to invoke various additional arguments. We recall that this system connects the coefficients of the first and second fundamental forms of the surface

$$I = ds^2 = E\,du^2 + 2F\,du\,dv + G\,dv^2$$

$$II = (\mathbf{n}d^2 r) = -(d\mathbf{n}, dr) = L\,du^2 + 2M\,du\,dv + N\,dv^2.$$

It contains the Gauss equation—a non-linear algebraic equation

$$LN - M^2 = (EG - F^2)K, \tag{4.1}$$

where $K = K_{int}$ is expressed in terms of E, F, G and their first and second order derivatives according to the well-known Gauss formula (see Finikov (1952) or Blaschke (1930), for example), and the two Peterson-Codazzi equations, which for a given metric tensor $g_{ij} = \begin{pmatrix} E & F \\ F & G \end{pmatrix}$ are linear first-order partial differential equations in L, M, N.

When $K < 0$ we can eliminate one of the three functions L, M, N, using (4.1); we then obtain a hyperbolic quasilinear system of two first-order equations, which after reduction to the so-called *Riemann invariants* $r(u, v)$, $s(u, v)$ (see Rozhdestvenskij and Yanenko (1978)) takes the form

$$r'_u + sr'_v = \mathscr{P}(r, s), \qquad s'_u + rs'_v = \mathscr{P}(s, r), \qquad (4.2)$$

where $\mathscr{P}$ is a polynomial of the third degree in the new unknown functions r and s, whose coefficients are linear combinations of the Christoffel symbols and the quantities

$$Q = \tfrac{1}{2}\ln k, \quad Q'_u, Q'_v \qquad (4.3)$$

(see Rozhdestvenskij (1962) and Poznyak (1966), and also Poznyak and Shikin (1986) and Poznyak (1973) in Part I); $k = \sqrt{(-K)}$. The system of equations (4.2) is equivalent to the original Gauss-Peterson-Codazzi system with the additional condition

$$r \neq s. \qquad (4.4)$$

If it is satisfied, then L, M, N can be expressed algebraically in terms of r and s. We note that r and s have a simple geometrical meaning: they are the slopes of the images of the asymptotic lines of the unknown surface on the plane of the parameters (u, v). The asymptotic lines serve as characteristics of the quasi-linear system (4.2), so condition (4.4) signifies the requirement that the system is non-degenerate on a given solution of it.

4.1.2. Following Efimov (1966b), in asymptotic coordinates we shall write $E = e^2$, $G = g^2$; then $F = eg \cos \omega$, where ω is the angle between the asymptotic lines on the surface. In these coordinates $L = N = 0$ identically, and if we introduce the normalized coefficients of II, dividing L, M, N by $\sqrt{(EG - F^2)}$, and putting $m = (EG - F^2)^{-1/2}M$, then the Gauss equation takes the form $m^2 = K_{\text{int}}$. However, if the curvature K_{int} is variable, and we discuss the question of an immersion of the given metric in E^3, then in asymptotic coordinates we cannot regard $E = e^2$, $F = eg \cos \omega$, $G = g^2$ and K_{int} as known functions, since the asymptotic lines themselves are an object of the extrinsic geometry of the surface. Nevertheless, as we shall see below, the equations of surface theory, written in asymptotic coordinates, play an important role. In these coordinates the Peterson-Codazzi equations take the form

$$\begin{cases} (Ek)'_u + Fk'_v = 0, \\ Fk'_u + (Gk)'_v = 0. \end{cases} \qquad (4.5)$$

Regarding as the first family of asymptotic lines the one that has right-handed screw torsion, we go over in (4.5) from derivatives with respect to u, v to derivatives $\dfrac{\partial}{\partial s_i}$ with respect to arc length of asymptotic lines of the i-th family and to derivatives $\dfrac{\partial}{\partial s_i^*}$ with respect to arc length of curves orthogonal to them; then (4.5) can be reduced to the form (see Efimov and Poznyak (1961))

$$\frac{\partial \ln(ek)}{\partial s_2} = \frac{\partial Q}{\partial s_1^*} \sin \omega, \qquad \frac{\partial \ln(gk)}{\partial s_1} = -\frac{\partial Q}{\partial s_2^*} \sin \omega. \qquad (4.6)$$

Next, from the Gauss-Bonnet formula, applied to a quadrangle $\mathscr{D}$ of the asymptotic net, after certain transformations and a limiting process when $\mathscr{D}$ contracts to a point, we obtain an equation of the form

$$\mathscr{L}_1(\mathscr{L}_2(\omega)) = \mathscr{A} \sin \omega \tag{4.7}$$

for the angle ω, where $\mathscr{L}_1$ and $\mathscr{L}_2$ are quasilinear first-order differential operators. For the specific form of them and an expression for the multiplier $\mathscr{A}$, see Efimov (1966b). When $k = 1$ (4.7) goes over to (1.6).

From (4.7) by means of (4.6), introducing the auxiliary quantities

$$\mathscr{W}_i = (-1)^{i+1}\left(\left(\frac{\partial^2}{\partial s_i^2}\sqrt{k}\right)_\Gamma - k^{3/2}\right)k^2, \tag{4.8}$$

where the symbol $\left(\dfrac{\partial^2}{\partial s_i^2}\cdots\right)_\Gamma$ denotes the second-order derivative with respect to arc length of the geodesic touching an asymptotic line of the i-th family, we can obtain the following relations for the derivatives of the geodesic curvatures $\varkappa_i$ of the asymptotic lines:

$$\begin{cases} \dfrac{\partial}{\partial s_2}(e^3 k^{3/2}\varkappa_1) = \mathscr{W}_1 e^3 \sin \omega, \\[2mm] \dfrac{\partial}{\partial s_1}(g^3 k^{3/2}\varkappa_2) = \mathscr{W}_2 g^3 \sin \omega. \end{cases} \tag{4.9}$$

The right-hand sides of (4.6) and the quantities (4.8) admit estimates in terms of the intrinsic metric of the surface. Hence, in equations (4.6)–(4.7) (or (4.6) and (4.9)) there is revealed a connection between the intrinsic and extrinsic geometries of the surface: these equations, together with the Beltrami-Enneper theorem on the torsion of the asymptotic lines, gives information about the influence of the intrinsic Gaussian curvature on the asymptotic net of the surface, and hence on the possibility of positioning the surface in space, since a surface of negative curvature is woven from its asymptotic lines.

4.1.3. The classical derivation of the Gauss-Peterson-Codazzi equations is based on the equality of mixed third-order derivatives ($r'''_{uuv} = r'''_{uvu}$, and so on) and so it assumes that r is C^3-smooth as a function of u and v. However, in specific cases when the coordinate lines are constructed on the basis of certain geometrical requirements, a higher smoothness of the surface itself is sometimes needed. Thus, for example, on transition from arbitrary coordinates to asymptotic coordinates, there is, generally speaking, a loss of smoothness of the vector-function r. This loss may be two units (from C^n to C^{n-2}, see Hartman and Wintner (1953)), so to derive equations (4.5)–(4.7) and (4.9) it was originally necessary (Efimov and Poznyak (1961)) to assume that $\mathscr{F} \in C^5$; see also Efimov (1966b). However, there is another way of deriving them, in which we need significantly less smoothness of the surface. Before presenting it, we mention certain facts.

Suppose that the metric ds^2 is considered in semigeodesic coordinates ξ, η, in which the geodesics are the curves $\eta = 0$ and $\xi = $ const; then it can be written like (2.8), but with ρ and φ replaced by ξ and η with the additional condition

$$B(0, \eta) = 1, \qquad B'_\xi(0, \eta) = 0. \qquad (4.10)$$

Such coordinates (ξ, η) are uniquely determined by the choice of the initial point $\xi = \eta = 0$ and the direction at this point of one of the coordinate curves, say the ξ-curve. We can prove (see Rozendorn (1966)) that the order of smoothness of $K_{\mathrm{int}}(\xi, \eta)$ is invariant under transition from one coordinate system (ξ, η) to another similar one.

Below, when we speak about the C^r-smoothness of K_{int}, we shall have in mind the intrinsic curvature $K_{\mathrm{int}} = K$, specified in semigeodesic coordinates ξ, η satisfying (4.10).

We shall say that a surface $\mathscr{F}$ belongs to the class $\mathfrak{A}^n$, $n = 2, 3, \ldots, \infty$, if on $\mathscr{F}$ there is a no more than countable set (possibly empty) of points $\{O_j\}$ isolated in its intrinsic metric and such that $\mathscr{F} \backslash \bigcup O_j \in C^n$, and $\mathscr{F} \in C^1$ in the large. If, moreover, each point O_j on $\mathscr{F}$ has a neighbourhood in which *the mean curvature* $H = \frac{1}{2}(k_1 + k_2)$ *is bounded*, we shall say that $\mathscr{F} \in \mathfrak{B}^n$. We have the following inclusions of these classes:

$$\begin{cases} \mathfrak{A}^m \subset \mathfrak{A}^n, \qquad \mathfrak{B}^m \subset \mathfrak{B}^n \quad \text{when } m > n \geqslant 2; \\[2mm] C^n \subset \mathfrak{B}^n \subset \mathfrak{A}^n \quad \text{when } n \geqslant 3; \\[2mm] C^2 \subset \mathfrak{B}^2 \subset C^{1,1}; \qquad \mathfrak{B}^2 \subset \mathfrak{A}^2 \subset C^1. \end{cases} \qquad (4.11)$$

Example. The surfaces of which we spoke in 2.1.8 and 2.3.3 (see Figs. 5 and 29) can be constructed so that they belong to the class $\mathfrak{B}^\infty$. This tells us that Theorem B (see 1.4) not only cannot be extended from the class C^2 to the class $C^{1,1}$, but even to the narrower class $\mathfrak{B}^2$.

We can prove that equations (4.5) and (4.6) remain true if $\mathscr{F} \in \mathfrak{A}^2$, $K < 0$, $K \in C^1$. The plan of the proof is as follows. First, on the assumption that $\mathscr{F} \in C^2$, $K < 0$, $K \in C^1$, on the basis of an analysis of the relations between the elements of an infinitely small net quadrangle $\mathscr{D}$ we can establish the existence in asymptotic coordinates of a mixed derivative of the radius vector r, understood as the limit

$$r''_{uv} = \lim_{\Delta u, \Delta v \to 0} \frac{1}{\Delta u \Delta v} [r]_{\mathscr{D}}, \qquad (4.12)$$

where the symbol $[\ldots]_{\mathscr{D}}$, just as in (1.2), denotes the alternating sum. After this we can establish the equality

$$r''_{uv} = keg\mathbf{n} \sin \omega - (Q'_v r'_u + Q'_u r'_v). \qquad (4.13)$$

With higher smoothness of the surface, when $\mathscr{F} \in C^4$, (4.13) is none other than the well-known Gauss derivational formula for the mixed derivative r''_{uv}, written in asymptotic coordinates. From (4.13) by identity transformations, *without additional differentiations*, we obtain the equalities (4.5), and from them (4.6).

Next we carry out an investigation of the structure of the asymptotic net in a neighborhood of the points O_j and establish that (4.13), (4.5) and (4.6) can also be extended to these points.

The equalities (4.9) and (4.7) are valid if $\mathscr{F} \in \mathfrak{A}^3$, $K < 0$, $K \in C^2$. They were also first established by means of an analysis of the relations between elements in $\mathscr{D}$ on the assumption of higher smoothness: $\mathscr{F} \in C^3$ when $K < 0$, $K \in C^2$, and then by a limiting process they were extended to the points O_j.

Simultaneously with the deduction of the formulae (4.5), (4.6), (4.9) and (4.7) the existence of all the derivatives occurring in them can be established. For the details see Rozendorn (1966) and Rozendorn (1966) in Part I.

4.2. $(h, \varDelta)$-Metrics. In this and the next section we consider the classes singled out by Efimov (Efimov (1966b)) of surfaces for which K varies so slowly from point to point that the properties appear to be similar to those of surfaces of constant curvature.

One of these classes (a more restricted one) is defined as follows. Suppose we are given numbers h and $\varDelta$ such that

$$\varDelta > 0, \qquad h \in (0, 1), \qquad h\varDelta^2 > \tfrac{3}{4} \tag{4.14}$$

and suppose that

$$|(\ln R)'| \leqslant \frac{1}{\varDelta}, \qquad (\sqrt{R})'' \leqslant (1 - h)|K|^{3/4}, \tag{4.15}$$

where (as above) $R = |K|^{-1/2}$, and primes denote derivatives with respect to the arc length of an arbitrary geodesic.

Definition (Efimov (1966b)). The surface of negative curvature (1.7) has an $(h, \varDelta)$-metric if the inequalities (4.14)–(4.15) are satisfied under the normalization $K \leqslant -1$.

Definition (Efimov (1966b)). A domain $\mathscr{U}$ on a plane (or on a surface homeomorphic to a plane) with a metric specified in it is called a *simple zone* if its completion $\bar{\mathscr{U}}$ in this metric is non-compact, and the metrical boundary $\partial\mathscr{U} = \bar{\mathscr{U}} \backslash \mathscr{U}$ is either empty or consists of at most two non-compact connected components.

Examples. The branches of a hyperbola divide E^2 into three simple zones. L^2 in its metric and a half-plane of L^2 are simple zones, and the conditions (4.15) on L^2 are satisfied for any $\varDelta > 0$ and $h \in (0, 1)$.

Theorem (Efimov (1966b)). *A simple zone with $(h, \varDelta)$-metric that is isometrically and regularly immersed in E^3 cannot contain more than one complete asymptotic line of each of two of its families.*

The plan of the proof is as follows. If a simple zone $\mathscr{U}$ contains two complete asymptotic lines, then they bound a simple zone $\mathscr{U}_1 \subset \mathscr{U}$. It can be established that $\mathscr{U}_1$ contains a simple zone $\mathscr{U}_2$ homeomorphic to a half-plane whose boundary $\partial\mathscr{U}_2$ has a special structure: $\partial\mathscr{U}_2 = l_- \cup l_0 \cup l_+$, where l_-, l_0, l_+ are arcs of

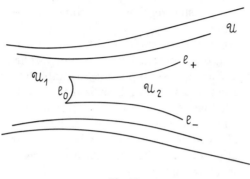

Fig. 43

asymptotic lines, and l_0 is homeomorphic to an interval and belongs to one family of asymptotic lines, and l_- and l_+ are homeomorphic to a ray and belong to another family; in addition, on l_0 the geodesic curvature $\varkappa \neq 0$ and it has a sign such that l_0 is converted to $\mathscr{U}_2$ by its "convexity", Fig. 43. In asymptotic coordinates (u, v) we carry out a two-sided estimate of the integral

$$I = \iint_{\mathscr{U}_2} \sin \omega \, du \, dv, \qquad \sin \omega > 0. \tag{4.16}$$

In addition we discover the following contradiction: according to the upper estimate the integral (4.16) converges, $I < +\infty$, and according to the lower estimate it diverges, $I = +\infty$; for the details see Efimov (1966b). In the proof one uses the conditions (4.14)–(4.15) and the analytic apparatus considered in 4.1.2–4.1.3. The proof goes through when $K \in C^2$, $\mathscr{F} \in \mathfrak{A}^3$.

Theorem A (see 1.2) can be obtained as a corollary of the theorem about a simple zone.

On surfaces in E^3 there occur simple zones with $(h, \varDelta)$-metric and even with constant curvature that contain a complete asymptotic line. As an example we can mention the Bianchi-Amsler surface; see 2.1.5 and Fig. 19 above.

4.3. q-Metrics. A surface has a q-metric if on it

$$|R'| \leqslant q < +\infty, \tag{4.17}$$

where $q > 0$; this term was introduced in Efimov (1966b).

Special cases of q-metrics are the $(h, \varDelta)$-metrics: they have $q = 1/\varDelta$ if $K \leqslant -1$. When $q = 0$ we obtain a metric of constant curvature.

On a surface with a q-metric the asymptotic net has the generalized Chebyshev property. Namely, the difference of the lengths of the opposite sides of the net quadrangle $\mathscr{D}$, measured in the l-metric (3.25), is estimated above by half the product of the number q and the area of the quadrangle $\mathscr{D}$, measured in the same l-metric (Efimov (1966b)).

An *asymptotic quadrant* is a simple zone on a surface of negative curvature, homeomorphic to a half-plane and bounded by the arcs of two asymptotic lines of different families, starting from one point and indefinitely extended, each on one side (Efimov (1966b)).

Asymptotic quadrants exist on surfaces with a q-metric in E^3, for example, on the universal covering of a smooth half of a pseudosphere. The generalized Chebyshev property of a net makes it possible to establish (Efimov (1966b)) that if $q < \sqrt{2}$, then the net of asymptotic lines is regular in the large in a whole asymptotic quadrant, that is, it is homeomorphic to a Cartesian net in an ordinary quadrant on E^2. The simultaneous use of the structure of an asymptotic net in quadrants, the methods of arguments with concave supports and the additional investigation of the spherical image of a geodesic curve make it possible to prove the following theorem.

Theorem (Perel'man (1988b) in Part I). *There exists a $q_0 > 0$ such that when $q \in [0, q_0]$ a half-plane with q-metric does not admit a C^∞-smooth isometric immersion in E^3.*

Here we understand a half-plane as a simple zone whose boundary is a geodesic curve. Hence the result of Efimov (Efimov (1975) in Part I) about the non-immersibility of a Lobachevskij half-plane in E^3 is generalized to a q-metric with small q.

As to complete q-metrics, for any $q \geqslant 0$ they are non-immersible into E^3 in the class of C^2-smooth surfaces according to Corollary 2 of Theorem B (see 1.4).

Although Theorem B itself cannot be extended to the class $\mathfrak{B}^2$, the assertion of its Corollary 2 remains in force in the classes $\mathfrak{B}^2$ and $\mathfrak{A}^2$. The point is that the application of the analytic apparatus that was in question in 4.1.3 enables us to carry out such an investigation of an asymptotic net in the neighbourhood of isolated points of violation of C^2-smoothness of a surface, which shows that at these points there cannot arise a saddle order $m > 2$ so long as the intrinsic metric of the surface is a q-metric; see Rozendorn (1966) in Part I. Hence there follows the local homeomorphism of the Gaussian spherical map. This in turn makes it possible (Rozendorn (1972)) to apply the methods of arguments with concave supports and to extend to the class $\mathfrak{A}^2$ the theorem of 3.3.3 under the additional assumption that $K \in C^1$, and also the result about non-immersibility of a complete q-metric in E^3.

4.4. Immersion of Metrics of Negative Curvature in E^3

4.4.1. The local aspect of the given question is considered in detail in Part III, so we discuss here problems connected with geometry in the large.

Examples of well-known surfaces such as the helicoid, hyperbolic paraboloid, catenoid, hyperboloid of one sheet, show that under a rapid decrease of K at infinity many complete metrics of negative curvature are immersible in E^3, but general theorems of this type are so far unknown.[4] Examples of a pseudosphere

---

[4] Shikin (see Shikin (1990)) and Perel'man (see Perel'man (1990a) in Part I) have obtained some sufficient conditions for immersibility in E^3 for a metric of a rotation of the form (5.7) (see above) with $K < 0$.

and a "flabby" pseudosphere have forced us to assume that compact domains of two-dimensional metrics with curvature of the form (1.7) are immersible in E^3, but for a long time this problem has not yielded a solution. Success here was attained by Poznyak (Poznyak (1966)), who constructed an immersion of a geodesic disc of arbitrarily large radius with any negative variable curvature of the form (1.7). This paper served as a stimulus for the development of research on immersibility in the form of surfaces with a boundary of incomplete but non-compact metrics and negative curvature. This problem is closely connected with the theory of Monge-Ampère hyperbolic equations and quasilinear systems. An important contribution to this was made by Shikin: he obtained a series of results about immersibility in E^3 when $K < 0$ of non-compact manifolds with boundary and extended the methods he developed to certain cases of the vanishing of the curvature, which corresponds to hyperbolic equations with degeneration.

Referring an interested reader to the details in the surveys Aminov (1982), Poznyak and Shikin (1974), Poznyak and Sokolov (1977) and Poznyak (1973) in Part I and the papers Shikin (1975, 1980, 1982), we mention two concrete results which can be formulated particularly transparently.

4.4.2. Theorem (Poznyak (1977b)). *Any proper polygon in L^2 admits a C^∞-smooth isometric immersion into E^3.*

We recall (Alekseevskij, Vinberg and Solodovnikov (1988)) that a proper polygon in L^2 can have both ordinary and infinitely distant vertices, but it does not contain any half-plane. In Fig. 44 in the conformal model of L^2 on a disc there is represented a proper polygon $\mathcal{M}$ with five vertices, of which four are infinitely distant.

The proof of the theorem is based on the fact that the compact part of a polygon admits an immersion in E^3 with great arbitrariness, and this arbitrariness can be used so that then to the compact part we "add" in E^3 immersed neighbourhoods of each of the infinitely distant vertices, preserving the necessary smoothness of the whole surface.

4.4.3. Let $w(\eta)$ be a positive continuous function, specified on the whole η-axis. We denote by Π_w the simple zone on the $\xi\eta$-plane specified by the inequalities

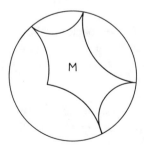

Fig. 44

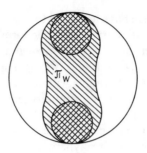

Fig. 45

$$|\xi| < w(\eta), \qquad -\infty < \eta < +\infty. \tag{4.18}$$

Suppose that in Π_w there is specified a C^4-smooth metric

$$ds^2 = d\xi^2 + B(\xi, \eta)^2 \, d\eta^2, \qquad K = -B^{-1}B''_{\xi\xi} < 0 \tag{4.19}$$

and that (4.10) is satisfied. The following theorem holds.

Theorem (Tunitskij (1987)). *The simple zone Π_w with metric (4.10), (4.19) can be isometrically immersed in E^3 in the form of a C^3-smooth surface.*

Hence it follows, in particular, that in E^3 there are C^3-smooth surfaces of constant negative curvature on which there are distributed two non-intersecting horocycles of L^2. This geometrical fact was unknown until quite recently, although surfaces of constant negative curvature had been the subject of much research (see, in particular, the survey in Steuerwald (1936)), and independently of Tunitskij's theorem it was established in Kaidasov and Shikin (1986). A convex simple zone $\Pi_w \subset L^2$ which according to the theorem in Tunitskij (1987) admits an immersion in E^3 and contains two horocycles is schematically shown in Fig. 45 and on it is marked a hatching, and on the horocycles a double hatching.

The proof of the theorem is analytic, and is based on the methods of the theory of differential equations (see Bellman and Kalaba (1965) and Rozhdestvenskij and Yanenko (1978)). By means of auxiliary methods, which can be shown to be artificial, in the whole simple zone (4.18) one can construct a solution of the system (4.2) that satisfies the requirement (4.4), so it follows that the metric (4.19) specified in Π_w can be immersed in E^3. In this proof there is a geometric background that goes back to the paper of Poznyak (1966): the solution of the posed problem about immersion is sought in the form of a surface contracted into a narrow roll—narrower than a simple zone which is widened rapidly in its metric ds^2 as $\eta \to \pm\infty$.

4.5. Study of the Boundary of a Surface. From Theorem B formulated in 1.4 it follows that there are no complete C^2-smooth surfaces in E^3 with negative curvature separated from zero. Therefore there naturally arises the question of studying the boundary of a C^n-smooth surface with curvature as in (1.7), $n \geqslant 2$.

For this we need to assume that the surface under consideration is inextensible over its intrinsic geometric boundary, preserving its former C^n-smoothness and the fact that K is negative. Clearly, if inextensibility over the boundary is not required, then the diversity of singularities that can appear on the boundary becomes boundless.

Amsler proved in Amsler (1955) that an inextensible surface in E^3 of curvature as in (1.1) has a smooth arc in the structure of its intrinsic-geometric boundary. In addition he assumed that the surface is either analytic or C^3-smooth with the additional condition that the set of its branch points is empty or finite. The fact that there can be branch points was shown in the same paper (Amsler (1955)). The importance of the requirement that the set of branch points is finite was shown by Wissler (Wissler (1972)): he gave a construction of a surface with constant curvature $K = -1$ whose intrinsic-geometric boundary contains a countable set of branch points and does not contain any smooth arc.

Amsler's result on the existence of a smooth arc of the boundary in the case of (h, Δ)-metrics of variable curvature was extended by Vinogradskij (Vinogradskij (1970)), who imposed additional conditions on the behaviour of asymptotic lines close to the boundary.

Without additional requirements on the derivatives of the curvature the result is certainly incorrect. This can be seen from the example of the surface shown in Fig. 5a and repeatedly discussed above: its C^∞-smooth part represents a surface with a boundary consisting of four points.

4.6. Surfaces with Slowly Varying Curvature in a Riemannian Space

4.6.1. So as not to divert attention to additional questions about smoothness, we shall assume here that all the objects belong to the class C^∞. On a two-dimensional surface $\mathcal{F}$ in a three-dimensional Riemannian space $\mathfrak{M}^3$ there are defined the intrinsic curvature K_{int} according to (2.7), and the extrinsic curvature K_{ext} as the ratio of the discriminants of the second and first quadratic forms analogous to (4.1). They are not equal to each other, and Gauss's theorem instead of (2.18) takes the form of the equality

$$K_{\text{int}} = K_{\text{ext}} + K_{\mathfrak{M}}, \qquad (4.20)$$

where $K_{\mathfrak{M}}$ is the curvature of the space $\mathfrak{M}^3$ in the direction of two-dimensional area element tangent to $\mathcal{F}$ at a given point of it; it is also defined by analogy with (2.7).

The following analogue of Hilbert's theorem is known (Spivak (1975)). In a three-dimensional space of constant curvature there are no complete regular surfaces with constant negative intrinsic and extrinsic curvatures.

Of course, constancy and negativity of just the intrinsic curvature is insufficient, since in L^3 there are planes L^2. Constancy and negativity of the extrinsic curvature is also insufficient: in three-dimensional elliptic space there is the *Clifford torus* with locally Euclidean intrinsic metric and therefore, according to (4.20), constant negative extrinsic curvature.

4.6.2. Surfaces with slowly varying negative curvature in a three-dimensional Riemannian space of negative curvature have been studied by Brandt. He showed that the situation is more complicated there than in the Euclidean case: the generalized Chebyshev property and the regularity of the structure in the large of an asymptotic net is ensured by the smallness of a quantity that depends not only on the rapidity of the change from point to point of the curvature of the intrinsic metric of the surface itself, but also on the curvature of the ambient space at points of the surface; see Brandt (1970a), Brandt (1970b); see also Poznyak (1991).

4.7. Influence of the Metric on the Regularity of a Surface

4.7.1. The problem of immersion in E^3 of a metric of positive curvature $K > 0$ is elliptic, and therefore the high regularity of the metric necessarily implies the high regularity of the surface; see Pogorelov (1969) in Part I. When $K < 0$ the situation is very different because the problem is hyperbolic (Holmgren (1902)). As an illustration it is again convenient to take surfaces of constant curvature. Locally they can be obtained by solving Darboux's problem for the equation (1.6). The regularity of the solution $\omega(u, v)$, and together with it the regularity of the surface, will depend on the boundary conditions—the data of Darboux's problem—although the metric is analytic.

However, if we impose additional conditions of regularity of boundary type, then from the smoothness of the metric we can obtain deductions about the smoothness of the surface when $K < 0$.

Let $\mathscr{G}$ be a domain homeomorphic to a disc on a surface of the class $\mathfrak{B}^2$, and let $\mathscr{G}_\varepsilon$ be a subdomain of it homeomorphic to a circular ring and consisting of those points of it that are at a distance less than ε from its boundary $\partial\mathscr{G}$ (Fig. 46). Then if $\mathscr{F}$ is an immersion in E^3 of the metric with negative C^r-smooth curvature $(1 \leqslant r \leqslant +\infty)$ and $\mathscr{F} \in C^{r+1}$ in $\mathscr{G}_\varepsilon$, then $\mathscr{F} \in C^{r+1}$ everywhere inside $\mathscr{G}$.

If $\mathscr{G} = \mathscr{D}$ is a quadrangle of an asymptotic net, then instead of conditions in $\mathscr{D}_\varepsilon$ it is sufficient to assume that of two adjacent sides of the quadrangle $\mathscr{D}$ their curvature as curves in E^3 is a C^{r-2}-smooth function of arc length, $2 \leqslant r < +\infty$.

The proofs (see Rozendorn (1967) in Part I) rely on the analytic apparatus described in 4.1. The a priori smoothness of $\mathfrak{B}^2$ cannot be reduced to that of $\mathfrak{A}^2$, as examples constructed by Cohn-Vossen show (Cohn-Vossen (1928)).

Corollary. *Let $\mathscr{F}$ be a C^∞-smooth surface in E^3 homeomorphic to a closed disc and having curvature $K < 0$. We assume that $\mathscr{F}$ is bendable (see §6 of Part III)*

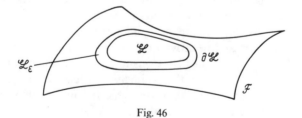

Fig. 46

and it is subjected to a bending under which its C^2-smoothness is preserved and its C^∞-smoothness in some domain $\mathscr{F}_\varepsilon$ situated along its boundary $\partial\mathscr{F}$ and homeomorphic to a circular ring is preserved too. Then on the bent surface its C^∞-smoothness is preserved everywhere.

4.7.2. We continue the discussion of questions touching on the regularity of a surface of negative curvature. We recall that from Aleksandrov's theorem on the smoothness of a convex surface with bounded specific curvature (see Aleksandrov (1942) and Aleksandrov (1948), both in Part I) it follows that an edge, that is, a curve of violation of C^1-smoothness, if it exists on such a surface, must go out to its boundary (in particular, on a closed convex surface with bounded specific curvature an edge cannot exist). We consider an example which shows that when $K < 0$ there can exist a singular curve that does not violate the C^1-smoothness of a surface even if the intrinsic metric of the surface is analytic.

Example. We take the Bianchi-Amsler surface (see 2.1.5 and Fig. 19) with curvature $K = -1$ and such that its rectilinear generator L is intersected on some interval by asymptotic lines of another family at an angle $\omega \neq \pi/2$ (this can certainly be done, Amsler (1955)). An arbitrary point of this interval can be taken as the centre of a geodesic disc of sufficiently small radius $\rho > 0$. Clearly the rectilinear generator splits a geodesic disc with centre $P \in L$ into two semi-discs. We denote one of them by V, joining its boundary to it (Fig. 47a). Let V' be the image of the geodesic semidisc V under symmetry in E^3 with respect to the line L. Then the surface $V \cup V'$ (with boundary) represents an isometric $C^{1,1}$-smooth *piecewise-analytic embedding* in E^3 of a closed disc (of radius ρ) taken from L^2. On a diameter of it the C^2-smoothness of the surface is violated. The fact is that there occurs here a break of the asymptotic lines that intersect a diameter $I \subset L$ (Fig. 47b).

As is obvious from what follows, the interval $I \subset L$ in this example can be regarded as an analogue of the edge in the theorem of Aleksandrov: here the intrinsic metric has high smoothness, and the singular curve I goes out to the boundary of the surface $\mathscr{F} = V \cup V'$ at its ends.

4.7.3. We can continue the analogy in question in 4.7.2. On an ovaloid an edge is possible of course (Fig. 48), but by Aleksandrov's theorem in a neighbour-

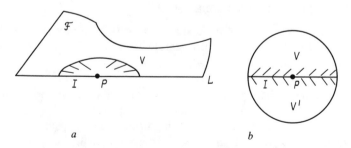

Fig. 47

Fig. 48

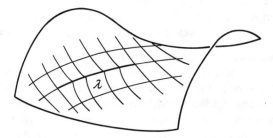

Fig. 49

hood of each of the end-points of the edge the specific curvature is unbounded, and so the loss of high smoothness of the intrinsic metric is inevitable. On a surface of negative curvature such a singular arc λ is also possible, which does not go out to the boundary of the surface $\mathscr{F}$ under consideration, and on λ the C^1-smoothness of the surface and the local homeomorphism of its spherical image are preserved, and at all points of the curve λ there exists $K_{\text{int}} < 0$ but the C^2-smoothness of the surface is violated. Such an arc together with the asymptotic lines intersecting it is shown schematically in Fig. 49. In addition, $\mathscr{F} \setminus \lambda$ can have arbitrarily high smoothness. Surfaces with the described singularity on an arc exist in E^3 in an arbitrarily small neighbourhood of an arbitrary C^∞-smooth surface of negative curvature (Rozendorn (1985)). By analogy with convex surfaces, here it is also natural to pose the problem of the loss of smoothness of the intrinsic metric at the ends of a singular arc λ. The answer to this (at least, a partial one) is presented in 4.7.4.

4.7.4. Suppose we are given in E^3 a surface $\mathscr{F}$ (with boundary $\partial \mathscr{F}$) homeomorphic to a closed disc and on it a simple arc λ ($\lambda \subset \mathscr{F}$), and the following conditions are satisfied.

a) $\mathscr{F} \in C^1$, $\mathscr{F} \setminus \lambda \in C^n$, $3 \leqslant n \leqslant \infty$.

b) The Gaussian spherical map on the surface $\mathscr{F}$ is a local homeomorphism.

c) $\lambda \in C^2$ and as a curve in E^3 it has bounded curvature.

d) On $\mathscr{F} \setminus \lambda$ the mean curvature H of the surface under consideration and the geodesic curvatures $\varkappa_j$ of its asymptotic lines are bounded.

Then we have the following result.

Theorem (Rozendorn (1988)). *If under the conditions* a, b, c, d *the surface* $\mathcal{F}$ *has* $K_{int} \in C^m$, $2 \leqslant m \leqslant n - 1$, $K_{int} < 0$, *then either both ends of the arc* λ *are situated on the boundary of the surface* $(\partial\lambda \subset \partial\mathcal{F})$ *or* $\mathcal{F} \in C^{m+1}$.

Remarks. 1. In the example in 4.7.2 all the conditions a, b, c, d are satisfied, $\lambda = I$, $n = \infty$. In the examples of Rozendorn (1985) mentioned in 4.7.3 we can also ensure that all the conditions a, b, c, d are satisfied.

2. If we assume that the admissible singularity is one-point (supposing that $\mathcal{F} \in \mathfrak{B}^n$), then the removability of the singularity is guaranteed under much weaker conditions; see Rozendorn (1966) in Part I.

The proof of the theorem in 4.7.4 (published in detail in the Transactions of the Moscow Mathematical Society) uses as auxiliary means the analytic apparatus from 4.1, the result on regularity of a surface inside an asymptotic quadrangle formulated in 4.7.1, and relies on approximation of the normal map of the surface $\mathcal{F}$ in a neighbourhood of a supposed end-point of the arc λ by some special C^2-smooth non-potential maps.

§ 5. On Surfaces with a Metric of Negative Curvature in Multidimensional Euclidean Spaces

5.1. Bieberbach's Theorem. In connection with Hilbert's theorem discussed in § 1, it is natural to pose the question of isometric immersions of L^p in E^N. A local embedding of L^p in E^{2p-1} was constructed by Schur (Schur (1886)), and in E^{2p-2} it is impossible to embed L^p even locally: see Cartan (1919–1920) and Liber (1938).

The first global result about the immersibility of L^2 in E^N, but when $N = \infty$, was obtained by Bieberbach (Bieberbach (1932)): he constructed an analytic embedding of L^2 in a Hilbert space. This embedding is given by explicit formulae. Before writing them out, we introduce an auxiliary notation, which will be useful in what follows. Let $F_m(z)$ be a finite or infinite collection of functions of the complex argument $z = u + iv$, possibly not analytic. We put

$$x_{2m-1} = \operatorname{Re} F_m(z), \qquad x_{2m} = \operatorname{Im} F_m(z), \qquad m = 1, \dots, p, \tag{5.1}$$

assuming that x_j are Cartesian coordinates in E^N, $N \geqslant 2p$.

Bieberbach's construction seems very simple: in (5.1) we take $F_m(z) = m^{-1/2}z^m$, $p = N = \infty$. Calculation shows that as a result when $\rho = |z| < 1$ we obtain an embedding in E^∞ of the metric $ds^2 = (1 - \rho^2)^{-2}(du^2 + dv^2)$, and this, as we know (Alekseevskij, Vinberg and Solodovnikov (1988)), is one of the forms of writing the metric of L^2. In addition, $x_1 = u$, $x_2 = v$, (2.2) is satisfied, and the surface is projected one-to-one onto a disc $\rho < 1$ of the $x_1 x_2$-plane.

Bieberbach proved that the surface of the form (5.1) that he constructed does not lie in any subspace of finite dimension and has the following interesting property: the whole group of motions is induced on it by motions of the ambient space. In the same paper (Bieberbach (1932)) he presented the proof due to

Schmidt of the fact that in E^N ($N < \infty$) on an immersed L^2 the group of all its motions cannot be induced by motions of E^N, although it was not known then whether such immersions were possible. In this connection see also Kadomtsev (1978).

5.2. Embedding and Immersion of L^p in E^N. An isometric embedding of L^p, $p \geqslant 3$, in E^∞ was constructed by Blanuša (Blanuša (1953)). The first embeddings of L^p in E^N for finite N were also constructed by him (Blanuša (1955)). Thus a result of Beltrami was generalized: not only for $p = 2$, but also for any $p \geqslant 3$, the geometry of a p-dimensional Lobachevskij space is realized in the form of the intrinsic geometry of some surfaces of Euclidean space of sufficiently large dimension. In Blanuša (1955)

$$N = 6p - 5 \quad \text{for } p \geqslant 3, \qquad N = 6 \quad \text{for } p = 2, \tag{5.2}$$

and the embeddings are C^∞-smooth and given by explicit formulae. We note that the general constructions of Nash and Gromov for an isometric embedding in E^N of a non-compact Riemannian manifold M^p require dimensions $N = N(p)$ considerably higher than (5.2); see Nash (1956), Gromov and Rokhlin (1970) and Gromov (1987), all in Part I.

We consider the embedding found by Blanuša in detail for $p = 2$, $N = 6$, in order to illustrate in this special case the methods that he developed. An important role in his construction is played by two pairs of auxiliary functions. One of them we denote below by $\varphi_j = \varphi_j(u) \geqslant 0$ ($j = 1, 2$); these are C^∞-smooth functions. The other one is $\psi_j(u) > 0$; these are step functions (that is, piecewise-constant functions).

For what follows it is important that the step functions on parts of constancy for differentiation behave as constants, but in the large on the number axis as $u \to \pm\infty$ they can increase arbitrarily quickly.

The functions φ_j are subject to the following conditions: $\varphi_1(u)^2 + \varphi_2(u)^2 = 1$ and each of the φ_j has zeros of infinite order at points of discontinuity of the function ψ_j, $j = 1, 2$. Clearly, φ_j with such properties can be chosen with great arbitrariness. In addition, we use one auxiliary function $f = f(u) \in C^\infty$, $f > 0$, whose concrete form will be stated below and depends on the metric ds^2 that is to be embedded in E^6. From these functions we compose the products

$$g_j(u) = \psi_j(u)^{-1}\varphi_j(u)f(u), \tag{5.3}$$

and the expression $g = (g_1')^2 + (g_2')^2 = g(u) > 0$.

For fixed f and φ_j the step functions ψ_j are chosen so that

$$g(u) < 1 - \varepsilon \quad (\varepsilon > 0) \quad \text{when } -\infty < u < +\infty, \tag{5.4}$$

and then in (5.1) we take

$$F_m = g_m(u)e^{iv\psi_m(u)}, \qquad m = 1, 2, \tag{5.5}$$

and put

$$x_5 = \int_0^u \sqrt{1 - g(\tau)}\, d\tau, \qquad x_6 = u. \tag{5.6}$$

Thanks to the fact that the jumps of the step functions ψ_j are as it were "cancelled" by the zeros of the functions φ_j, all the $x_j(u, v)$, $1 \leqslant j \leqslant 5$, are continuous here. The additional research in Blanuša (1955) shows that their partial derivatives of all orders are also continuous. The condition (2.2) is ensured on account of (5.4) and (5.6). Thus the resulting surface turns out to be C^∞-smooth (but certainly not analytic!), and it is projected one-to-one on the $x_5 x_6$-plane.

Where both the ψ_j are constants we have

$$ds^2 = du^2 + B(u)^2\, dv^2, \tag{5.7}$$

where $B(u) = \sqrt{1 + f(u)^2}$. At those points where the ψ_j are discontinuous, the metric (5.7) preserves continuity. In particular, if we take $f(u) = \operatorname{sh} u$, then the formulae (5.1), (5.5) and (5.6) give a C^∞-smooth embedding of L^2 in E^6.

By analogy with 2.2 we can assume that here the $x_5 x_6$-plane plays the role of a short embedding, which then undergoes a corrugation. In addition, the range of the dimension of the ambient space enables us to manage with a unique corrugation described by the formulae (5.5) and straight away in all directions at all points with tightening the metric of the short embedding to the necessary value.

In this construction, for a given $B(u) \in C^n$, $n \geqslant 2$, $B(u) \geqslant 1$, we can take $f(u) = \sqrt{B(u)^2 - 1}$, obtaining C^n-smooth embeddings of metrics of the form (5.7) in E^6.

We can modify the construction, substituting $f(u) = B(u)$ in (5.6) and also putting $x_6 = 0$. Then self-intersections arise in the surface, and the dimension of the ambient space is reduced by one. The form of the metric (5.7) is preserved. In particular, in this way, when $f(u) = \operatorname{ch} u$, and also when $f(u) = e^{\pm u}$, we obtain C^∞-smooth immersions of L^2 in E^5.

The last construction can be generalized in dimension and gives a C^∞-smooth immersion of L^p in E^N when $N = 4p - 3$ (Henke (1981)).

As for embeddings, using Blanuša's methods, his result (5.2) in dimension N when $p > 2$ has recently been improved by Azov (Azov (1985)). He established that L^p admits a C^∞-smooth embedding in E^N for

$$N = \begin{cases} 5p - 3 & \text{when } p = 2l + 1, \\ 5p - 4 & \text{when } p = 2l. \end{cases} \tag{5.8}$$

It is not known up to now whether we can further improve the result (5.8).

Immersions of domains from L^p to E^N, where $N \geqslant 2p - 1$, have been studied by Aminov; see Aminov (1980), (1983), (1988). He showed (Aminov (1983)) that when $p = 3$, $N = 5$ with such immersions there is associated a system of equations that generalizes the equations known in mechanics for the motion of a rigid body with a fixed point, and when $p = 4$, $N = 7$ with an immersion there is associated a tensor field (see Aminov (1988)) analogous to the tensor of an electromagnetic field, known in relativistic electrodynamics.

Xavier stated a supposition in Xavier (1985) about the non-immersibility of L^p in the large in E^{2p-1} and in the same paper for some special class of complete Riemannian manifolds of constant negative curvature obtained by factorizing L^p he proved their non-immersibility in E^{2p-1}.

5.3. Piecewise-Analytic Immersion of L^2 in E^4. We now give up the requirement of smoothness of the surface and consider a somewhat more general problem than in the title of this section: we shall look for an immersion in E^4 of the metric of a rotation of the form (5.7), assuming that $B(u) \geqslant 0$, $-\infty < u < +\infty$. The construction presented here is due to Sabitov (Sabitov (1989a)). Following Sabitov (1989a) we take in (5.1) $F_m = \rho_m(u)e^{i\alpha_m(u,v)}$, $m = 1$, 2, putting $\alpha_m = \psi_m(u)v + \beta_m(u)$, where ψ_m are step functions, as in Blanuša's method. Putting ds^2 equal to (5.7), calculated from (5.1), we obtain the system of equations

$$\begin{cases} \sum_{k=1}^{2} ((\rho_k\beta_k')^2 + (\rho_k')^2) = 1, \\ \sum_{k=1}^{2} (\psi_k\rho_k^2\beta_k')^2 = 0, \qquad \sum_{k=1}^{2} (\psi_k\rho_k)^2 = B^2. \end{cases} \tag{5.9}$$

Here and below derivatives with respect to the argument u are denoted by a prime.

Next we apply an artificial method: we put

$$\rho_1 = \frac{1}{\psi_1} B \cos \Theta, \qquad \rho_2 = \frac{1}{\psi_2} B \sin \Theta \tag{5.10}$$

and

$$\beta_1' = \frac{1}{\psi_2} A \sin^2 \Theta, \qquad \beta_2' = -\frac{1}{\psi_1} A \cos^2 \Theta, \tag{5.11}$$

where $A = A(u)$ and $\Theta = \Theta(u)$ are new unknown functions. By (5.10) and (5.11) the last two of the equations (5.9) are satisfied identically. For what follows it is convenient to introduce one more auxiliary function:

$$\Phi(u) = (\psi_1\psi_2)^{-2}(A \sin \Theta \cos \Theta)^2, \tag{5.12}$$

so that

$$A = \frac{2\psi_1\psi_2}{\sin 2\Theta} \sqrt{\Phi}. \tag{5.13}$$

In addition, we put $b = (\ln B)'$. Then, taking account of (5.12), the first equation of the system (5.9) takes the form

$$\Phi(u) = 1 - (b \cos \Theta - \Theta' \sin \Theta)^2\psi_1^{-2} - (b \sin \Theta + \Theta' \cos \Theta)^2\psi_2^{-2}. \tag{5.14}$$

The subsequent plan of action is as follows. For $\Theta = \Theta(u)$ we take an arbitrary continuous monotonic piecewise-analytic function, specified on the whole axis $-\infty < u < \infty$, admitting discontinuities of the first kind in its first order derivative at those points where $\Theta = m\pi/2$, m being an integer. Then we construct step functions $\psi_k(u) > 0$ so that the discontinuities of $\psi_1(u)$ fall at the zeros of cos $\Theta(u)$;

the discontinuities of $\psi_2(u)$ fall at the zeros of sin $\Theta(u)$;

as $u \to \pm\infty$ the functions $\psi_k(u)$ increase so rapidly that in (5.14) the inequality $\Phi(u) \geqslant 0$ is guaranteed.

At the points of discontinuity the functions $\psi_k(u)$ are defined either by right continuity or by left continuity – for what follows it makes no difference.

After this from (5.10) the functions $\rho_k(u)$ are defined uniquely, and from (5.13) and (5.11) the functions $\beta_k(u)$ are defined to within constants of integration; as a result we obtain the unknown immersion. The constants of integration in β_k in fact turn out as piecewise-constant functions of the argument u, but they have no effect on the metric ds^2. The vector function $r(u, v)$ is continuous, and its C^1-smoothness is violated only where $\psi_k(u)$ or Θ' are discontinuous; everywhere where they are continuous, the condition (2.2) is satisfied when $p = 2$. The derivatives $\rho'_k(u)$ are bounded in each compact part of the u-axis. Generally speaking, the derivatives $\beta'_k(u)$ are unbounded, but in the expression for r'_v they do not occur, and in r'_u they occur only in the composition of products

$$\rho_1\beta'_1 = B\sqrt{\Phi}\sin\Theta, \qquad \rho_2\beta'_2 = -B\sqrt{\Phi}\cos\Theta. \qquad (5.15)$$

A simple calculation shows that on any finite interval of the u-axis the quantities (5.15) are bounded, so $r \in C^{0,1}$. Further investigation shows that on singular curves, which are possible when $\Theta = m\pi/2$, local homeomorphism of the immersion is also satisfied.

5.4. Some Results on Non-Immersibility in the Multidimensional Case. The Riemannian manifolds considered in this and the following subsection and their immersions are assumed for simplicity of formulation to be C^∞-smooth.

5.4.1. The problem of Hilbert and Cohn-Vossen discussed in 1.4 admits various generalizations under transition to the multidimensional case. First of all, we can pose the question of the smallest dimension N for which in E^N an isometric immersion of a complete non-compact Riemannian manifold $\mathfrak{M}^p$ is possible, for which at all points in all two-dimensional directions the sectional curvature is negative (such Riemannian spaces under the condition of simple connectedness are an important special case of Hadamard manifolds (see Eberlein (1985) and Shiga (1984)), which requires additionally that it is uniformly separated from zero. A more special problem is to prove for $\mathfrak{M}^p = L^p$ that $N > 2p - 1$ (as Xavier assumed in Xavier (1985), see above at the end of section 5.2). From what was stated above it is obvious that for L^p we obtain what is so far only a two-sided estimate of $N(p)$. The lower estimate $N \geqslant 2p - 1$ is based only on local arguments (see Liber (1938), Cartan (1919–1920) and Schur (1886)), and the upper estimate $N \leqslant 4p - 3$ is based on the presentation of concrete immesions (Henke (1981)). A deeper investigation of this problem is a matter for the future, and it is natural to expect that in the case of Hadamard manifolds the unknown N may depend not only on their dimension p, but also on other properties, particularly if on the immersions we impose additional conditions, for example the requirement of saddle-shape; in this connection see Chapter 3 of Part I.

However, we can pose the problem differently: assume that the sectional curvature on $\mathfrak{M}^p$ is negative and separted from zero not in all directions but only on certain families of submanifolds on which $\mathfrak{M}^p$ is fibered. In this direction there are interesting results. The point is that in Riemannian geometry there are known conditions under which an isometric immersion $F: \mathfrak{M}^p \to E^N$ of a Riemannian product

$$\mathfrak{M} = \mathfrak{M}^p = \mathfrak{M}_0 \times \mathfrak{M}_1 \times \cdots \times \mathfrak{M}_l \qquad (5.16)$$

in a Euclidean space E^N represents a product of immersions, that is,

$$\begin{cases} \mathscr{F}(X) = \mathscr{F}(X_0, X_1, \ldots, X_l) = \{f(X_0), \ldots, f(X_l)\}, \\ X_j \in \mathfrak{M}_j, \quad f_j: \mathfrak{M}_j \to E^{n_j}, \quad N = \sum_{j=0}^{l} n_j \end{cases} \qquad (5.17)$$

and E^N is the direct sum of its pairwise orthogonal subspaces E^{n_j}. Therefore under suitable additional requirements on the factors in (5.16) and the dimension N from here we can extract results about non-immersibility, using Theorem B for example (see 1.4).

Let us formulate one of these conditions. Suppose that in the decomposition (5.16):

(A) $\mathfrak{M}_0$ is a connected flat Riemannian manifold of dimension $p_0 \geqslant 0$;

(B) $\mathfrak{M}_j$, $1 \leqslant j \leqslant l$, are connected Riemannian manifolds of dimension $p_j \geqslant 2$ on each of which the set of those of its points where one of the sectional curvatures vanishes does not have interior points in $\mathfrak{M}_j$. Then the following result holds.

Theorem (Moore (1976)). *If*

$$N = p_0 + l + \sum_{j=1}^{l} p_j \qquad (5.18)$$

and conditions (A) *and* (B) *are satisfied, then any isometric immersion of the Riemannian manifold* (5.16) *in* E^N *represents a product of immersions of the form* (5.17).

Corollary. *Suppose that the Riemannian space* $\mathfrak{M}$ *is a product of the form* (5.16), *and that conditions* (A) *and* (B) *are satisfied and there is a number* i, $1 \leqslant i \leqslant l$, *such that the dimension* $p_i = 2$, *and the two-dimensional factor* $\mathfrak{M}_i$ *is a complete space and its intrinsic metric has negative curvature satisfying the conditions of Theorem B* (p. 95). *Then the Riemannian product of* $\mathfrak{M}$ *in the large is non-immersible in a Euclidean space whose dimension does not exceed the quantity* (5.18).

For a proof and other details we refer the reader to the papers Alexander and Maltz (1976) and Moore (1971).

5.4.2. Apart from what we have said, there is another way of generalizing the problem of Hilbert and Cohn-Vossen to the multidimensional case. Namely, as an analogue of the intrinsic Gaussian curvature K_{int} when $p \geqslant 3$ we can take the *Ricci curvature*. In this way we obtain the following result.

Theorem (Smyth and Xavier (1987)). *Suppose that a complete p-dimensional Riemannian manifold $\mathfrak{M}_p$ with negative Ricci curvature is immersed in E^N in the form of a hypersurface ($N = p + 1$). Then the upper bound of the Ricci curvature on $\mathfrak{M}^p$ is equal to zero*
a) *if $p = 3$, or*
b) *if $p > 3$, and the sectional curvatures on $\mathfrak{M}^p$ take not all real values.*

In its proof, by the way, the investigation of the properties of the spherical map of the hypersurface under consideration plays an essential role.

5.5. On Closed Surfaces of Negative Curvature. The result about the impossibility in E^3 of a closed surface of non-positive Gaussian curvature can be generalized in dimension. Namely, we have the following result.

Theorem (Chern and Kuiper). *A compact p-dimensional Riemannian manifold of non-positive sectional curvature does not admit an isometric immersion in E^{2p-1}* (Chern and Kuiper (1952)).

Of course, from this and Moore's theorem we can also extract corollaries about the non-immersibility of Riemannian products. The problems connected with further development of the result of Chern and Kuiper are reflected in the surveys Aminov (1982) and Poznyak and Sokolov (1977). Here we return to the case $p = 2$. It is known that any compact two-dimensional Riemannian manifold (including one of negative curvature) can be isometrically immersed in E^6; see Gromov (1987) in Part I. Thus, between the lower and upper estimates for N there remains the interval $4 \leqslant N \leqslant 6$, and the theorem of Chern and Kuiper does not prohibit the existence of two-dimensional closed surfaces of negative intrinsic curvature even in E^4. Examples show that they really exist there. From a visual point of view their construction seems fairly simple. Regarding E^3 as a hyperplane in E^4, we take in E^3 the surface constructed in 2.1.8 and shown in Fig. 5a, denoting it by $\mathscr{F}_1$. We then take a second copy $\mathscr{F}_2$ of the same surface, obtained from $\mathscr{F}_1$ by a parallel translation in E^4 in the direction orthogonal to the original E^3. We then bend towards each other the narrowed tubes of the surfaces $\mathscr{F}_1$ and $\mathscr{F}_2$, smoothing the edges arising at their meeting place. After this we cut out on $\mathscr{F}_1$ and on $\mathscr{F}_2$ the neighbourhoods of their singular points (those where C^2-smoothness is violated, and the saddle order $m = 3$). Suitable calculation shows (see Rozendorn (1962)) that the boundaries of the perforations thus formed can also be joined to each other in E^4 by "tubes" of negative curvature. As a result we obtain an unknown closed surface, a topological sphere with seven handles, whose intrinsic metric has negative curvature. The construction can be carried out so that this surface has smoothness C^∞. One of the intermediate stages of its construction is shown schematically in Fig. 50. In view

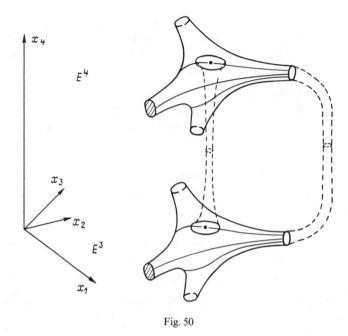

Fig. 50

of the compactness its curvature is separated from zero and the supremum in (1.11) is finite, so its intrinsic metric ds^2 is a q-metric. In this connection Efimov stated the supposition, at present not confirmed but also not disproved, that this surface can be deformed, preserving the sign of its curvature and reducing the value of q to zero, and we obtain in the limit a closed two-dimensional surface situated in E^4 of constant negative curvature.

Commentary on the References

The most important steps in setting up and developing the theory of surfaces of negative curvature were taken in the papers Beltrami (1868a), Hilbert (1901), Bieberbach (1932) and Efimov (1964) and are closely connected with the problem of interpreting non-Euclidean geometry.

The methods developed in the study of surfaces of negative curvature have been applied in the theory of mappings (Efimov (1964), Klotz-Milnor (1972), Efimov (1968)) and in the investigation of non-linear hyperbolic partial differential equations; in this connection see Efimov (1976), Geisberg (1970), Azov (1983), Azov (1984), Brys'ev (1985), Shikin (1980), and also Hartman and Wintner (1952), Bäcklund (1905), Steuerwald (1936), Galeeva and Sokolov (1984a), Gribkov (1977), Kantor (1981), Poznyak (1979), Popov (1989), and Tenenblat and Terng (1980), Terng (1980).

The equation $z''_{xy} = \sin z$, which appeared and was first used in the geometrical papers Chebyshëv (1878), Hazzidakis (1879), and Hilbert (1901), turned out to be one of the typical representatives of non-linear equations of mathematical physics (see Barone, Esposito, Magee and Scott (1971)), and many current problems are connected with it. One can become acquainted with them in Barone and Paternò (1982), Enz (1964), Kosevich (1972) and Lamb (1971), and in Barone and Paternò (1982) the reader will find an extensive bibliography.

The theory of surfaces plays a major role in mechanics in the study of the properties and the calculation of thin elastic shells (see Gol'denveizer (1976), Vekua (1982), Pogorelov (1967), and also Gol'denveizer (1979) and Gol'denveizer, Lidskij and Tovstik (1979)). However, the application of specific properties of surfaces of negative Gaussian curvature to these questions is still only in its initial stages (Klabukova (1983)).

In recent years new interesting connections have been revealed between geometrical problems concerned with surfaces of negative curvature and various questions of mechanics and physics. In this connection, see in particular Aminov (1971), Aminov (1983), Aminov (1988), Rozendorn (1980), and Poznyak and Popov (1991).

Despite the fact that in recent decades the theory of surfaces of negative curvature has been significantly developed and has actually become an independent branch of geometry, in the current world literature there is not a single monograph specially devoted to it.

There are the well-known surveys Efimov (1966a) and Poznyak and Shikin (1974). In addition, there are the quite detailed (partly survey) journal articles Efimov (1966b), Klotz-Milnor (1972), Shikin (1975), Eberlein (1979), Hartman and Wintner (1951), and also Rozendorn (1966) in Part I, and the surveys Poznyak (1977a), Poznyak and Shikin (1980) and Poznyak and Shikin (1986) are devoted to various aspects of the problem. Here we can refer to the last chapter of the monograph Bakel'man, Verner and Kantor (1973) in Part I. In addition, a number of questions relating to the geometry of surfaces of negative curvature are elucidated in the surveys Aminov (1982), Poznyak and Sokolov (1977) and Poznyak (1973) in Part I.

The bibliography given below does not cover all journal articles on surfaces of negative curvature. However, it enables us to trace the main directions of this branch of geometry and its interconnections with other branches of mathematics and its applications. The reader can find additional references in the surveys listed here.

References*

Adel'son-Vel'skij, G.M. (1945): Generalization of a geometrical theorem of S.N. Bernstein. Dokl. Akad. Nauk SSSR 49, 399–401 (Russian), Zbl.61,373

Aleksandrov, A.D. (1938): On a class of closed surfaces. Mat. Sb. 4, 69–77 (Russian), Zbl.20,261

Aleksandrov, V.A. (1990): Efimov's theorem on differential criteria for homeomorphism. Mat. Sb. 181, 183–188. Engl. transl.: Math. USSR. Sb. 69, No. 1, 197–202 (1991).

Alekseevskij, D.V., Vinberg, E.B., Solodovnikov, A.G. (1988): The geometry of spaces of constant curvature. Itogi Nauki Tekh. Ser. Sovrem. Probl. Mat., Fundam. Napravleniya 29, 5–146, Zbl.699.53001. Engl. transl. in: Encycl. Math. Sc. 29, Springer-Verlag, Heidelberg (in preparation)

Alekseevskij, D.V., Vinogradov, A.M., Lychagin, V.V. (1988): The basic ideas and concepts of differential geometry. Itogi Nauki Tekh., Ser. Sovrem. Probl. Mat., Fundam. Napravleniya 28, 5–297, Zbl.675.53001. Engl. transl.: Encycl. Math. Sc. 28, Springer-Verlag, Heidelberg (1991)

Alexander, S., Maltz, R. (1976): Isometric immersions of Riemannian products in Euclidean space. J. Differ. Geom. 11, 47–57, Zbl.334.53053

Aminov, Yu.A. (1968): n-dimensional analogues of S.N. Bernstein's integral formula. Mat. Sb., Nov. Ser. 75, 375–399. Engl. transl.: Math. USSR, Sb. 4, 343–367 (1968), Zbl.176,187

Aminov, Yu.A. (1971): An energy condition for the existence of a vortex. Mat. Sb., Nov. Ser. 86, 325–334. Engl. transl.: Math. USSR, Sb. 15, 325–334 (1972), Zbl.221.52005

Aminov, Yu.A. (1980): Isometric immersions of domains of n-dimensional Lobachevskij space in $(2n - 1)$-dimensional Euclidean space. Mat. Sb., Nov. Ser. 111, 402–433. Engl. transl.: Math. USSR, Sb. 39, 359–386 (1981), Zbl.431.53023

---

*For the convenience of the reader, references to reviews in Zentralblatt für Mathematik (Zbl.), compiled using the MATH database, and Jahrbuch über die Fortschritte der Mathematik (Jbuch) have, as far as possible, been included in this bibliography.

Aminov, Yu.A. (1982): Embedding problems: geometrical and topological aspects. Itogi Nauki Tekh., Ser. Sovrem. Probl. Geom. *13*, 119–156. Engl. transl.: J. Sov. Math. *25*, 1308–1331 (1984), Zbl. 499.53018

Aminov, Yu.A. (1983): Isometric immersions of domains of three-dimensional Lobachevskij space in five-dimensional Euclidean space and the motion of a rigid body. Mat. Sb., Nov. Ser. *122*, No. 1, 12–30. Engl. transl.: Math. USSR, Sb. *50*, 11–30 (1985), Zbl.539.53006

Aminov, Yu.A. (1988): Isometric immersions of domains of n-dimensional Lobachevskij space in Euclidean spaces with a flat normal connection. A model of a gauge field. Mat. Sb., Nov. Ser. *137*, 275–299. Engl. transl.: Math. USSR, Sb. *65*, No. 2, 279–303 (1990), Zbl.663.53018

Amsler, M.H. (1955): Des surfaces à courbure négative constante dans l'espace à trois dimensions et de leurs singularités. Math. Ann. *130*, 234–256, Zbl.68,351

Arnol'd, V.I. (1990): Catastrophe Theory (3rd edition). Nauka, Moscow. English transl. (of the 2nd. edition): Springer-Verlag, Berlin-Heidelberg-New York, 1984, Zbl.517.58002

Azov, D.G. (1983): On a class of hyperbolic Monge-Ampère equations. Usp. Mat. Nauk *38*, No. 1, 153–154. Engl. transl.: Russ. Math. Surv. *38*, No. 1, 170–171 (1983), Zbl.524.35072

Azov, D.G. (1984): Some generalizations of a theorem of N.V. Efimov on hyperbolic Monge-Ampère equations. In: Differential equations and their applications, Collect. Artic., Moscow 1984, 60–64 (Russian), Zbl.595.35088

Azov, D.G. (1985): Immersion by D. Blanuša's method of some classes of complete n-dimensional Riemannian metrics in Euclidean spaces. Vestn. Mosk. Univ., Ser. I, No. 5, 72–74. English transl.: Mosc. Univ. Math. Bull. *40*, No. 5, 64–66 (1985), Zbl.581.53040

Bäcklund, A.V. (1905): Concerning surfaces with constant negative curvature. New Era Printing Co., Lancaster

Baikoussis, C., Koufogiorgos, T. (1980): Isometric immersions of complete Riemannian manifolds into Euclidean space. Proc. Amer. Math. Soc. *79*, 87–88, Zbl.466.53003

Bakievich, N.I. (1960): Some boundary-value problems for equations of mixed type that arise in the study of infinitesimal bendings of surfaces of revolution. Usp. Mat. Nauk *15*, No. 1, 171–176 (Russian), Zbl.91,341

Barone, A., Esposito, F., Magee, C.J., Scott, A.C. (1971): Theory and application of the sine-Gordon equation. Riv. Nuovo Cimento *1*, 227–267

Barone, A., Paternò, G. (1982): Physics and Applications of the Josephson Effect. Wiley, Chichester-New York

Bellman, R., Kalaba, R. (1965): Quasilinearization and Nonlinear Boundary-Value Problems. Elsevier, New York, Zbl.139,107

Beltrami, E. (1868a): Saggio di interpretazione della geometria non-euclidea. Giorn. di Mat. Napoli *6*, 285–315, Jbuch 1, 275

Beltrami, E. (1868b): Teoria fondamentale degli spazii di curvature constante. Ann. Mat. Pura Appl. *2*, 232–255, Jbuch 1, 209

Beltrami, E. (1872): Sulla superficie di rotazione che serve di tipo alle superficie pseudosferiche. Giorn. di Mat. Napoli *10*, 147–159

Bernstein, S.N. (1960a): On a geometrical theorem and its applications to partial differential equations of elliptic type. Collected Works, Vol. 3, 251–258, Zbl.178,447

Bernstein, S.N. (1960b): Strengthening of a theorem on surfaces of negative curvature. Collected Works, Vol. 3, 361–370, Zbl.178,447

Bianchi, L. (1927): Lezioni di geometria differenziale. Vol. 1, parte 1–2, 4. ed., Zanichelli, Bologna, Jbuch 48, 784

Bieberbach, L. (1932): Eine singularitätenfreie Fläche konstanter negativer Krümmung im Hilbertschen Raum. Comment. Math. Helv. *4*, 248–255, Zbl.5,82

Blanuša, D. (1953): Eine isometrisch singularitätenfreie Einbettung des n-dimensionalen hyperbolischer Räume im Hilbertschen Raum. Monatsh. Math. *57*, 102–108, Zbl.53,109

Blanuša, D. (1955): Über die Einbettung hyperbolischer Räume in Euklidische Räume. Monatsh. Math. *59*, 217–229, Zbl.67,144

Blaschke, W. (1930): Vorlesungen über Differentialgeometrie und Geometrische Grundlagen Einsteins Relativitätstheorie. I. Elementare Differentialgeometrie. 3rd. ed., Springer-Verlag, Berlin, Jbuch 56, 588

Bolin, B. (1956): An improved barotropic model and some aspects of using the balance equation on three-dimensional flows. Tellus 8, 61–75

Brandt, I.S. (1970a): Surfaces of negative extrinsic curvature in a Riemannian space with non-positive Riemannian curvature. Dokl. Akad. Nauk SSSR 194, 747–749. Engl. transl.: Sov. Math. Dokl. 11, 1270–1272 (1970), Zbl.213,481

Brandt, I.S. (1970b): Some properties of surfaces with slowly changing negative extrinsic curvarture in a Riemannian space. Mat. Sb., Nov. Ser. 83, 313–324. Engl. transl.: Math. USSR, Sb. 12, 313–324 (1971), Zbl.203,243

Brys'ev, A.B. (1985): An estimate of the domain of regularity of solutions of some non-linear differential inequalities. Ukr. Geom. Sb. 28, 19–21. Engl. transl.: J. Sov. Math. 48, No. 1, 15–17 (1990), Zbl.584.34006

Burago, Yu.D. (1989), Shefel', S.Z.: The geometry of surfaces in Euclidean spaces. Itogi Nauki Tekh., Ser. Sovrem. Probl. Mat., Fundam. Napravleniya 48, 5–97. Engl. transl. in: Encycl. Math. Sc. 48, Springer-Verlag, Heidelberg, 1–85, 1992 (Part I of this volume)

Burago, Yu.D., Zalgaller, V.A. (1974): Sufficient criteria for convexity. Zap. Nauchn. Semin. Leningr., Otd. Mat. Inst. Steklova 45, 3–52. Engl. transl.: J. Sov. Math. 8, 395–435 (1978), Zbl. 348.52003

Cartan, E. (1919–1920): Sur les variétés de courbure constante d'un espace euclidien ou non-euclidien. Bull. Soc. Math. France 47, 125–160; 48, 132–208, Zbuch.47,692

Chebyshëv, P.L. (1878): Sur la coupe des habits. Lecture to the Association française pour l'avancement des sciences, 28 August 1878

Chern, S.S., Kuiper, N.H. (1952): Some theorems on the isometric imbedding of compact Riemannian manifolds in Euclidean space. Ann. Math., II. Ser. 56, 422–430, Zbl.52,176

Cohn-Vossen, S.E. (1928): Die parabolische Kurve. Math. Ann. 99, 273–308, Jbuch 54, 498

Dini, V. (1865): Sulle superficie nelle quali la somma due raggi di curvature principale è constante. Ann. Mat. Pura Appl. 7, 5–18.

Dubrovin, B.A., Novikov, S.P., Fomenko, A.T. (1979): Modern Geometry. Methods and Applications. Nauka, Moscow, Zbl.433.53001. Engl. transl.: Grad. Texts Math. 93, Part I (1984) and 104, Part II (1985)

Eberlein, P. (1979): Surfaces of nonpositive curvature. Mem. Amer. Math. Soc. 218, 90 pp., Zbl.497.53043

Eberlein, P. (1985): Structure of manifolds of nonpositive curvature. Lect. Notes Math. 1156, 86–153, Zbl.569.53020

Efimov, N.V. (1953): Investigation of a one-to-one projection of a surface of negative curvature. Dokl. Akad. Nauk SSSR 93, 609–611, Zbl.52,172

Efimov, N.V. (1963): The impossibility in three-dimensional Euclidean space of a complete regular surface with a negative upper bound of the Gaussian curvature. Dokl. Akad. Nauk SSSR 150, 1206–1209. Engl. transl.: Sov. Math. Dokl. 4, 843–846 (1963), Zbl.135,400

Efimov, N.V. (1964): The origin of singularities on surfaces of negative curvature. Mat. Sb., Nov. Ser. 64, 286–320 (Russian), Zbl.126,374

Efimov, N.V. (1966a): Hyperbolic problems in the theory of surfaces. Proc. Int. Congr. Math., Moscow 1966, 177–188. Engl. transl.: Transl., II. Ser., Am. Math. Sec. 70, 26–38 (1968), Zbl.188,536

Efimov, N.V. (1966b): Surfaces with slowly changing negative curvature. Usp. Mat. Nauk 21, No. 5, 3–58. Engl. transl.: Russ. Math. Surv. 21, No. 5, 1–55 (1966), Zbl.171,199

Efimov, N.V. (1968): Differential criteria for homeomorphism of certain mappings with application to the theory of surfaces. Mat. Sb., Nov. Ser. 76, 499–512. Engl. transl.: Math. USSR, Sb. 5, 475–488 (1968), Zbl. 164,215

Efimov, N.V. (1976): Estimates of the dimensions of the domain of regularity of certain Monge-Ampère equations. Mat. Sb. Nov. Ser. 100, 356–363. Engl. transl.: Math. USSR, Sb. 29, 319–326 (1978), Zbl.332.35018

Efimov, N.V. (1984): Surfaces of negative curvature. Math. Encyclopaedia, vol. 4, 163–173

Efimov, N.V., Poznyak, E.G. (1961): Some transformations of the basic equations of the theory of surfaces. Dokl. Akad. Nauk SSSR 137, 25–27. Engl. transl.: Sov. Math. Dokl. 2, 225–227 (1961), Zbl.108,339

Enz, U. (1964): Die Dynamik der Blochschen Wand. Helv. Phys. Acta 37, 245–251

Finikov, S.P. (1950): Theory of Congruences. GTTI, Moscow-Leningrad. German transl.: Akademie-Verlag, Berlin (1959), Zbl.85,367

Finikov, S.P. (1952): A Course of Differential Geometry. GTTI, Moscow (Russian) Zbl.48,149

Fomichëva, Yu.G. (1978): On the existence of a surface of negative curvature in E^3 with a given product of the principal curvatures defined as a function of the normal. In: Modern geometry. Repub. Collect. Sci. Works, Leningrad 1978, 136–139 (Russian), Zbl.423.53004

Fomichëva, Yu.G. (1979a): Stability of the solution of the problem of constructing in E^3 a surface with a given spherical mapping from its negative curvature defined as a function of the normal. In: Questions of global geometry. Collect. Sci. Works, Leningrad 1979 (Russian), Zbl.477.53004

Fomichëva, Yu.G. (1979b): On the existence in E^3 of an asymptotic quadrangle with a univalent spherical mapping and given negative Gaussian curvature. In: Studies in the geometry of immersed manifolds and in projective geometry. Collect. Sci. Works, Leningrad 1979, 109–121 (Russian), Zbl.532.53003

Galchenkova, R.I., Lumiste, Yu.G., Ozhigova, E.P., Pogrebysskij, I.B. (1970): Ferdinand Minding. Nauka, Leningrad (Russian), Zbl.225.01007

Galeeva, R.F., Sokolov, D.D. (1984a): On the geometrical interpretation of solutions of some non-linear equations of mathematical physics. In: Investigations on the theory of surfaces in Riemannian spaces. Interuniv. Collect. Sci. Works, Leningrad 1984, 8–22 (Russian), Zbl.616.35074

Galeeva, R.F., Sokolov, D.D. (1984b): Infinitesimal deformations of a hyperboloid of one sheet. ibid., 41–44 (Russian), Zbl.601.53004

Gauss, D.F. (1823–1827): Disquisitiones generales circa superficies curvas. Comm. Soc. Reg. Scient. Göttingensis Recentiores 6, 99–146. Also in: Gauss, C.F., Werke, Band 4, Gesellschaft der Wissenschaft, Göttingen, 1873, pp. 217–258

Geisberg, S.P. (1970): On the properties of the normal mapping generated by the equation $rt - s^2 = -f^2(x, y)$. Mat. Sb., Nov. Ser. 82, 224–232. Engl. transl.: Math. USSR, Sb. 11, 201–208 (1970), Zbl.194,525

Gisina, F.A. et al. (1976): Dynamic Meteorology. Gidrometeoizdat, Leningrad (Russian)

Gol'denveizer, A.L. (1976): The Theory of Thin Elastic Shells. 2nd ed., Nauka, Moscow (Russian), Zbl.461.73052

Gol'denveizer, A.L. (1979): Mathematical rigidity of surfaces and physical rigidity of shells. Izv. Akad. Nauk SSSR Mekh. Tverd. Tela 1979, No. 6, 66–67. Engl. transl.: Mech. Solids 14, No. 6, 53–63 (1979)

Gol'denveizer, A.L., Lidskij, B.V., Tovstik, P.E. (1979): Free Oscillations of Thin Elastic Shells. Nauka, Moscow (Russian)

Gribkov, I.V. (1977): The construction of some regular solutions of the "sine-Gordon" equation by means of surfaces of constant curvature. Vestn. Mosk. Univ., Ser. I., No. 4, 78–83. Engl. transl.: Mosc. Univ. Math. Bull. 32, No. 4, 63–67 (1977), Zbl.359.53003

Hartman, P., Wintner, A. (1951): On the asymptotic curves of a surface. Am. J. Math. 73, 149–172, Zbl.42,157

Hartman, P., Wintner, A. (1952): On hyperbolic partial differential equations. Am. J. Math. 74, 834–864, Zbl.48,333

Hartman, P., Wintner, A. (1953): On asymptotic parametrisation. Am. J. Math. 75, 488–496, Zbl.50,378

Hazzidakis, J.N. (1879): Über einige Eigenschaften der Flächen mit konstante Krümmungsmass. J. Reine Angew. Math. 88, 68–73, Jbuch 11, 527

Heinz, E. (1955): Über Flächen mit eineindeutiger Projektion auf der Ebene, deren Krümmungen durch Ungleichungen eingeschränkt sind. Math. Ann. 129, 451–454, Zbl.65,372

Henke, W. (1981): Isometrische Immersionen des n-dimensional hyperbolischen Räumes H^n im E^{4n-3}. Manuscr. Math. 34, 265–278, Zbl.458.53035

Hilbert, D. (1901): Flächen von konstanter Gauss'schen Krümmung. Trans. Amer. Math. Soc. No. 2, 87–99, Jbuch 32, 608

Hilbert, D. (1903): Über Flächen von konstanter Gauss'sche Krümmung. In: Grundlagen der Geometrie. 2nd. ed., Teubner, Leipzig, pp. 162–175, Jbuch 34, 223

Hilbert, D., Cohn-Vossen, S.E. (1932): Anschauliche Geometrie. Springer Verlag, Berlin. Engl. transl.: Geometry and the Imagination. Chelsea, New York, 1952, Zbl.5,112

Holmgren, E. (1902): Sur les surfaces à courbure constante négative. C. R. Acad. Sci. Paris *134*, 740–743, Jbuch 33, 643

Ivanova-Karatopraklieva, I. (1983): Sufficient conditions for rigidity of some classes of surfaces that project one-to-one on a plane. Banach Cent. Publ. *12*, 83–93 (Russian), Zbl.561.53006

Ivanova-Karatopraklieva, I. (1984a): Non-rigidity of some classes of surfaces of revolution of mixed curvature. C. R. Akad. Bulg. Sci. *37*, 569–572 (Russian), Zbl.546.53003

Ivanova-Karatopraklieva, I. (1984b): Conditions for rigidity of some classes of surfaces of mixed type. I. Serdica *10*, 287–302 (Russian), Zbl.567.53002

Ivanova-Karatopraklieva, I. (1985): Rigidity of some classes of surfaces of revolution of mixed curvature whose boundary is not a parallel. Serdica *11*, 330–340 (Russian), Zbl.611.53006

John, F. (1968): On quasi-isometric mappings. I. Commun. Pure Appl. Math. *21*, 77–110, Zbl.157,458

Kadomtsev, S.B. (1978): The impossibility of some special immersions of Lobachevskij space. Mat. Sb., Nov. Ser. *107*, 175–198. Engl. transl.: Math. USSR, Sb. *35*, 461–480 (1979), Zbl.394.53032

Kaidasov, Zh., Shikin, E.V. (1986): An isometric immersion in E^3 of a convex domain of the Lobachevskij plane containing two horocycles. Mat. Zametki *39*, 612–617. Engl. transl.: Math. Notes *39*, 335–338 (1986), Zbl.609.53005

Kantor, B.E. (1970): On the question of the normal image of a complete surface of negative curvature. Mat. Sb., Nov. Ser. *82*, 220–223. Engl. transl.: Math. USSR, Sb. *11*, 197–200 (1970) Zbl.194,525

Kantor, B.E. (1976): Rigidity of surfaces of negative curvature. Sib. Mat. Zh. *17*, 1052–1057. Engl. transl.: Sib. Math. J. *17*, 777–781 (1977), Zbl.358.53003

Kantor, B.E. (1978a): On the rigidity of some surfaces of negative curvature. In: Questions of global geometry. 31st Herzen lecture. Leningrad State Ped. Inst., Leningrad, 16–18 (Russian)

Kantor, B.E. (1978b): On the unique determination of some classes of surfaces of negative curvature. In: Modern geometry. Republ. Collect. Sci. Works, Leningrad 1978, 67–73 (Russian), Zbl.421.53002

Kantor, B.E. (1980): The absence of closed asymptotic lines on a class of tubes of negative curvature. Sib. Mat. Zh. *21*, No. 6, 21–27. Engl. transl.: Sib. Math. J. *21*, 768–773 (1980), Zbl.467.53002

Kantor, B.E. (1981): On the uniqueness of the solution of a Monge-Ampère equation of hyperbolic type with two fixed characteristics. In: Modern geometry (Investigations on differential geometry). Interuniv. Collect. Sci. Works, Leningrad 1981, 78–81 (Russian), Zbl.548.53004

Khineva, S. (1977): Infinitesimal bendings of surfaces of negative Gaussian curvature. God. Sofij. Univ., Fak. Mat. Mekh. 68 (1973/1974), 295–309 (Russian), Zbl.393.53001

Klabukova, L.S. (1983): On the differential operator of problems of the theory of moment-free elastic shells with negative Gaussian curvature. Zh. Vychisl. Mat. Mat. Fiz. *23*, 1477–1486. Engl. transl.: USSR Comput. Math. Math. Phys. *23*, No. 6, 120–126 (1983), Zbl.563.73081

Klotz-Milnor, T. (1972): Efimov's theorem about complete immersed surfaces of negative curvature. Adv. Math. *8*, 474–543, Zbl.236.53055

Kosevich, A.M. (1972): The Foundations of the Mechanics of a Crystal Lattice. Nauka, Moscow (Russian)

Kovaleva, G.A. (1968): An example of a homotopic tube of a surface with closed asymptotic lines. Mat. Zametki *3*, 403–413. Engl. transl.: Math. Notes *3*, 257–263 (1968), Zbl.169,237

Kuiper, N.H. (1955): On C^1-isometric imbeddings. I, II. Nederl. Acad. Wetensch. Proc. Ser. A *58* (Indagationes Math. *17*), 545–556, 683–689, Zbl.67,396

Lamb, G.L. (1971): Analytical description of ultrashort optical pulse propagation in the resonant medium. Rev. Mod. Phys. *43*, 99–124

Lavrant'ev M.A., Shabat, B.V. (1973): Problems of Hydrodynamics and Their Mathematical Models. Nauka, Moscow (Russian)

Liber, A.E. (1938): On a class of Riemannian spaces of constant negative curvature. Uch. Zap. Saratov. Univ. Ser. Fiz.-Mat. *1*, No. 2, 105–122 (Russian)

Mikhailovskij, V.I. (1962a): Infinitesimal bendings of surfaces of revolution of negative curvature with conical sleeve-like constraints. Dopov. Akad. Nauk Ukr. SSR 1962, No. 8, 990–993 (Russian), Zbl.139,149

Mikhailovskij, V.I. (1962b): Infinitesimal "gliding" deformations of surfaces of revolution of negative curvature. Ukr. Mat. Zh. *14*, 18–29 (Russian), Zbl.141,188

Mikhailovskij, V.I. (1962c): Infinitesimal bendings of the lateral surface of a glued tube of revolution of negative curvature. Vestn. Kiev. Univ. Ser. Mat. Mekh., No. 5, Part 1, 58–64 (Russian)

Mikhailovskij, V.I. (1962d): Infinitesimal bendings of piecewise regular surfaces of revolution of negative curvature. Ukr. Mat. Zh. *14*, 422–426 (Russian), Zbl.146,432

Mikhailovskij, V.I. (1988): On some boundary-value problems of the theory of infinitesimal bendings of surfaces of negative curvature. 9th All-Union Geometry Conference, Kishinev, pp. 218–219

Miller, J.D. (1984): Some global inverse function theorems. J. Math. Anal. Appl. *100*, 375–384, Zbl.549.58006

Minagawa, T., Rado, T. (1952): On the infinitesimal rigidity of surfaces. Osaka Math. J. *4*, 241–285, Zbl.48,153

Minding, F. (1839): Wie sich entscheiden lässt, ob zwei gegebene krumme Flächen auf einander abwickelbar sind oder nicht; nebst Bemerkungen über die Flächen von unveränderlichen Krümmungsmasse. J. Reine Angew. Math. *19*, 370–387

Minding, F. (1840): Beiträge zur Theorie der kürzesten Linien und krummen Flächen. J. Reine Angew. Math. *20*, 323–327

Moore, J.D. (1971): Isometric immersions of Riemannian products. J. Differ. Geom. *5*, 159–168, Zbl.213,238

Nirenberg, L. (1963): Rigidity of a class of closed surfaces. In: Nonlinear problems. Proc. Symp. Madison 1962, 177–193, Zbl.111,344

Norden, A.P. (ed.) (1956): On the foundations of geometry. In: Classic Works on Lobachevskij Geometry and the Development of its Ideas. GITTL, Moscow (Russian), Zbl.72,156

Novikov, S.P. (1986): Topology. Itogi Nauki Tekh., Ser. Sovrem. Probl. Mat., Fundam. Napravleniya *12*, 5–552, Zbl.668.55001. Engl. transl. in: Encycl. Math. Sc. *12*, Springer-Verlag, Heidelberg (1992).

Novikov, S.P., Fomenko, A.T. (1987): Elements of Differential Geometry and Topology. Nauka, Moscow (Russian), Zbl.628.53002

Pogorelov, A.V. (1967): Geometrical Methods in the Nonlinear Theory of Elastic Shells. Nauka, Moscow (Russian), Zbl.168,452

Popov, A.G. (1989): An analogue of the phase space for the sine-Gordon equation. Vestn. Mosk. Univ., Ser. Fiz. Astron. *30*, No. 4, 19–22. Engl. transl.: Mosc. Univ. Phys. Bull. *44*, No. 4, 20–23 (1989)

Pourciau, B. (1988): Global invertibility of nonsmooth mappings. J. Math. Anal. Appl. *131*, 170–179, Zbl.666.49004

Poznyak, E.G. (1966): On the regular realization in the large of two-dimensional metrics of negative curvature. Dokl. Akad. Nauk SSSR *170*, 786–789. Engl. transl.: Sov. Math. Dokl. 7, 1288–1291 (1966), Zbl.168,195

Poznyak, E.G. (1977a): Geometrical investigations connected with the equation $Z_{xy} = \sin Z$. Itogi Nauki Tekh., Ser. Sovrem. Probl. Geom. *8*, 225–241. English transl.: J. Sov. Math. *13*, 677–686 (1980), Zbl.432.53003

Poznyak, E.G. (1977b): Isometric immersion in E^3 of some non-compact domains of the Lobachevskij plane. Mat. Sb., Nov. Ser. *102*, 3–12. Engl. transl.: Math. USSR, Sb. *31*, 1–8 (1977), Zbl.344.53008

Poznyak, E.G. (1979): Geometrical interpretation of regular solutions of the equation $Z_{xy} = \sin Z$. Differ. Uravn. *15*, 1332–1336. Engl. transl.: Differ. Equations *15*, 948–951 (1980), Zbl.413.35078

Poznyak, E.G. (1991): Isometric immersions of Riemannian spaces. Differential geometry structures on manifolds and their applications. Collection of materials of the All-Union geometry school, Chernovtsy Univ., pp. 177–182. (Deposited at VINITI, No. 562-B91)

Poznyak, E.G., Popov, A.G. (1991): The geometry of the sine-Gordon equation, Itogi Nauki Tekh., Ser. Sovrem. Problem. Geom. *23*, 99–130

Poznyak, E.G., Shikin, E.V. (1974): Surfaces of negative curvature. Itogi Nauki Tekh., Ser. Algebra, Topologija, Geom. *12*, 171–207. English transl.: J. Sov. Math. *5*, 865–887 (1976), Zbl.318.53050

Poznyak, E.G., Shikin, E.V. (1980): Isometric immersions of a domain of the Lobachevskij plane in Euclidean spaces. Tr. Tbilis Mat. Inst. Razmadze *64*, 82–93 (Russian), Zbl.497.53006

Poznyak, E.G., Shikin, E.V. (1986): Analytic tools of the theory of imbeddings of negative curvature two-dimensional manifolds·. Izv. Vyssh. Uchebn. Zaved., Mat. 1986, No. 1, 56–60. Engl. transl.: Soviet Math. *30*, 72–79 (1986), Zbl.607.53033

Poznyak, E.G., Sokolov, D.D. (1977): Isometric immersions of Riemannian spaces in Euclidean spaces. Itogi Nauki Tekh., Ser. Algebra, Topologija, Geom. *15*, 173–211. English transl.: J. Sov. Math. *14*, 1407–1428 (1980), Zbl.448.53040

Rozendorn, E.R. (1962): On complete surfaces of negative curvature $K \leqslant -1$ in the Euclidean spaces E^3 and E^4. Mat. Sb., Nov. Ser. *58*, 453–478 (Russian), Zbl.192,272

Rozendorn, E.R. (1966): Investigation of the main equations of the theory of surfaces in asymptotic coordinates. Mat. Sb., Nov. Ser. *70*, 490–507 (Russian), Zbl.145,415

Rozendorn, E.R. (1972): Non-immersibility of complete q-metrics of negative curvature in the class of weakly non-regular surfaces. Mat. Sb., Nov. Ser. *89*, 83–92. Engl. transl.: Math. USSR, Sb. *18*, 83–92 (1973), Zbl.241.53035

Rozendorn, E.R. (1980): Reduction of a problem of meteorology to a geometrical problem. Usp. Mat. Nauk 35, No. 6, 167–168. Engl. transl.: Russ. Math. Surv. *35*, No. 1, 101–102 (1980), Zbl.458.53013

Rozendorn, E.R. (1985): A class of surfaces of negative curvature with singularities on curves. Vestn. Mosk. Univ. Ser. I, No. 1, 50–52. Engl. transl.: Mosc. Univ. Math. Bull. *40*, No. 1, 75–79 (1985) Zbl.576.53004

Rozendorn, E.R. (1987): The integral form of writing the equations of infinitesimal bendings for surfaces of negative curvature. All-Union conference on geometry "in the large". Novosibirsk, p. 105 (Russian)

Rozendorn, E.R. (1988): Investigation of the regularity of a surface with a regular metric of negative curvature. All-Union School on optimal control, geometry and analysis. Kemerovo, p. 43 (Russian)

Rozhdestvenskij, B.L. (1962): A system of quasilinear equations of the theory of surfaces. Dokl. Akad. Nauk SSSR *143*, 50–52. Engl. transl.: Sov. Math. Dokl. *3*, 351–353 (1962), Zbl.137,407

Rozhdestvenskij, B.L., Yanenko, N.N. (1978): Systems of quasilinear equations and their applications to gas dynamics. 2nd. ed., Nauka, Moscow. Engl. transl.: Transl. Math. Monographs, Vol. *55*, Providence (1983), Zbl.513.35002, Zbl.177,140

Sabitov, I.Kh. (1989a): On isometric immersions of the Lobachevskij plane in E^4. Sib. Mat. Zh. 30, No. 5, 179–186. Engl. transl.: Sib. Math. J. *30*, No. 5, 805–811 (1989), Zbl.697.53014

Sabitov, I.Kh. (1989b): Local theory of bendings of surfaces. Itogi Nauki Tekh., Ser. Sovem. Probl. Mat., Fundam. Napravleniya 48, 196–270. Engl. transl. in: Encycl. Math. Sc. *48*, Springer-Verlag, Heidelberg, 179–250, 1992 (Part III of this volume)

Schur, F. (1886): Über die Deformation der Räume konstanten Riemannschen Krümmungsmasses. Math. Ann. 27, 163–176, Jbuch 18, 444

Seifert, H., Threlfall, W. (1934): Lehrbuch der Topologie. Teubner, Leipzig-Berlin. Reprint: Chelsea Publ. Co., New York, 1947, Zbl.9,86

Seminar on geometry in the large (1986): Scientific seminar of the department of mathematical analysis. Joint enlarged meeetings devoted to the 75th birthday of N.V. Efimov (26–30 September 1985). Vestn. Mosk. Univ. Ser. Mat. Mekh., No. 5, 92–98

Shiga, K. (1984): Hadamard manifolds. In: Geometry of geodesics and related topics. Proc. Symp., Tokyo 1982, Adv. Stud. Pure Math. *3*, 239–281, Zbl.565.53025

Shikin, E.V. (1975): Isometric imbeddings in E^3 of noncompact domains of nonpositive curvature. Itogi Nauki Tekh., Ser. Probl. Geom. 7, 249–266 (Russian), Zbl.551.53006

Shikin, E.V. (1980): Isometric immersion of two-dimensional manifolds of negative curvature by the method of Darboux. Mat. Zametki 27, 779–794. Engl. transl.: Math. Notes *27*, 373–382 (1980), Zbl.438.53004

Shikin, E.V. (1982): Isometric imbeddings in three-dimensional Euclidean space of two-dimensional manifolds of negative curvature. Mat. Zametki *31*, 601–612. Engl. transl.: Math. Notes *31*, 305–312 (1982), Zbl.491.53002

Shikin, E.V. (1990): On the parabolicity of immersible and the hyperbolicity of non-immersible two-dimensional manifolds of negative curvature. Vestn. Mosk. Univ., Ser. I Mat. Mekh. No. 5, 42–45 (Russian). Engl. transl.: Mosc. Univ. Math. Bull. *45*, No. 5, 40–42 (1991), Zbl.718.53004

Smyth, B. and Xavier, F. (1987): Efimov's theorem in dimension greater than two. Invent Math. *90*, 443–450, Zbl.642.53007

Sokolov, D.D. (1980): Surfaces in pseudo-Euclidean space. Itogi Nauki Tekh., Ser. Probl. Geom. *11*, 177–201. Engl. transl.: J. Sov. Math. *17*, 1676–1688 (1981), Zbl.469.53053

Spivak, M. (1975): Some left-over problems from classical differential geometry. Proc. Symp. Pure Math. *27*, Part 1, 245–252, Zbl.306.53004

Steuerwald, R. (1936): Über Enneper'sche Flächen und Bäcklund'sche Transformation. Abh. Bayer. Akad. Wiss. *40*, 1–106, Zbl.16,224

Ten, L.V. (1980): Rigidity of complete surfaces of negative curvature that coincide with a hyperbolic paraboloid outside a compact domain. Usp. Mat. Nauk *35*, No. 6, 175–176. Engl. transl.: Russ. Math. Surv. *35*, No. 6, 111–112 (1980), Zbl.458.53037

Tenenblat, K., Terng, C.L. (1980): Bäcklund's theorem for n-dimensional submanifolds of R^{2n-1}. Ann. Math., II. Ser. *111*, 477–490, Zbl.462.35079

Terng, C.L. (1980): A higher dimensional generalization of the sine-Gordon equation and its soliton theory. Ann. Math., II. Ser. *111*, 491–510, Zbl.447.53001

Tunitskij, D.V. (1987): On a regular isometric immersion in E^3 of unbounded domains of negative curvature. Mat. Sb., Nov. Ser. *134*, 119–134. Engl. transl.: Math. USSR, Sb. *62*, No. 1, 121–138 (1989), Zbl.636.53003

Vekua, I.N. (1982): Some General Methods of Constructing Different Versions of the Theory of Shells. Nauka, Moscow, Zbl.598.73100. Engl. transl.: Pitman, Boston etc. (1985)

Vinberg, E.B., Shvartsman, O.V. (1988): Discrete groups of motions of spaces of constant curvature. Itogi Nauki Tekh., Ser. Sovrem. Probl. Mat., Fundam. Napravleniya *29*, 147–259, Zbl.699.20017. Engl. transl. in: Encycl. Math. Sc. *29*, Springer-Verlag, Heidelberg (in preparation)

Vinogradskij, A.S. (1970): Boundary properties of surfaces with slowly changing negative curvature. Mat. Sb., Nov. Ser. *82*, 285–299. Engl. transl.: Math. USSR, Sb. *11*, 257–271 (1970), Zbl.195,230

Vorob'eva, L.I. (1976): The impossibility of a C^2-isometric immersion in E^3 of a Lobachevskij half-plane. Vestn. Mosk. Univ. Ser. I, No. 5, 42–46. Engl. transl.: Mosc. Univ. Math. Bull. *30*, No. 5/6, 32–35 (1975), Zbl.327.53036

Wintner, A. (1945): The nonlocal existence problem of ordinary differential equations. Am. J. Math. *67*, 277–284

Wissler, C. (1972): Global Tschebyscheff-Netze auf Riemannschen Mannigfaltigkeiten und Fortsetzung von Flächen konstanter negativer Krümmung. Comment. Math. Helv. *47*, 348–372, Zbl.257.53003

Xavier, F. (1984): Convex hulls of complete minimal surfaces. Math. Ann. *269*, 179–182, Zbl.528.53009

Xavier, F. (1985): A non-immersion theorem for hyperbolic manifolds. Comment. Math. Helv. *60*, 280–283, Zbl.566.53046

III. Local Theory of Bendings of Surfaces

I. Kh. Sabitov

Translated from the Russian
by E. Primrose

Contents

Preface

The origin of the theory of bendings as one of the basic problems of metrical geometry is associated with the names of Euler, Lagrange, Legendre, Cauchy and Gauss. After it was discovered that on surfaces there is an "intrinsic geometry" that does not depend on the external form of the surface, there naturally arose the question of the possibility of deforming the surface, preserving its intrinsic geometry. Consideration of isometric immersions (or, as we say, realizations) of abstractly given Riemannian metrics also leads to the problem of bendings of surfaces as to some problem about the uniqueness or non-uniqueness of an immersion.

In this article we are concerned mainly with local questions of the theory of bendings of two-dimensional surfaces in three-dimensional Euclidean space. The main complication of the problem turns out to be connected with the question of whether the point for which we wish to study the bendings of its neighbourhood is singular. The specific meaning of singular points is well known in many branches of mathematics, and geometry, of course, is no exception – the question is only which points should be regarded as singular, and which are generic points or, in another way, are points of general position. In the problem of bendings of surfaces "in the small", which we consider, the *singular points* are the so-called *flat points*, at which the surface has contact with its tangent plane of more than the first order. In complete correspondence with the well-known thought of the great sage Tolstoy "All happy families resemble one another, but each unhappy family is unhappy in its own way" it turns out that in a neighbourhood of points of general position all surfaces resemble each other – they are all bendable, and the arbitrariness of their bendings, or in other words the character of non-uniqueness of surfaces with a given metric, can be described quite precisely, while in a neighbourhood of flat points the bendability of a surface is in some sense an individual property of it, which depends on many parameters that characterize the behaviour of the surface near the flat point.

In the historical setting of the 20th century the most active period for the study of local questions of the theory of bendings occurred in the 40s. After the comparative "calm" in the 50s and 60s, from the beginning of the 70s there again began to appear papers in which there were proposed new statements and new methods of solving problems of the local theory of bendings. In this survey we attempt to set forth with sufficient completeness the main results involved here, not forgetting to mention unsolved questions.

§1. Definitions and Terminologies

1.1. A Surface and Its Metric. Since in this article discussions will be conducted only "in the small", in a neighbourhood of a marked point, in the definitions we restrict ourselves to the case of simply-connected surfaces. Furthermore, all surfaces and their deformations will be assumed to be in a sufficiently regular class; each time the precise class of their smoothness will be indicated, but they will all have smoothness at least C^1.

By a $C^{n,\alpha}$-smooth, $n \geqslant 1, 0 \leqslant \alpha \leqslant 1$, ($C^\infty$-smooth and C^A-smooth or analytic) surface we shall understand a $C^{n,\alpha}$-smooth (respectively, C^∞-smooth and C^A-smooth) map f of the disc $D: u^2 + v^2 < R^2$ into a three-dimensional Euclidean space E^3, and we shall assume that $df \neq 0$, which guarantees the local injectiveness of the map f. We shall denote the pair (D, f) defining a surface by one letter S; obviously, in our conditions S can be identified locally with $f(D) \subset E^3$, that is, we can assume that $S = f(D)$, not forgetting, however, that each point $M \in S$ has coordinates (x, y, z) defined by the map f with components

$$x = x(u, v), \quad y = y(u, v), \quad z = z(u, v); \quad (u, v) \in D. \tag{1}$$

The parameters (u, v) ae called the *intrinsic coordinates* of points of the surface. Curves on S are defined as images of curves in D under the map f.

The ambient Euclidean Space E^3 induces on S a rule for measuring the lengths $s = \int \sqrt{dx^2 + dy^2 + dz^2}$ of curves $x = x(t), y = y(t), z = z(t)$ (here $x(t) = x(u(t), v(t))$ and so on), which leads to the appearance on the disc D of the Riemannian metric

$$ds^2 = E\, du^2 + 2F\, du\, dv + G\, dv^2, \tag{2}$$

where

$$E = x_u^2 + y_u^2 + z_u^2,$$
$$F = x_u x_v + y_u y_v + z_u z_v, \tag{3}$$
$$G = x_v^2 + y_v^2 + z_v^2.$$

Having the lengths of curves, we can convert S in a known way to a metric space, taking as the distance between points on S the lower bound of lengths of curves that join these points.

1.2. Isometric Surfaces and Isometric Immersions. Now let $f^*: D^* \to E^3$ be another surface S^* with metric ds^{*2}, where $D^*: u^{*2} + v^{*2} < R^{*2}$. We say that the surfaces S and S^* are *isometric*, or that S^* is obtained by an *isometric transformation* of S, if there is an isometry $g: S \to S^*$ (as between metric spaces). Obviously, g preserves the lengths of all rectifiable curves, that is, any rectifiable curve γ on S must go over, under the isometry g, to some curve γ^* on S^* with the same length as γ on S. In terms of metrics this means that between the discs D and D^* there is established a homeomorphism $h = f^{*-1} \circ g \circ f$ that preserves the lengths of curves, measured in the metrics ds^2 and ds^{*2} respectively. One comparatively recent result of the theory of isometric maps is that the homeomorphism h is automatically a diffeomorphism if the metrics ds^2 and ds^{*2} are sufficiently regular. Namely, if the coefficients of both metrics belong to the Hölder class $C^{n,\alpha}$, $n \geqslant 0, 0 \leqslant \alpha \leqslant 1, n + \alpha \neq 0$ or 1, then $h \in C^{n+1,\alpha}$; see Reshetnyak (1982), Ch. 6, and Calabi and Hartman (1970).

Consequently, the composition $F = f^* \circ h$ enables us to express the coordinates of points on $S^* = f^*(D^*)$ in terms of functions of (u, v), and points on S and S^* that correspond to each other under the isometry g will have identical intrinsic coordinates (u, v), and in addition we shall have $df^2 = dF^2$, that is, isometric surfaces of class $C^{1,\alpha}$, $\alpha > 0$, can be regarded as given by various maps of the form (1) with general metric of the form (2).

This fact naturally leads to another way of approaching the study of isometric maps of surfaces into each other. Suppose that in the disc $D: u^2 + v^2 < R^2$ there is given a metric (2) with known coefficients E, F, G. If a map f of the form (1) defines a surface S on which the metric induced by the space E^3 coincides with the metric (2) already existing in D, that is, the equalities (3) are satisfied, we say that the map $f: D \to E^3$ gives an *isometric immersion* in E^3 of the disc D with metric (2), or as we said earlier, the metric (2) is realized in E^3 (on the surface S). Hence two isometric surfaces S and S^* from the previous discussion are none other than two isometric immersions f and F of the disc D with metric (2).

Remark 1. A homeomorphism g between isometric surfaces S and S^* of class C^1 or between Riemannian manifolds with metric of class C^0 need not belong to the class C^1 (but only to the *Sobolev class* W_p^1 with any $1 \leqslant p \leqslant \infty$, or more simply to the Lipschitz class $C^{0,1}$; see the example in Reshetnyak (1982) and Calabi and Hartman (1970)). Therefore, if S and S^* are two C^1-smooth isometric surfaces, then the establishment of an isometry between them by the equality of intrinsic coordinates, that is, the representation of S and S^* as two isometric immersions of one domain in E^3, say of a disc D with metric (2), can lead to the fact that one of these maps is not of class C^1 (but its image as before is a C^1-smooth surface!); similarly there is an observation of Borisov in Zalgaller (1962): a C^1-smooth surface can be the image of a non-C^1-smooth isometric immersion of a continuous metric (2). An example of Calabi and Hartman (1970) gives precisely this situation: a certain locally Euclidean continuous metric of the form (2) under a non-C^1-smooth map $x = x(u, v)$, $y = y(u, v)$, $z = 0$ is isometrically immersed in E^3 by an analytic surface (plane) $S^*: z = 0$, which can simultaneously be regarded as an immersion $S = S^*$ of the metric $ds^2 = dx^2 + dy^2$. An example with different S and S^* can be constructed as follows: the locally Euclidean continuous metric ds^{*2} in Calabi and Hartman (1970), which cannot be mapped C^1-smoothly onto the metric $ds^2 = dx^2 + dy^2$, can be immersed isometrically in E^3 by a C^1-smooth map in the form of some surface S^* according to Kuiper (Kuiper (1955)). Then S^* and $S = (x, y)$-plane are isometric, but the establishment of isometry between them by the equality of intrinsic coordinates leads to violation of C^1-smoothness of one of the maps that determine S and S^* in common coordinates. Consequently, the method frequently encountered in academic literature of reducing isometric surfaces to general intrinsic coordinates ("isometry by the equality of intrinsic coordinates") holds only in the class of surfaces of smoothness $C^{1,\alpha}$, $\alpha > 0$.

Among surfaces S^* isometric to S there are surfaces congruent to S, that is, obtainable from S by a motion of it in E^3 (without the inclusion of a mirror reflection!) and called *trivially isometric* to S. If an isometry between surfaces does not reduce to a motion, then it is called *non-trivial*, and the corresponding surfaces are called non-trivially isometric. Consequently, non-trivially isometric surfaces are either mirror symmetric or have different external forms but with a common metric.

In order to exclude from consideration surfaces trivially isometric to a given surface S, on all surfaces isometric to S we fix the position of some specific point, say the image of the point $(0, 0) \in D$, the tangent plane to the surface, and the directions of the tangent vectors to images of the curves $u = 0$ and $v = 0$ passing through this point. In turn, applying an affine transformation of the plane of intrinsic coordinates and contracting the neighbourhood of the point $(0, 0)$ if necessary, we can assume that for coefficients of the form (2) given in D we have $E(0, 0) = G(0, 0) = 1$, $F(0, 0) = 0$; such a specification of the metric will be called *standard*. Then by a choice of the axes x, y, z in E^3 we can arrange that the equations of S and all the surfaces isometric to it have the form

$$x = u + o(r), \quad y = v + o(r), \quad z = o(r),$$
$$r^2 = u^2 + v^2 \to 0. \tag{4}$$

If the equation of some surface has the form (4), we shall say that the surface is *correctly situated*. Thus, if two correctly situated surfaces are isometric, then the isometry between them is either identical or non-trivial (just because in the set of correctly situated surfaces two mirror symmetric surfaces have different equations (4), we do not suppose that such surfaces are trivially isometric; if, as is usually done, we suppose they are trivially isometric, then almost every time it will be necessary to make special stipulations or refinements, since in many bending problems mirror symmetric surfaces occur as non-identifiable objects; bearing in mind that such surfaces are usually supposed to be trivially isometric, maybe we ought to use the term "non-triviality in a wide sense of the word").

Thus, the study of a set of non-trivially isometric surfaces reduces to the study of the structure of the set $\mathfrak{M}$ of all correctly situated surfaces with a given standard metric (2) common to them all. Since we are interested only in local questions of isometry, in $\mathfrak{M}$ we can consider the following variants of the introduction of topology.

1) We restrict ourselves to the consideration of a subset of $\mathfrak{M}$ consisting of isometric maps f of the form (4) of the disc D_ε: $u^2 + v^2 < \varepsilon^2 \leqslant R^2$ in E^3 belonging to a given smoothness class $C^{n,\alpha}$, C^∞ or C^A; such a subset will be denoted by $\mathfrak{M}_\varepsilon$, and the topology in it can be introduced differently depending on the specific statement of the problem, for example, starting from the pointwise convergence of maps together with their derivatives up to a certain order, and so on.

2) We can consider the set $\widehat{\mathfrak{M}}_\varepsilon$ of isometric immersions of a closed disc $\bar{D}_\varepsilon$ in E^3 of a given smoothness class; the topology in $\widehat{\mathfrak{M}}_\varepsilon$ can be introduced,

for example, as in the set of maps of a compactum, taking account of the corresponding smoothness, of course.

3) In $\mathfrak{M}_\varepsilon$ it is natural to introduce the following equivalence relation: two maps f_1 and f_2 from $\mathfrak{M}_\varepsilon$ are regarded as *equivalent*, $f_1 \sim f_2$, if there is a disc $D_\delta : u^2 + v^2 < \delta^2, \delta \leqslant \varepsilon$, in which f_1 and f_2 coincide identically. The set obtained from $\mathfrak{M}_\varepsilon$ by factorization (identification) of elements with respect to a stated equivalence relation is denoted by $\mathfrak{M}_\varepsilon^0$; the topology in it is induced by the usual rules from the topology in $\mathfrak{M}_\varepsilon$ (that is, in $\mathfrak{M}_\varepsilon^0$ we introduce the so-called factor topology). Speaking differently, $\mathfrak{M}_\varepsilon^0$ is the *set of germs* at the point $(0,0)$ of *isometric maps* $f : D_\varepsilon \to E^3$.

In the more general statement of the problem the set $\mathfrak{M}$ consists of all isometric immersions in E^3 of a neighbourhood of the point $(0,0)$ of D, and since this neighbourhood is not fixed in advance, the topology in $\mathfrak{M}$ must be introduced as in the set of germs of maps of the form (4), defined in open neighbourhoods of the point $(0,0)$ and having each time a specifically stated smoothness class. In this topology the set $\mathfrak{M}$ is naturally denoted by $\mathfrak{M}_0^0$; the structure of the set $\mathfrak{M}$ has not been studied either in the form $\mathfrak{M}_0^0$ or in the form $\mathfrak{M}_\varepsilon^0$.

1.3. Bendings of Surfaces. If for any $\varepsilon > 0$ the set $\mathfrak{M}_\varepsilon$ of immersions of a given smoothness class consists only of $S(\varepsilon)$, the restriction of S to D_ε (and its mirror image), we say that the surface S is locally uniquely determined by its metric (in a given smoothness class). If $S(\varepsilon)$ for some $\varepsilon > 0$ enters into a linearly connected component $\mathfrak{N}_\varepsilon$ of the set $\mathfrak{M}_\varepsilon$, we say that S is *locally bendable*, and any continuous path in $\mathfrak{N}_\varepsilon$ beginning in $S(\varepsilon)$ determines a local bending of S. Speaking differently, a *bending* of a surface S is a continuous deformation of it under which the lengths of curves on the surface remain constant. If a deformation of S is symbolically denoted by

$$S \to S_t,$$

then for bendings the family of surfaces S_t is continuous with respect to the deformation parameter t, $t \in [0, 1)$, with $S_0 = S$. Since henceforth it will always be a question of local bendings, that is, bendings of a surface $S(\varepsilon)$ with some small $\varepsilon > 0$, henceforth we shall obviously not point to the dependence of S, $\mathfrak{M}$ and $\mathfrak{N}$ on ε.

If the isometric surfaces S and S^* belong to one connected component $\mathfrak{N} \subset \mathfrak{M}$, they are called *applicable* to each other, and the process of application itself can be represented as a bending S_t, $t \in [0, 1]$, with $S_0 = S$ and $S_1 = S^*$. If the surface S is applicable either to S^* or to $\bar{S}^*$, the mirror image of S^*, we shall say that S is applicable to S^* in a wide sense.

Any motion of the surface S in E^3,

$$S \to S_t = A(t)S + B(t) \qquad \text{(symbolic notation)}, \tag{5}$$

where $A(t)$ is an orthogonal matrix (with determinant $+1$) and $B(t)$ is a translation vector depending continuously on t with $A(0) = E$, $B(0) = 0$, determines a bending S_t, which is called *trivial*. Obviously, if the bending $S_t \neq S$ takes place in a family of correctly situated surfaces, it is necessarily non-trivial. A surface that admits only trivial bendings is called *non-bendable*. A connected component of a locally non-bendable surface S in any $\mathfrak{M}_\varepsilon$ consists only of the surface S itself.

Remark 2. Sometimes in the literature the term "bending" is understood to mean any isometric correspondence between two surfaces, and a "bending" in the sense applicable in the given article is used in combination with the word "continuous". We prefer to include the property of continuity in the concept of bending itself as intuitively corresponding to the conventional meaning of the word "bending" not as a mathematical term, but as a word of the living Russian language (as to the terminology in English, see Spivak (1979), Ch. 12).

In the topological terminology a bending of a surface $S = f(D)$ with $f: D \to E^3$ is defined as a homotopy $f_t: D \times I \to E^3$, $t \in I = [0, 1]$, of the map $f = f_0$ with the condition that every map f_t with fixed t is an isometric immersion of D in E^3. Henceforth on the whole we shall need to be concerned with bendings for which smoothness classes of surfaces involved in the homotopy and the smoothness of the homotopy with respect to the deformation parameter will be essentially different and independent of each other, for example, bendings can take place in the class of analytic surfaces only with continuous dependence of them on a parameter, or vice versa the surfaces themselves can be only C^1-smooth, and the dependence on the parameter is analytic. Therefore we shall describe the *smoothness of bendings* in a special notation. Generalizing the definitions of Efimov (1948b), let us agree to say that bendings take place in a smoothness class or bendings have a smoothness class $C_{k;m}^{n,\alpha}$ ($C_{k;m}^\infty$ or $C_{k;m}^A$), where $0 \leqslant k \leqslant n\,(k < \infty)$, $0 \leqslant m \leqslant \infty$ or $m = A$, if:

1) a bendable surface S is included in a family of surfaces S_t, $t \in T$, where T is some interval $[a, b]$ containing the point $t = 0$, for which $S_0 = S$;

2) each surface S_t is a $C^{n,\alpha}$ (C^∞ or C^A)-smooth isometric immersion f_t of the form (4) of some disc $D_\varepsilon: u^2 + v^2 < \varepsilon^2$ in E^3, all of whose derivatives (when $m \leqslant \infty$)

$$\frac{\partial^{i+j} f_t}{\partial u^l \partial v^{i-l} \partial t^j}, \qquad 0 \leqslant l \leqslant i \leqslant k, \qquad 0 \leqslant j \leqslant m, \qquad (6)$$

are continuous with respect to $(u, v) \in D_\varepsilon$ for each fixed $t \in T$ and continuous with respect to t at each fixed point $(u, v) \in D_\varepsilon$; when $m = A$ the derivatives in (6) with any $j < \infty$ are continuous with respect to $(u, v) \in D_\varepsilon$ for each fixed $t \in T$ and analytic with respect to t at each point $(u, v) \in D_\varepsilon$.

In relation to the set $\mathfrak{M}$ the class of bendings we have introduced has the following meaning: we consider the space $\mathfrak{M}_\varepsilon$ of isometric immersions $f: D_\varepsilon \to E^3$ of the form (4) and of smoothness $C^{n,\alpha}$ (C^∞ or C^A) with topology of pointwise convergence of position vectors of surfaces and their derivatives up to order $k \leqslant n$, and in this space we seek a path S_t of smoothness C^m, beginning in the surface $S = S_0$ or passing through it.

Further, we shall say that bendings have uniform smoothness of class $\hat{C}_{k;m}^{n,\alpha}$ ($\hat{C}_{k;m}^\infty$ or $\hat{C}_{k;m}^A$), $0 \leqslant k \leqslant n\,(0 \leqslant k < \infty)$, $0 \leqslant m \leqslant \infty$ or $m = A$, if:

1) the bendable surface S is included in a family of surfaces S_t, $t \in T$, where T is an interval $[a, b]$ containing the point $t = 0$ for which $S_0 = S$;

2) each surface S_t is a $C^{n,\alpha}$ (C^∞ or C^A)-smooth isometric immersion f_t of the form (4) of the closed disc $\bar{D}_\varepsilon: u^2 + v^2 \leqslant \varepsilon^2 < R^2$ in E^3, and all the derivatives (when $m \leqslant \infty$) in (6) are continuous with respect to (u, v) for each fixed $t \in T$ and are continuous with respect to t uniformly with respect to $(u, v) \in \bar{D}_\varepsilon$; when $m = A$ the derivatives in (6) with any $j < \infty$ are continuous with respect to (u, v) for each fixed $t \in T$ and analytic with respect to t uniformly with respect to $(u, v) \in \bar{D}_\varepsilon$, that is, for radii of convergence of series with centre at an arbitrary point $t \in T$ there is a positive lower estimate that does not depend on $(u, v) \in \bar{D}_\varepsilon$ (we observe that when we speak of maps f_t of class C^A of a closed disc $\bar{D}_\varepsilon$ in E^3 we have in mind that the f_t are analytic in the open disc D_ε and belong at least to the class C^k in $\bar{D}_\varepsilon$).

In relation to the set $\mathfrak{M}$ the class of bendings we have introduced has the following meaning: we consider the set $\hat{\mathfrak{M}}_\varepsilon$ of isometric immersions $\bar{D}_\varepsilon$ in E^3 of smoothness $C^{n,\alpha}$ (C^∞ or $C^A(D_\varepsilon) \cap C^k(\bar{D}_\varepsilon)$) with topology of uniform convergence in the class $C^k(\bar{D}_\varepsilon)$ and in this space we seek a path S_t of smoothness C^m, beginning in $S = S_0$ or passing through it.

In certain cases the metric in $\hat{\mathfrak{M}}_\varepsilon$ is taken with respect to the norm of the space $C^{k,\beta}(\bar{D}_\varepsilon)$; then bendings of the class $\hat{C}^q_{k,\beta;m}$ ($q = n, \alpha$ or ∞ or A, $0 \leqslant m \leqslant A$) will signify that the path S_t in $\hat{\mathfrak{M}}_\varepsilon \subset C^q(\bar{D}_\varepsilon)$ has C^m-smoothness in the metric $C^{k,\beta}(\bar{D}_\varepsilon)$.

Finally, bendings of the class $C^A_{A;A}$ signify that the maps $f_t: D_\varepsilon \to E^3$ are analytic as functions of three arguments, and in this case T is assumed to be an open interval.

With some complication of the topology in $\mathfrak{M}_\varepsilon$ or in $\hat{\mathfrak{M}}_\varepsilon$ we can introduce bendings of the classes $C^\infty_{\infty;m}$, $C^A_{\infty;m}$ and so on, $0 \leqslant m \leqslant \infty$ or $m = A$, but we shall not consider such bendings.

Remark 3. In all situations that we know the component $\mathfrak{N}$ is an open set in the sense that a bendable surface S can be included as an internal point in the family of bendings S_t, $-\delta < t < \delta$, where $S = S_0$; speaking differently, in $\mathfrak{N}$ there is a path passing *through* S. It is easy to give examples when bendings of the surface S for $-\delta < t \leqslant 0$ and $0 \leqslant t < \delta$ have different smoothness, but it is not known whether it can happen that *any* bendings of the given surface S constitute a family S_t, $-\delta < t < \delta$, $S = S_0$, which is piecewise smooth on the whole with respect to t and smooth separately in $(-\delta, 0]$ and $[0, \delta)$ or there is generally only one path beginning in S so that S is, as it were, an end-point at which there "comes to a stop" any bending of S_t for which $S = \lim_{t \to 0+} S_t$; in other words, it is not known whether the set $\mathfrak{N}$ can be homeomorphic in some small neighbourhood of S to the half-segment $[0, \delta)$ or, more generally, whether S can be a boundary point of the set $\mathfrak{N}$ (for polyhedra this situation does not occur; see §9).

Remark 4. The concept of a bending "in the small" can be extended as follows: a surface S is assumed to be included in some family of isometric surfaces S_t for which continuity with respect to the parameter t of the maps f_t and/or their derivatives up to some order is assumed to occur only at the point $(0, 0)$. In this general formulation the bendability of surfaces has not been studied in detail; we note, however, that some theorems in this formulation are true in Hopf and Schilt (1938), also in Efimov (1948b).

1.4. Infinitesimal Bendings of Surfaces. We now consider deformations of a surface when on it there is admitted a variation of lengths of curves, but this variation in some sense is small in comparison with the variation of spatial distances between points of the surface. Namely, suppose that the deformation has the form

$$f_t = f + tz_1 + \cdots + t^n z_n + o(t^n), \qquad t \to 0$$
$$f, z_i \in C^1(D), \qquad 1 \leqslant i \leqslant n. \tag{7}$$

We call the deformation (7) an *infinitesimal bending of order n* of the surface S with position vector f if

$$ds_t^2 - ds^2 = df_t^2 - df^2 = o(t^n), \qquad t \to 0. \tag{8}$$

Consequently, for infinitesimal bendings of order n the spatial distances between points of the surface vary, generally speaking, by order $O(t)$, and the lengths of curves on the surface vary by order $o(t^n)$, $t \to 0$. The vector-valued functions z_i that determine the infinitesimal bending (7) will be called *fields of infinitesimal bendings* of the corresponding order. The *smoothness of an infinitesimal bending* is defined as the smallest smoothness of the fields $z_i(u, v)$, $1 \leqslant i \leqslant n$.

If in the motion (5) the matrices $A(t)$ and $B(t)$ have derivatives up to order n inclusive at the point $t = 0$, then replacing $A(t)$ and $B(t)$ by their Taylor polynomials of degree n we obtain a deformation of the form (7) for which (8) is satisfied; such infinitesimal bendings, which can be regarded as induced by initial Taylor expansions of some sufficiently smooth motion, are called *trivial infinitesimal bendings*. Under trivial infinitesimal bendings of order n the spatial distances between points of a surface also vary by order $o(t^n)$. The converse is also true: if under some deformation of the form (7) the spatial distances between points of a surface (not containing a plane domain) vary by order $o(t^n)$, then this deformation can be represented as a Taylor expansion of some motion of smoothness C^n. Therefore trivial infinitesimal bendings (7) of order n can be characterized by the condition of varying spatial distances (consequently also lengths of curves on the surface) by order $o(t^n)$. A surface that admits only trivial infinitesimal bendings of order n is called *rigid with respect to infinitesimal bendings of order n*.

An infinitesimal bending (7), non-trivial as a whole, can contain an initial part of a trivial infinitesimal bending of some order less than n. In order to exclude from consideration trivial components of infinitesimal bendings, we introduce a certain normalization of them. For this we observe that a composition with the infinitesimal bending (7) of the motion (5) of smoothness C^n again leads to a certain infinitesimal bending of order n. If in (7) the initial terms up to order $k - 1 < n$ inclusive represent a trivial infinitesimal bending of order $k - 1$, then by applying the corresponding motion to (7) we can arrange that the infinitesimal bending takes the form

$$f_t = f + t^k z_k + \cdots + t^n z_n + o(t^n), \qquad t \to 0, \tag{9}$$

of course with z_i other than in (7).

If $k - 1 = n$, then the infinitesimal bending (7) is trivial as a whole and then in (9) we have $f_t = f + o(t^n)$. If $k \leqslant n$, then again by applying some motion with matrix of the form $A = E + t^k A_k + \cdots$ we can arrange that the non-trivial infinitesimal bending (7) has components of the form

$$z_i = o(r), \qquad r^2 = u^2 + v^2 \to 0, \qquad k \leqslant i \leqslant n \tag{9'}$$

(that is, in order that the deformed surfaces should also be in correct situation). Conversely, if at the very beginning we assume that in (7) the deformed surfaces are correctly situated, that is, $z_i = o(r)$, $r \to 0$, then k is equal to the number of the first field z_i not identically equal to zero.

Thus, *normalized* infinitesimal bendings of order n have the form (9) with z_i of the form (9'). We observe that the definition of a non-trivial infinitesimal bending of order n mentioned here differs from the usually adopted condition of non-triviality in which it is assumed that a deformation begins with a non-trivial field z_1 (see Efimov (1948a), for example). Therefore, generally speaking, we need some refinements in the definitions. We call the infinitesimal bending (7) (k, n)-*non-trivial*, $1 \leqslant k \leqslant n$, if it contains a maximal trivial initial part of infinitesimal bendings of order $k - 1 \geqslant 0$; the normalized representation of such an infinitesimal bending has the form (9), where the z_i have the form (9'). For a (k, n)-non-trivial infinitesimal bending the spatial distances between points of the surface vary by the precise order $O(t^k)$, while distances on the surface vary by order $o(t^n)$, $t \to 0$. We note that a $(1, n)$-non-trivial infinitesimal bending coincides with a non-trivial (in the sense of Efimov (1948a)) infinitesimal bending of order n, and the definition of a (k, n)-non-trivial infinitesimal bending given here is close to other versions of the definition of n-th order rigidity mentioned in Efimov (1952) in a footnote. The class of deformations defined according to (9) as non-trivial infinitesimal bendings is wider than non-trivial infinitesimal bendings in the "usual" sense; for example, if a surface has m linearly independent fields of infinitesimal bendings of the 1st order $z_1, \ldots, z_m$, then the deformation

$$f_t = f + t^k z_1 + \cdots + t^n z_{n-k+1}, \qquad n < 2k, \qquad n \leqslant m + k - 1,$$

is an infinitesimal bending of order (k, n), but it is not an infinitesimal bending of any high order in the classical sense.

Suppose that in a (k, m)-non-trivial infinitesimal bending

$$f_t = f + t^k z_k + \cdots + t^n z_n + t^{n+1} z_{n+1} + \cdots + t^m z_m + o(t^m) \tag{10}$$

the fields z_i, $k \leqslant i \leqslant n$, are the same as in (9). We then say that the infinitesimal bending (9) of order (k, n) is *extended* to the infinitesimal bending (10) of order (k, m). Any (k, n)-non-trivial infinitesimal bending is an extension of each "truncated" initial (k, n')-field of it, $k \leqslant n' < n$.

Not every field of an infinitesimal bending of some order can be extended to an infinitesimal bending of higher order. In a non-local situation an example of such an infinitesimal bending is constructed as follows. Suppose that a surface S is a flat plate (membrane) with a fixed boundary, situated on the (x, y)-plane in the form of a domain ω; then the deformation $f_t = (x, y, 0) + t(0, 0, h(x, y))$, $h|_{\partial\omega} = 0$ is an infinitesimal bending of order $(1, 1)$, but there is no infinitesimal bending of order $(1, 2)$ of the form

$$f_t = (x, y, 0) + t(0, 0, h(x, y)) + t^2(\xi(x, y), \eta(x, y), \zeta(x, y)).$$

In a local setting an example of a non-extendable field of infinitesimal bendings of the 1st order will be given in § 8.

If a surface is bent in the class $C^1_{1,A}$, then any initial part of the expansion of the bending f_t in a Taylor series with respect to the parameter t is a non-trivial infinitesimal bending of some order (k, n). This fact is one of the motivations for introducing the concept of an *infinitesimal bending of order (k, n)*: it is naturally required that any initial part of a bending which is analytic in the parameter can be regarded as a deformation that does not preserve but slowly changes the lengths of curves on the surface, and if the expansion of an analytic bending begins with a term t^k, $k > 1$, then the usual definition of a non-trivial infinitesimal bending does not enable us to consider the part of the expansion from t^k to t^n, $n > k$, as some non-trivial infinitesimal bending.

1.5. Bendings of Surfaces and the Theory of Elastic Shells. The choice of the term *rigidity* has a real physical meaning: in the theory of thin elastic shells it has been proved that a rigid surface (relative to infinitesimal bendings of the 1st order) in the above-mentioned geometrical sense represents a stable form of equilibrium of a shell under the action of internal stress forces caused by a given load. For non-rigid surfaces their infinitesimal bendings (of the 1st order) have the following mechanical interpretation: the shifts of a median surface under its infinitesimal bendings represent its virtual movements for the system of external forces, under the action of which in the shell there is realized a momentless stress state of equilibrium, and at least for convex shells the converse is also true: if for a given system of external forces any infinitesimal bending of the median surface can be interpreted as a virtual movement of the shell, admissible under these forces, then under the action of these forces in the shell there is realized a momentless stress state of equilibrium; see the monographs Vekua (1959), Vekua (1982). More clearly this can be represented as follows: a non-rigid surface (closed or with fixed boundary) reacts to an exterior load by the appearance of internal stresses that give to the shell an unstable form of equilibrium. In the mathematical scheme it turns out that a system of equations of momentless stress state of equilibrium of the shell is closely connected with a system of equations of infinitesimal bendings of the median surface of the shell: under one choice of fundamental characteristics of the stress state of a shell and field of infinitesimal bendings these systems coincide, and for other choices of such characteristics the solutions of the corresponding systems are in some sense mutually determined: see Vekua (1959), Vekua (1982) and other references in these books. Isometric deformations of a surface can also be associated with forms of equilibrium of a thin elastic shell admitted by it under the action of various loads; see Pogorelov (1967), Pogorelov (1986). In particular, a bendable surface represents an unstable form of equilibrium of a shell and under the influence of external loads it changes its form (it is bent), tending to minimize the internal stresses; and if the surface is non-bendable, then under the influence of external loads that exceed some critical value it can sharply "with a flick" change its form, remaining isometric to the original form, but forming, as a rule, an edge contributing to its resistance to the increasing load. Starting from what we have said, in principle one can assert that any theorem on the theory of bendings has some analogue or some interpretation in the theory of elastic shells, at least in the case of convex surfaces.

For shells that have parts of negative curvature or substantial thickness the connections between mechanical and geometrical rigidity are more complicated than for thin convex shells. Some indications of the situations that arise here can be found in Part II of this book by Rozendorn (see p. 105).

1.6. Areal Deformations. Of other questions close to the theory of bendings and infinitesimal bendings we mention the investigation of deformations under which the element of area of a surface either does not vary (the so-called *equiareal* or *A-deformations*) or varies by order $o(t^n)$, $t \to 0$, under deformations of the form (7) (the so-called *A*-deformations of the *n*-th order). The class of *A*-deformations is wider than the class of isometric deformations. It turns out, however, that the equations of *A*-deformations of the *n*-th order are closely connected with the equations of infinitesimal bendings of order $(1, n)$.

The papers Boudet (1961) and Vincensini (1962) were apparently the first in which *A*-deformations were defined and studied; for modern results and literature see the survey Sinyukov (1986).

§2. Statement of Problems

The detailed list of problems of the local theory of bendings would be apparently very impressive, so we restrict ourselves to the enumeration of problems of fundamental character.

1) Does a given surface admit non-trivial isometric transformations?

We shall see that in all known cases a surface has another surface non-trivially isometric to it, so as an addition to 1) there appears the problem:

1') Is there a surface locally uniquely determined by its metric in a given smoothness class?

Similar questions can be posed for bendings:

2) Does a given surface admit non-trivial bendings?

2') Is there a locally non-bendable surface?

3) Does a surface admit infinitesimal bendings of given order (k, n)?

3′) Is there a surface that is locally rigid with respect to infinitesimal bendings of a given order (k, n)?

In 1.2 we saw that in local questions there is sometimes a sense of identifying surfaces that coincide in some neighbourhood of the point under consideration. Then it is natural to pose the following question:

4) Suppose that two isometric immersions $f_1: D_\varepsilon \to E^3$ and $f_2: D_\varepsilon \to E^3$ coincide in some disc $D_\delta \subset D_\varepsilon$. Will they coincide in the whole of D_ε? In other words, in the notation of 1.2 the question is formulated as follows: do the spaces $\mathfrak{M}_\varepsilon$ and $\mathfrak{M}_\varepsilon^0$ coincide?

Finally, there are questions touching on connections between all deformations of a surface that we have introduced:

5) Suppose that two surfaces are isometric to each other. Are they applicable to each other, or, in other words, do the properties of the two surfaces of being isometric and being applicable to each other coincide?

6) To describe the set of connected components in the set of all isometric transformations of a given surface – is this set finite, and by which characteristics are the surfaces of each component determined?

7) To describe a connected component, or in other words the configuration space of a given surface under all possible bendings of it. In particular, for such surfaces is their configuration space finite-dimensional? See also Remark 3 at the end of 1.3.

8) What is the connection between bendings and infinitesimal bendings or between the non-bendability of a surface and its rigidity of order (k, n)? For example, can a given (k, n)-non-trivial infinitesimal bending be extended to an infinitesimal bending of higher order (k, m) and, in particular, to a bending analytic with respect to the parameter?

A pair consisting of a surface and the field of its infinitesimal bendings of order $(1, 1)$ has a number of interesting properties not of differential but rather of algebraic character: projective invariance, symmetry and so on. We shall therefore formulate this circle of questions separately as:

9) Algebraic properties of fields of infinitesimal bendings.

Answers to all these questions depend on which smoothness class the surfaces under consideration and their deformations belong to; in addition, there emerges the dependence on the behaviour of the curvature of the surface in a neighbourhood of a given point. Therefore the answer to each question is not settled by one theorem, and it is necessary to examine many different cases.

Later in the text the order of presentation is the following: first in §3 we discuss the questions 8) and 9), in §6 questions 4), 5), 6) for the cases of curvature $K > 0$ and $K < 0$, with a simultaneous positive answer to question 2), in §7 these same questions in a neighbourhood of a point with $K = 0$, and in the same place there is information on the question 4), and §8 is devoted to questions of rigidity and/or non-rigidity of surfaces with respect to infinitesimal bendings. The general answer to question 1) is given in the same place as it is formulated, the answer to question 1′) in the class C^1 is negative (see §4), and in smoothness classes C^2 and higher it is not known.

§3. Connection Between Bendings and Infinitesimal Bendings of Surfaces

Since a series of results of the theory of bendings is obtained with the use of infinitesimal bendings, we begin with question 8) from §2. In turn, this question requires "works" with a system of equations defining infinitesimal bendings of order (k, n).

3.1. General Equations of Infinitesimal Bendings of Arbitrary Order. Representing infinitesimal bendings of order (k, n) in the form of a deformation

$$f_t = f + 2t^k z_k + \cdots + 2t^n z_n, \tag{9''}$$

we have the following equations from condition (8):

$$df\, dz_k = 0,$$

$$df\, dz_{k+1} = 0,$$

$$\cdots \cdots \cdots$$

$$df\, dz_{2k-1} = 0.$$

$$df\, dz_{2k} + dz_k^2 = 0, \tag{11}$$

$$df\, dz_{2k+1} + 2dz_k\, dz_{k+1} = 0,$$

$$\cdots \cdots \cdots \cdots \cdots$$

$$df\, dz_{2i} + 2dz_k\, dz_{2i-k} + \cdots + 2dz_{i-1}\, dz_{i+1} + dz_i^2 = 0,$$

$$df\, dz_{2i+1} + 2dz_k\, dz_{2i-k+1} + \cdots + 2dz_i\, dz_{i+1} = 0,$$

$$\cdots \cdots \cdots \cdots \cdots \cdots \cdots$$

$$(2i \text{ and } 2i + 1 \leqslant n)$$

This system can be replaced by the following: there are vectors $y_i(u, v)$ such that

$$dz_k = [y_k \times df], \qquad \text{(vector product)}$$

$$dz_{k+1} = [y_{k+1} \times df],$$

$$\cdots \cdots \cdots \cdots$$

$$dz_{2k-1} = [y_{2k-1} \times df],$$

$$dz_{2k} = [y_{2k} \times df] + [y_k \times dz_k], \tag{11'}$$

$$\cdots \cdots \cdots \cdots \cdots$$

$$dz_{2i} = [y_{2i} \times df] + [y_k \times dz_{2i-k}] + \cdots + [y_{2i-k} \times dz_k],$$

$$dz_{2i+1} = [y_{2i+1} \times df] + [y_k \times dz_{2i-k+1}] + \cdots + [y_{2i+1-k} \times dz_k],$$

$$\cdots \cdots \cdots \cdots \cdots \cdots \cdots \cdots \cdots \cdots$$

Hence it follows in particular that trivial infinitesimal bendings of order (k, n) have the form (we regard the point $(0, 0, 0)$ as fixed)

$$z_i = [y_i \times f], \quad k \leqslant i < 2k, \quad z_i = [y_i \times f] + \sum_{j=k}^{i-k} [y_j \times z_{i-j}], \quad 2k \leqslant i \leqslant n,$$

where y_j are constant vectors, $k \leqslant j \leqslant n$.

3.2. Transition from Infinitesimal Bendings of High Order to Infinitesimal Bendings of Low Order.
This question, in this or a different form, was discussed long ago and by Cohn-Vossen (Cohn-Vossen (1936)), for example, it was formulated as the question of the relation between non-bendability and infinitesimal rigidity.

The meaning of the general answer to the question is that if we do not make the smoothness class of deformations more precise, then these properties of surfaces are different, generally speaking – there are surfaces that are non-bendable but non-infinitesimally rigid, and bendable but infinitesimally rigid. But nevertheless between them there are very close general interconnections; the starting point for searches of such connections is the following observation: the initial speed of a smooth (with respect to t) bending can be interpreted as the field of an infinitesimal bending of the 1st order. In fact, if $df_t^2 = df^2$, then $\dfrac{d}{dt}(df_t^2) = 0$, in particular, when $t = 0$ we have $2df\, dz = 0$, where $z = \dfrac{d}{dt}(f_t)$, $t = 0$. Consequently, infinitesimal bendings appear as the result of "linearization" of the problem of determining bendings, which is essentially a non-linear problem. However, here there is a certain complication, which consists in the fact that the infinitesimal bending determined by the field $z = f_t'(0)$ can be trivial. Nevertheless, under very general assumptions, with bendings we can associate a certain non-trivial infinitesimal bending of the 1st and even the 2nd order.

Theorem 3.1 (compare Efimov (1948a), p. 121). *Suppose that the surface S admits a non-trivial infinitesimal bending (9″) of order (k, n), $k > 1$. Then it admits non-trivial infinitesimal bendings of order $(1, 1)$, and when $n \geqslant 2k$, of order $(1, 2)$, of the same smoothness as the fields z_k and z_{2k} in (9″). This is true if S admits a bending analytic with respect to the parameter t.*

The proof is short and so we give it. By hypothesis, there is a deformation (9″) that is an infinitesimal bending of order (k, n), and the fields z_i, $k \leqslant i \leqslant n$, satisfy the equations (11). If a deformation is a bending of at least class $C^1_{1;A}$, then the initial part of the expansion of this deformation in the degrees of t also satisfies the same equations (11). Among them we have two equations

$$df\, dz_k = 0,$$

$$df\, dz_{2k} + dz_k^2 = 0 \qquad \text{(when } n \geqslant 2k\text{)},$$

of which it is clear that the deformations

$$f_\tau = f + 2\tau z_k \quad \text{and} \quad f_\tau = f + 2\tau z_k + 2\tau^2 z_{2k}$$

give respectively infinitesimal bendings of the order $(1, 1)$ and $(1, 2)$ of smoothness asserted in the theorem.

Corollary 3.2. *If a surface $S \in C^1$ has C^1-rigidity of the 1st or 2nd order, then it is non-bendable in the class of $C^1_{1;A}$-smooth deformations (that is, in the class of C^1-smooth surfaces with analytic dependence on t of the first derivatives of the position vector of the surface) and is rigid with respect to C^1-smooth infinitesimal bendings of any order (k, n), $k \geqslant 1$, or of order (k, n), $n \geqslant 2k \geqslant 2$, respectively.*

Thus we can say that between bendings and infinitesimal bendings of the 1st and 2nd orders some connection is established. Let us go over to infinitesimal bendings of higher order. The unconditional influence of analytic bendability or non-rigidity of high order (k, n), $1 < k < n$, on the availability of non-rigidity of order (l, m), $1 < l < k$, $m < n$, has not been established up to now. Of course, there are trivial cases when, say, the initial degrees in the expansion of the deformation f with respect to t are proportional to one number:

$$f_t = f + t^{k_1} z_1 + \cdots + t^{k_n} z_n + \cdots,$$

$$k_1 = p l_1, \ldots, k_n = p l_n, \ldots$$

then the substitution $\tau = t^p$ gives a representation of the same deformation in the form $f_\tau = f + \tau^{l_1} z_1 + \cdots + \tau^{l_n} z_n$, that is, as an infinitesimal bending of order (l_1, l_n). In the general case, generally speaking, the statement of the problem requires a substantial refinement, since several versions of the original information and the required answer are possible. For example, suppose that for a surface there is known the existence of a non-trivial infinitesimal bending of a given high order (k, n) (or an analytic bending); we need to investigate the non-rigidity with respect to infinitesimal bendings of low orders, and for the required fields of infinitesimal bendings of low orders we can require or establish only the fact of their existence ("qualitative" statement of the question) or it is necessary to find these fields, indicating methods for constructing them depending or conversely not depending on the specific form of the original field of infinitesimal bendings of high order ("quantitative" or "constructive" statement of the problem). In the latter case we can make the following assertion (see Sabitov (1987)), which is a generalization of a result of Efimov in Efimov (1952).

Theorem 3.3. *Suppose that all the C^1-smooth infinitesimal bendings of the 1st order of a non-planar C^1-smooth surface S are parallel to one field of infinitesimal bendings z_1^0 and that S admits bendings of class $C^n_{1;A}$ with some n, $1 \leqslant n \leqslant A$. Then S has C^n-non-rigidity of any order $(1, m)$. This is still true if S admits C^n-smooth infinitesimal bendings of order (k, l), where $k > 1$, $km < l$.*

The idea of the proof consists in introducing a new parameter with respect to which the initial part of a given deformation f_t can be represented as an infinitesimal bending of order $(1, m)$, and for the field z_1 we take the first (by order) derivative of f_t with respect to t that is non-zero at the point $t = 0$.

We see that in the general case the field $z = z(u, v)$ that determines a non-trivial infinitesimal bending of the 1st order does not have to be the field of initial speeds of some bending that generates it. Moreover, if we are given some bending field without the assumption that it is analytic with respect to the parameter, then in the general case it is impossible to connect with it some field of an infinitesimal bending: 1) for example, all smooth surfaces are bendable in the class of C^1-smooth surfaces (see §4), but there are examples of C^∞-smooth surfaces that are rigid with respect to C^1-smooth infinitesimal bendings of the 1st order, see Sabitov (1973); 2) for surfaces of positive curvature and any smoothness C^q, $3 \leqslant q \leqslant A$, we can find a bending of smoothness $C^q_{p;\infty}$, $p \leqslant q$, that is not the extension of any infinitesimal bending of order (k, n); for an example see Klimentov (1984) (but of course there are infinitesimal bendings of any order that are in no way connected with the given bending).

Nevertheless we can state some connection of a general form between the bendability of a surface and its non-rigidity of order $(1, m)$ in the form of the following assertion.

Theorem 3.4. *Suppose that a surface $S \in C^n$, $1 \leqslant n \leqslant A$, admits non-trivial bendings S_t in the smoothness class $C^k_{l;m}$, $1 \leqslant m \leqslant \infty$, $l \leqslant k \leqslant n$, $l < \infty$. Then in the family of isometric surfaces S_t, $0 \leqslant t \leqslant 1$, arbitrarily close to $S = S_0$ there is a surface that admits non-trivial C^l-smooth infinitesimal bendings of order $(1, m)$.*

We give the proof that we need for future discussion. If the surfaces S_t: $f_t = f + U(u, v; t)$ in correct situation are obtained by a bending of the surface S_0: $f_0 = f$, then the field $z(u, v; \tau) = U(u, v; t) - U(u, v; t_0)$, $\tau = t - t_0$, gives a bending of the surface S_{t_0}. Since $\dfrac{\partial U}{\partial t}\bigg|_{t=t_0}$ cannot be identically equal to zero (with respect to u, v) for all t_0 close to $t = 0$, there is a value $t = t_0$ arbitrarily close to $t = 0$ for which $\dfrac{\partial z}{\partial \tau}\bigg|_{\tau=0} \neq 0$. Hence the segment of the Taylor polynomial of the field $z \in C^l(u, v)$ with respect to τ from τ to τ^m is a non-trivial infinitesimal bending of order $(1, m)$ for S_{t_0}.

From the reasoning above we can observe that bendings of surfaces take place "in the main" in the set $\mathfrak{R}$ of non-rigid surfaces with the field of an infinitesimal bending $z_t = \dfrac{\partial f_t(u, v; t)}{\partial t}$. Consequently, we can regard the trajectory S_t of an isometric $C^1_{1;1}$-smooth deformation of the surface S_0 as an integral curve of the "vector" field of infinitesimal bendings $z(S)$ defined at each point $S \in \mathfrak{R}$ (the surface S_0 itself may not occur in $\mathfrak{R}$ or it may be that $z_0 = 0$).

Here it is relevant to set up the converse question: given some trajectory in $\mathfrak{R}$ beginning at $S_0 \in \mathfrak{M}$; in which case is it a bending of the surface S_0? The answer is as follows: suppose that the family $f_t = f + F(u, v; t) \in C^1_{1;1}$ consists of non-rigid surfaces so that there is a field z_t with $df_t\, dz_t = 0$; then in order that the deformation f_t should be a bending it is sufficient that $z_t = \dfrac{\partial F}{\partial t}$ and this condition is necessary if f_t has a unique field of an infinitesimal bending of the 1st order (here this is most likely, and is the first cause of the condition of finiteness of the number of fields of an infinitesimal bending of the 1st order, which ensures that their surfaces of rotation are non-rigid of the 2nd order; see Poznyak (1959)).

Thus, by Theorem 3.4 we can say that if a surface S is bent in the class $C_{1;m}^k$, $m \geqslant 1$, then it is either non-rigid or the limit of non-rigid surfaces S_t. In the case of the presence of an analytic bending (with respect to the parameter) the surface is necessarily non-rigid; apparently a more general fact is true: a surface that admits $C_{1;1}^1$-smooth bendings is non-rigid with respect to infinitesimal bendings of the 1st order, although the field of its non-trivial infinitesimal bendings is not necessarily the initial speed of the given bending.

As to the influence of non-bendability of a surface "in the small" on its rigidity of some order, here there is not one result known to us. In this connection we mention that the proof, which is available for polyhedra, of the assertion that from the non-bendability of a polyhedron there follows its rigidity of some order (k, n), see §9, cannot be extended to smooth surfaces, since the problem of investigating the rigidity of a polyhedron reduces to the solution of some linear algebraic system, and in the smooth (analytic!) case, although we can represent the corresponding system of differential equations formally in the form of an analytic functional equation, all the same there remains the problem of the feasibility of differential relations between the components of the solution obtained in the form of a formal power series.

3.3. Transition from Infinitesimal Bendings of Low Order to Infinitesimal Bendings of High Order.

There is no simple answer to the possibility of extending infinitesimal bendings of low order to infinitesimal bendings of higher order in the general case – there are examples where a surface locally has a field of an infinitesimal bending of the 1st order that is not extendable to an infinitesimal bending of the 2nd order, see Sabitov (1979a) (consequently, it is not extendable to an infinitesimal bending of any high order, in particular, to an analytic bending with respect to the parameter), and conversely, there are surfaces for which each non-trivial infinitesimal bending of any order $(1, n)$ can be extended to an infinitesimal bending of any order $(1, m)$, $m > n$, and even into an analytic bending with respect to the parameter. This assertion holds for surfaces of positive curvature and it has been proved in an article of Isanov (Isanov (1977)) for surfaces and deformations of class $C^{l,\alpha}$, $l \geqslant 2$, in the case $n = 1$. For another version of the proof, for any n, see Klimentov (1984); the method of this paper of Klimentov gives an example of the use of a representation of a family S_t of bendings as "curves" touching fields of infinitesimal bendings at each point S_t: infinitesimal bendings are regarded in Klimentov (1984) as jets of a map that assigns a bending of the surface S.

3.4. Algebraic Properties of Fields of Infinitesimal Bendings of the 1st Order.

Besides properties that depend on the specific form of a surface, infinitesimal bendings have a series of properties of algebraic geometry nature, valid for infinitesimal bendings of all surfaces and based on formal operations over fields of infinitesimal bendings.

3.4.1. Darboux Surfaces. The field z_1 of an infinitesimal bending of the 1st order of a surface F_1 with position vector $f_1: D \to E^3$ satisfies the equation

$$df_1 \, dz_1 = 0. \tag{12}$$

The equation (12) is symmetric with respect to f_1 and z_1: we can assume that the field f_1 determines an infinitesimal bending of a surface Z_1 with position vector $z_1: D \to E^3$, called the *diagram of bendings*. This symmetry between F_1 and Z_1 leads to an unusually elegant construction known by the name of *Darboux wreath*; see Darboux (1986), Ch. 3.

First of all, having the condition (12), we can show that there is a field $z_2(x, y)$ such that

$$dz_1 = [z_2 \times df_1].$$

The field z_2 is called the *rotation field* for the field of infinitesimal bendings z_1, and the surface Z_2 with position vector $z_2: D \to E^3$ is called the *diagram of rotations*.

We now construct the field $f_2 = z_1 - [z_2 \times f_1]$. It is easy to verify that $df_2 \, dz_2 = 0$, so by analogy with (12) there is a field z_3 such that

$$dz_2 = [z_3 \times df_2].$$

We again construct the field $f_3 = z_2 - [z_3 \times f_2]$, and so on. Extending this process by the formulae

$$dz_i \, df_i = 0, \qquad dz_i = [z_{i+1} \times df_i], \qquad f_{i+1} = z_i - [z_{i+1} \times f_i],$$

we discover that $z_7 = z_1$, $f_7 = f_1$! Fig. 1 shows schematically the connections between the resulting 12 Darboux surfaces F_i and Z_i with the position vectors f_i

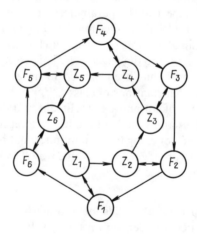

Fig. 1

and z_i respectively, $1 \leqslant i \leqslant 6$. The arrows between F_i and Z_i show their equal status, since $df_i \, dz_i = 0$ and each surface determines the field of infinitesimal bendings of the other. The arrow from Z_i to Z_{i+1} shows that Z_{i+1} is the diagram of rotations for Z_i, regarded as the diagram of bendings for F_i; the arrow from F_i to F_{i-1} shows that F_{i-1} is the diagram of rotations for F_i, regarded as the diagram of bendings for Z_i.

Apart from the relations just mentioned, between Darboux surfaces there is still a series of algebraic geometry connections, for example, the surfaces F_i and Z_{i+2} are in polar correspondence, F_i and Z_{i+1} have parallel tangent planes at corresponding points; for these and other connections see Voitsekhovskij (1979) and the literature mentioned there.

Differential geometry properties of Darboux surfaces depend mainly on the sign of the curvature of the original bendable surface F_1; they have been studied in detail only for the case when F_1 has positive curvature; see Sabitov (1965a), Sabitov (1965b). The diagram of rotations turned out to be more interesting and useful; by means of it two shorter proofs of the rigidity of regular ovaloids (closed surfaces of positive curvature) were given. We give one of them: we first establish that the diagram of rotations at an internal point does not have locally supporting planes (see Sabitov (1965a)), but for a closed surface Z_2 there must be a supporting plane – a contradiction!

In Sabitov (1965a) and Sabitov (1965b) there is a series of other tests for rigidity of surfaces of positive curvature, based on local properties of the Darboux surfaces Z_1, Z_2 and so on; see also Kann (1970).

The diagram of rotations for infinitesimal bendings of domains on a sphere represents a minimal surface, and conversely, with each minimal surface we can associate some infinitesimal bending of a spherical domain. This fact was already known to Liebmann in 1919; for a new proof and details see Sabitov (1967). In the general case of surfaces of positive curvature the diagrams of rotation for their infinitesimal bendings represent surfaces of negative curvature with isolated branch points; these surfaces, associated by W. Süsse in the 20s and 30s with so-called relative minimal surfaces, have many properties in common with minimal surfaces, and they give an excellent real example of immersions with branches, studied in Gulliver, Osserman and Royden (1973).

3.4.2. Projective Invariance of Infinitesimal Bendings. The property of projective invariance of infinitesimal bendings is that under projective transformations of space a non-rigid surface goes over to a non-rigid surface, and the field of infinitesimal bendings of the 1st order of the transformed surface is obtained explicitly in terms of the field of infinitesimal bendings of the 1st order of the original surface. The proof of this assertion, which is known as the *Darboux-Sauer theorem* (Darboux (1896), Sauer (1935)), is obtained as the composition of the following two easily verifiable relations (see Yanenko (1954) also for the multidimensional case):

1) if the transformation of space is affine:

$$\tilde{X} = XA + B, \qquad X = (x_1, x_2, x_3), \qquad \tilde{X} = (\tilde{x}_1, \tilde{x}_2, \tilde{x}_3) \qquad (13)$$

and S is a non-rigid surface with a field of infinitesimal bending of the 1st order

$$Z = \begin{bmatrix} z_1 \\ z_2 \\ z_3 \end{bmatrix},$$

then after the transformation (13) S goes over to some surface $\tilde{S}$ with a field of infinitesimal bending

$$\tilde{Z} = A^{-1}Z;$$

2) if the transformation of space is projective of special form

$$\tilde{x}_1 = \frac{x_1}{x_3}, \qquad \tilde{x}_2 = \frac{x_2}{x_3}, \qquad \tilde{x}_3 = \frac{1}{x_3}, \tag{14}$$

then after the transfomation (14) the non-rigid surface S goes over to a surface $\tilde{S}$ with a field of infinitesimal bending

$$\tilde{z}_1 = \frac{z_1}{x_3}, \qquad \tilde{z}_2 = \frac{z_2}{x_3}, \qquad \tilde{z}_3 = -\frac{x_1 z_1 + x_2 z_2 + x_3 z_3}{x_3}.$$

The main applications of the property of projective invariance of infinitesimal bendings are known not in the local but the global theory, and they consist of the following.

1) Many properties of infinitesimal bendings of the 1st order are true on surfaces of the form $z = f(x, y)$, that is, projected one-to-one onto some plane, and are untrue without this property; for the use of such properties the application of a projective transformation of the form (14) enables us to transfer a closed surface S to an infinite surface $\tilde{S}$ of the form $z = f(x, y)$, and then, obtaining the corresponding information for $\tilde{S}$, we can make deductions about infinitesimal bendings of the original surface S. In such a way there were obtained, for example, proofs of the infinitesimal rigidity of closed convex surfaces: regular and of strictly positive curvature in Vekua (1959), with isolated zeros of the curvature in Kann (1970), and the general case in Pogorelov (1969). In the same way, assertions about infinitesimal bendings, true with respect to surfaces of the form $z = f(x, y)$, were carried over to surfaces that are star-shaped with respect to some point separated from the surface by a plane.

2) In boundary-value problems conditions of a relation of infinitesimal bendings with respect to a point by a projective transformation can be carried over to conditions of a connection with respect to a plane, or conversely.

The extension of the results of these two sections to infinitesimal bendings of order (k, n), $1 < k < n < 2k$, is obvious. For an infinitesimal bending of order (k, n), $n \geqslant 2k$, we see from (11') that there are fields y_1, y_2 and so on that are analogues of rotation fields, but it is not known which fields can be connected with them to obtain an analogue of the Darboux wreath.

Under projective transformations there naturally takes place a corresponding change of Darboux surfaces; for the algebraic geometry interconnections between Darboux surfaces that arise here see Bol (1955) and Sauer (1948).

3.4.3. Algebraic Properties of the Set of Solutions of Equations of Infinitesimal Bendings.

The definition of infinitesimal bendings of order (k, n) in 1.4 relies on the representation of a deformation of a surface in the form (9); we call the vector functions $z_k(u, v), \ldots, z_n(u, v)$ fields defining the given infinitesimal bending. If we make a substitution of the parameter $t = t(\tau) \in C^n$ with

$$t = a_1 \tau + a_2 \tau^2 + \cdots + a_n \tau^n + o(\tau^n), \qquad a_1 \neq 0,$$

then with respect to τ the deformation (9) takes the form

$$\tilde{f}_\tau = f + \tau^k \tilde{z}_k + \cdots + \tau^n \tilde{z}_n + o(\tau^n), \qquad \tau \to 0.$$

This deformation is also an infinitesimal bending of order (k, n), however the vector functions that determine it are now different:

$$\tilde{z}_k = a_1^k z_k, \quad \tilde{z}_{k+1} = k a_2 a_1^{k-1} z_k + a_1^{k+1} z_{k+1}, \ldots,$$

$$\tilde{z}_n = a_1^n z_n + p_{n-1,n} z_{n-1} + \cdots + p_{k,n} z_k, \tag{15}$$

where $p_{i,j}$ are certain numerical coefficients. Obviously we need to assume that the deformations f_τ and $\tilde{f}_\tau$ determine the same infinitesimal bending. This identification means that in the correspondence between solutions of the system (11) and the infinitesimal bendings determined by them we also need to introduce necessary refinements, namely, in the set of solutions of the system (11) we need to introduce the condition of equivalence of solutions. So far, from the representation of the fields $\tilde{z}_i$ in terms of z_i it follows that if we know some solution $(z_k, \ldots, z_n)$ of the system (11), then there is a collection of functions $(\tilde{z}_k, \ldots, \tilde{z}_n)$ expressible linearly in terms of $z_k, \ldots, z_n$ by the formulae (15), which is also a solution of the system (11) and gives an infinitesimal bending equivalent to the infinitesimal bending determined by the fields $z_k, \ldots, z_n$. Starting from this, it is natural to raise the following questions:

1) Let $z^0 = (z_k^0, \ldots, z_n^0)$ be a solution of the system (11). Which are the linear transformations A for which Az^0 is also a solution of the system (11)?

2) When do the two solutions $(z_k^{(1)}, \ldots, z_n^{(1)})$ and $(z_k^{(2)}, \ldots, z_n^{(2)})$ of the system (11) determine one infinitesimal bending in the sense described above?

3) From the solution $z = (z_k, \ldots, z_n)$ of the system (11) can we "construct" by some linear transformation Az a field $(z_l, \ldots, z_m)$ that determines an infinitesimal bending of order (l, m), $l < k$?

All these questions are still unsolved. The discussion of Theorem 3.4 shows that the properties of the set of solutions of the system (11) depend essentially on whether the field of infinitesimal bendings of the 1st order is unique or not. In particular, on this there depends the uniqueness of an extension of infinitesimal bendings of low orders to infinitesimal bendings of high order.

§4. Bendings of Surfaces in the Class C^1

One of the unexpected discoveries in the theory of immersions of metrics and bendings of surfaces was the establishment of the fact that any metric (2) with continuous coefficients is isometrically immersed in E^3 in the form of a C^1-smooth surface and the freedom of the immersions is such that they constitute a $C_{1;0}^1$-smooth family; quite astonishing here was the bendability in the class $C_{1;0}^1$ of analytic surfaces "in the large", for example spheres, even without preserving the orientation: Gromov and Rokhlin (1970), Kuiper (1955), Nash (1954).

Subsequently in the paper Borisov (1965) the following extension of the smoothness class of bendings was proposed: an analytic surface of positive curvature that is homeomorphic to a disc is bendable in the class $C_{1;0}^{1;\alpha}$ with $\alpha < 1/7$ and there the hypothesis was stated that this is true with $\alpha < 1/5$. Of course, in the process of such a bending, on the surfaces we lose well-known classical connections between extrinsic and intrinsic geometry, for example, although these surfaces have positive intrinsic curvature, nevertheless at any point they are not locally convex. In the same way, we can bend a plane in a family of surfaces of smoothness $C_{1;0}^{1;\alpha}$, $\alpha < 1/7$, and the resulting surfaces of intrinsic curvature $K = 0$ will not contain rectilinear generators. However, the possibility of the existence of a surface isometric to a plane but not containing rectilinear generators was established by Lebesgue (Lebesgue (1902)), but the surfaces there were only smooth almost everywhere.

But we should say that here there are no sufficiently simple descriptive or analytic representations of such surfaces (like, say, for continuous but nowhere differentiable curves), and so it was interesting to study algorithmic approaches to the construction or description of such surfaces (with the possible use of a computer).

The best estimate of the values of those α under which bendability "in the small" exists for any surface of classical smoothness C^n, $2 \leqslant n \leqslant A$, is not known to the author, but it is known (Borisov (1965)) that when $\alpha > 2/3$ classical extrinsic geometry holds on surfaces of positive curvature. Together with this we see in §8 that there are C^∞-smooth surfaces, locally rigid in the class of C^1-smooth infinitesimal bendings of the 1st order; hence we should expect that in the class $C_{1;1}^1$ there is not always bendability.

It was shown in Pogorelov (1969) that if the spherical image of a smooth surface has bounded absolute variation (or bounded area taking account of the multiplicity of the covering – the so-called *surfaces of bounded extrinsic curvature*), then on such surfaces there remain the classical connections between their extrinsic and intrinsic geometries. Therefore, bearing in mind certain general properties of maps with small smoothness, for example the impossibility of filling a square by a Peano curve of class $C^{0,\alpha}$ when $\alpha > 1/2$ and the possibility of such a filling when $\alpha \leqslant 1/2$, and also special properties of a spherical map itself, we can expect that the border between universal "exotic" bendability and bendability depending on the local structure of the surface goes through to class $C^{1,1/2}$.

In Rozendorn's article II in the present volume we can find a more detailed description of the idea of immersing metrics in E^3 in the form of a surface of small smoothness; see also § 3 of Ch. 4 in the article I of Burago and Shefel'.

§ 5. Auxiliary Information: Classification and Integral Characteristics of Points of a Surface; Equations of Immersion and Bending

5.1. Four Types of Points on a Surface. In the smoothness classes C^2 and above, local bendability of a surface depends essentially on the structure of the surface near the point under consideration. We recall various characteristics of a "marked" point of a surface. First of all, one of the important characteristics of a point is the *order of contact* of the surface with its tangent plane at that point. Suppose there is a correctly situated surface S: $z = f(x, y) \in C^k, k \geqslant 2$; we have $f(0, 0) = f_x(0, 0) = f_y(0, 0) = 0$. Let n $(2 \leqslant n \leqslant k)$ be the first number for which at least one of the n-th order partial derivatives of $f(x, y)$ is non-zero at the point $(0, 0)$. Then from the local Taylor formula we have

$$z = f^{(n)}(x, y) + R(x, y), \qquad R = o(r^n), \qquad r^2 = x^2 + y^2 \to 0, \qquad (16)$$

where $f^{(n)}$ is a homogeneous form of degree n in x and y. The number $p = n - 1$ is called the *order of contact* of the surface with the tangent plane at the point $M_0(0, 0, 0) \in S$. When $p = 1$, with respect to the discriminant $\Delta(M_0)$ of the second fundamental form of the surface we have the following known classification of points of the surface:

a) $\Delta > 0$ – the point is *elliptic* $\Leftrightarrow$ the Gaussian curvature $K(0, 0) > 0$.

b) $\Delta < 0$ – the point is *hyperbolic* $\Leftrightarrow$ $K(0, 0) < 0$.

c) $\Delta = 0$ – the point is *parabolic* $\Leftrightarrow$ $K(0, 0) = 0$ (we draw attention to the fact that in the definition of a parabolic point adopted here at least one of the coefficients of the second fundamental form is non-zero). The points of these three types will be called *points of general position*.

When $p > 1$ the point $M_0 \in S$ is called a *flat* (or planar) point, and the number $q = p - 1 = n - 2$ is called the *order of flattening*. At a flat point $K = 0$, and in addition all the coefficients of the second fundamental form are zero. Points in general position will sometimes be called *points with flattening of zero order*.

If $f(x, y)$ has smoothness C^k and $f = o(r^k)$, then obviously one cannot determine the order of flattening in the general case, and in specific problems for the function $f(x, y)$ and its derivatives up to some order we need to assume the existence of a representation as in (16) with $R = o(|f^{(n)}|)$ and with additional refinement of the form of the principal part $f^{(n)}$, for example, as we need to do for a description of flattenings of exponential order.

5.2. Arithmetic Characteristics of a Regular Point of a Surface. In Efimov (1948b) further details of the structure of a surface in a neighbourhood of a marked point were given. If a point $M_0 \in S$ has some neighbourhood in which there are no normals to S parallel to the normal $n(M_0)$, then the point M_0 is called *regular*, and the corresponding neighbourhood of the point is called *canonical*. Suppose that the surface S: $z = f(x, y)$ is given in correct situation; then if the point $M_0(0, 0, 0) \in S$ is regular, in its canonical neighbourhood the field grad $f(x, y)$ has an isolated singular point $(0, 0)$. The index of this vector field is called the *index of the point* M_0 and denoted by $\mathrm{ind}(M_0)$.

Let us consider the section of the surface S by the tangent plane Π: $z = 0$. It has been proved (on the assumption only that $S \in C^1$!) that in a small canonical neighbourhood U of the point M_0 the intersection of S and Π consists either of just the point M_0 or of an even number of curves Γ_i, $1 \leqslant i \leqslant l$, of smoothness C^1 dividing U into l domains, just as on a plane l rays starting from the point $(0, 0)$ divide a neighbourhood of this point into sectors, but in the general case it is impossible to guarantee some regularity of the appearance of Γ_i at M_0, for example, with a definite direction. The number $m = l/2$ is called the *order of saddleness* of the surface at M_0.

The numbers $p(M_0)$, the order of contact of S with the tangent plane, $\mathrm{ind}(M_0)$ and $m(M_0)$ are connected by the relations

$$\mathrm{ind}(M_0) = 1 - m(M_0),$$
$$m(M_0) \leqslant 1 + p(M_0). \tag{17}$$

Elliptic and hyperbolic points on S are always regular; for them we have $m = 0$ and $m = 2$ respectively (see Figs. 2 and 3). Parabolic points and flat points can be both regular and non-regular. If, for example, the curvature in a deleted

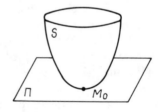

Fig. 2

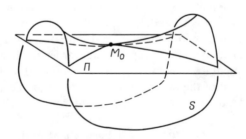

Fig. 3

neighbourhood of a point is of constant sign, then the point M_0 is regular, and when $K > 0$ we again have $m = 0$, but when $K < 0$ the order of saddleness can be any integer $\geqslant 2$, taking account of (17), of course. For the case of curvature of variable sign in the analytic class there is the following criterion (Efimov (1948b)): a point is non-regular if and only if through it there passes an arc of the curve $K = 0$ along which the surface has a stationary tangent plane. Such, for example, is each point of the parabolic curve on a circular torus.

Examples of surfaces with a regular parabolic point.

1) $S: z = y^2 + ax^n$, $n \geqslant 3$. Here for even $n = 2k$, $a > 0$ we have $K \sim n(n-1)ax^n \geqslant 0$. The intersection of S with the tangent plane $\Pi: z = 0$ consists of one point M_0, so $m = 0$; when $n = 2k$, $a < 0$ we have $K \leqslant 0$; $S \cap \Pi$ consists of four arcs with equations $y = \pm\sqrt{(-a)}x_k$, so $m = 2$; if $n = 2k + 1$, then K is of variable sign for any $a \neq 0$ and $S \cap \Pi$ consists of two arcs with equations $y = \pm\sqrt{(-ax)}x^k$, $ax < 0$, so $m = 1$. We see that for a parabolic point there are realized all the possible cases $m = 0, 1, 2$ that follow from the inequality (17) with $p = 1$.

2) $S: z = y^2 + a(x^{2n} + y^{2n})x^2$, $n \geqslant 1$, $a \neq 0$. This example is interesting in that the curvature $K \sim 4a(y^{2n} + (n+1)(2n+1)x^{2n})$ is of constant sign in a deleted neighbourhood of the point M_0. Here we have $m = 0$ when $a > 0$ and $m = 2$ when $a < 0$.

3) $S: z = y^2 + 2a \operatorname{Re}(x + iy)^{n+2}$, $n \geqslant 1$. The curvature $K \sim a(n+1) \times (n+2) \operatorname{Re}(x + iy)^n$ is of variable sign and it vanishes on $2n$ arcs that go into $(0, 0)$ with definite directions. If $n = 2k$, $a > 0$, then $m = 0$. If $n = 2k$, $a < 0$, then $m = 1$. Hence we see that the number of curves $K = 0$ is not connected with the order of saddleness.

Examples of surfaces with a regular flat point.

1) $S: z = (x^2 + y^2)^n$, $n \geqslant 2$. Here $K > 0$, $p = 2n - 1$, $m = 0$.

2) For the case $K < 0$ a typical example of a surface with a flat point is the graph of a harmonic function (see Fig. 22 in the article by Rozendorn in this volume). Suppose, for example, that $z = \operatorname{Re} w^n$, $w = x + iy$, $n \geqslant 3$. We have $p = n - 1$, $m = n$. Obviously equality in (17) is attained for any given p or m.

3) $S: z = (w\overline{w})^n \operatorname{Re} w^k$. Here $m = k$, $p = 2n + k - 1$, and the curvature is negative when $k^2 > 2n + k$ and of variable sign when $k^2 < 2n + k$. Hence for any m and p that have opposite parity and satisfy (17) there is an analytic surface that realizes m and p. Another more complicated example shows that m and p can be also of the same parity. We note that under the additional condition $K < 0$ such a surface may or may not exist (for example, we can show that when $m = 2$ and $p = 2$ there is no such surface). A description of all pairs (p, m) with the condition (17) for which there is a surface with curvature $K < 0$ that realizes them does not occur in the literature, and apparently nobody has formulated it as a meaningful problem. It is also not known whether there are surfaces with a flattening of infinite order at a point and with negative curvature in a deleted neighbourhood of the point (consequently, with finite order of saddleness). We recall that condition (17) connects m and p only in the analytic case, and in classes of finite smoothness this condition may be meaningless because of the non-existence of

p (but on a C^k-surface for which $p \leqslant k$ and m exist, (17) is valid). Interest in the order of flattening of a surface $z = f(x, y)$ under negative curvature of this surface in a neighbourhood of a flat point has been further induced by the following circumstance: many surfaces of negative curvature are obtained as solutions of elliptic equations of a definite class (with possible degeneracy at a point), consequently, if it is established that with the sign $K < 0$ there is no flattening of infinite order, then by the same token we shall obtain an assertion about the finiteness of the order of decrease of solutions of certain classes of elliptic equations in a neighbourhood of critical points of the solution.

As we shall see later, the sign of the curvature and the order of its vanishing will have an important significance for the structure of the set $\mathfrak{M}$.

5.3. Stability and Instability of Arithmetic Characteristics of a Point of a Surface. An arithmetic characteristic of the point under consideration is called *stable* (*unstable*) with respect to a given class of deformations if it is not changed (it is changed) under the indicated deformations of the surface. Here we are interested, of course, in the stability of the order of flattening and the index of a point under bendings and infinitesimal bendings of the surface.

Theorem 5.1. a) *For the instability* (*under bendings*) *of the order of flattening* $q = n - 2$ *at the point* $(0, 0)$ *of the surface*

$$z = f^{(n)}(x, y) + R(x, y), \qquad n \geqslant 3, \qquad f^{(n)} \not\equiv 0, \tag{18}$$

it is necessary that the Hessian of the form $f^{(n)}$,

$$H(f^{(n)}) = f^{(n)}_{xx} f^{(n)}_{yy} - f^{(n)2}_{xy},$$

is either identically zero or it contains a multiple linear factor $(ax + by)$.

b) *Suppose that the curvature* K *of a surface has the Taylor expansion*

$$K = K^{(k)}(x, y) + R(x, y), \qquad k \geqslant 1.$$

Then for q *to be stable it is sufficient that one of the following three conditions should be satisfied*:
1) $k = 1$,
2) $k = 2$ *and* $K^{(k)}$ *does not have the form* $\pm(ax + by)^2$,
3) $k > 2$ *and* $K^{(k)}$ *is a form of definite sign*.

c) *For* q *to be stable at the point* $(0, 0)$ *of the surface* (18) *it is sufficient that* $n \geqslant 4$ *and the form* $f^{(n)}$ *does not contain a multiple linear factor* $(ax + by)$.

d) *The index of a point in a deleted neighbourhood of which the curvature of the surface is of constant sign is stable with respect to bendings.*

The first two assertions of the theorem were proved in Hopf and Schilt (1938), and the rest in Efimov (1948b). The idea of the proof of a)–c) consists in a careful investigation of the connections between the coefficients of the initial forms in the expansions of f, the Hessian $H(f)$ and K by the Taylor formula and so these items are actually true in classes of surfaces and their bendings of some finite smoothness. Item d) is the result of a difficult lemma on the structure of the set of level curves of a function in a neighbourhood of a regular point, and it is true in the class of bendings of smoothness $\hat{C}_{1;0}^2$.

5.4. Equations of an Immersion and a Bending of a Surface. In equation (3) there are derivatives of only the first order, and so in them there is not detected the type of problem that depends on the sign of the curvature of the metric. In smoothness classes C^2 and above we can propose other equations of an immersion, and if their solutions are determined by some continuously variable conditions (for example, a variable boundary condition, a condition of Cauchy type, and so on), then the continuous totality of solutions of the equations gives a bending of the surface that enters into this totality.

A linear variation of the equations of an immersion leads to an equation of infinitesimal bendings of the 1st order; variations of higher orders give equations of infinitesimal bendings of the corresponding order $(1, n)$ (Belousova (1961), Klimentov (1984)).

The form of all these equations depends essentially on the choice of the system of intrinsic coordinates and basic unknown functions that determine a surface to be immersed. Here we give a series of specific forms of the equations of an immersion.

5.4.1. The Darboux Equation. Suppose that the metric (2) is immersed in E^3 by the surface (1). Since the metric $ds^2 - dz^2 = dx^2 + dy^2$ is locally Euclidean, this happens if and only if its Gaussian curvature is equal to zero, which gives for $z(u, v)$ the so-called Darboux equation

$$z_{11}z_{22} - z_{12}^2 = K(EG - F^2 - Ez_v^2 + 2Fz_uz_v - Gz_u^2), \tag{19}$$

where

$$z_{11} = z_{uu} - \Gamma_{11}^1 z_u - \Gamma_{11}^2 z_v,$$

$$z_{12} = z_{uv} - \Gamma_{12}^1 z_u - \Gamma_{12}^2 z_v,$$

$$z_{22} = z_{vv} - \Gamma_{22}^1 z_u - \Gamma_{22}^2 z_v,$$

and Γ_{ij}^k are the Christoffel symbols of the metric (2). If $z = z(u, v)$ is a solution of equation (19) with the condition $z_u(0, 0) = z_v(0, 0) = 0$, then in a small neighbourhood of the point $(0, 0)$ the form $ds^2 - dz^2$ will be positive definite, and then the maps $x = x(u, v)$, $y = y(u, v)$ that reduce the form $ds^2 - dz^2$ to the standard form $dx^2 + dy^2$ are found from the coefficients of this form by quadratures; an explicit form of this reduction can be found, for example, in Hartman and Wintner (1951) or in Sabitov (1988).

In questions of local bendings of a given surface $z = f(x, y)$ it is helpful to use the following form of the Darboux equation, composed on the basis of the metric of the surface $x = u$, $y = v$, $z = f(u, v)$:

$$(z_{uu}z_{vv} - z_{uv}^2)(1 + p^2 + q^2) - (rz_{uu} - 2sz_{uv} + tz_{vv})(pz_u + qz_v)$$

$$= (rt - s^2)(1 - z_u^2 - z_v^2), \tag{19'}$$

where $p = f_u$, $q = f_v$, $r = f_{uu}$, $s = f_{uv}$, $t = f_{vv}$, and $z(u, v)$ is the required solution.

5.4.2. Fundamental Equations of the Theory of Surfaces. Suppose that the required surface is determined by the coefficients L, M, N of its second fundamental form; then, as we know, the following equations for them are satisfied (the case of a surface of class C^3):

$$L_v - M_u = (\Gamma_{12}^2 - \Gamma_{11}^1)M + \Gamma_{12}^1 L - \Gamma_{11}^2 N,$$

$$M_v - N_u = (\Gamma_{22}^2 - \Gamma_{12}^1)M + \Gamma_{22}^1 L - \Gamma_{12}^2 N, \tag{20}$$

$$LN - M^2 = K(EG - F^2). \tag{21}$$

If the system (20), (21) is solved with respect to L, M, N, then from its solution the surface S in correct situation is uniquely restored. For $S \in C^2$ the system (20) needs to be replaced by its integral analogue; general coordinates are inconvenient for the work, but if we choose a system of isothermal coordinates, in which $E = G = \Lambda^2$, $F = 0$, we obtain a quite "efficient" integral equation in a domain D of change of coordinates (u, v) or $w = u + iv$:

$$L - N - 2iM = -\frac{2}{\pi} \iint\limits_D \frac{H\Lambda^2}{(\zeta - w)^2} d\xi \, d\eta + \frac{4}{\pi} \iint\limits_D \frac{\Lambda H \Lambda_\zeta}{\zeta - w} d\xi \, d\eta + F(w), \tag{20'}$$

where $F(w)$ is an arbitrary function holomorphic in D, and H is the mean curvature.

Equations for infinitesimal bendings will be described in §8.

§6. Bendings of Surfaces in a Neighbourhood of a Point of General Position

6.1. Analytic Case. With respect to bendings of smoothness C^4 in a neighbourhood of a point of general position all the surfaces behave identically (that is, as in another extreme case of smoothness – in bendings of class C^1). Namely, the following theorem is true.

Theorem 6.1. a) *The analytic metric (2) is locally realized in E^3 in the form of an analytic surface without a flat point, and any such realization is bendable in the analytic class.*

b) *Let S and S^* be two analytic realizations of the analytic metric (2), lying in correct situation, for which the point $M_0(0, 0, 0)$ is a point of general position. Then there is a bending of class $C_{A;A}^A$, with a possible addition of mirror reflection, of some neighbourhood of the point M_0 on S into the corresponding neighbourhood of the point M_0 on S^*.*

Part a) is a consequence of the applicability to (19) of the Cauchy-Kovalevskaya theorem (the assertion about the existence of a local realization of an analytic metric has been known for a long time; for the history and current state of this question see Gromov and Rokhlin (1970), Jacobowitz (1972a),

Spivak (1979)); part b) with $K \neq 0$ was proved by Levi in 1908, see for example Efimov (1948b), and in the case $K(M_0) = 0$ by Schilt (Schilt (1937)). The necessity of a mirror reflection in Theorem 6.1 depends, as it turns out, on the type of point, and correspondingly it is necessary to consider three cases, for which we can additionally make certain refinements on the order of smoothness of the bendings.

6.2. Surfaces of Positive Curvature. For a description of the set of regular isometric immersions of a given metric (2) with positive curvature we consider separately the Gauss equation (21). Introducing the functions

$$X = \frac{M}{\sqrt{\Delta K}}, \qquad Y = \frac{L - N}{2\sqrt{\Delta K}}, \qquad Z = \frac{L + N}{2\sqrt{\Delta K}}, \qquad \Delta = EG - F^2$$

we obtain

$$Z^2 = 1 + X^2 + Y^2. \tag{22}$$

Consequently, in the space of the variables (X, Y, Z) we obtain for S the map

$$h: S \to E^3 = (X, Y, Z), \tag{23}$$

which transforms S to some domain on a hyperboloid of two sheets. We note that to surfaces that are mirror symmetric with respect to the plane $z = 0$ after the map h there correspond domains on the hyperboloid that are centrally symmetric with respect to $(0, 0, 0)$. Therefore we can expect that the set of surfaces isometric to a given surface of positive curvature consists of two connected components. In fact, we have the following theorem.

Theorem 6.2. *Suppose that the metric (2) has smoothness of class $C^{n,\alpha}$, $n \geqslant 2$, $0 < \alpha < 1$, and that the curvature of the metric at the point $(0, 0)$ is positive. Next suppose that M is an arbitrarily given point on the hyperboloid of two sheets Γ with equation (22). Then for the point $(0, 0) \in D$ there exist a neighbourhood $U \subset D$ and a map $f: U \to E^3$ that gives an isometric immersion of the metric (2) in E^3 in the form of a correctly situated surface S and such that under the map (23) the point $M_0(0, 0, 0)$ goes into a point $M \in \Gamma$.*

The connection between the different immersions S and S^* of the metric (2) is given by the following theorem.

Theorem 6.3. *Suppose that the metric (2) of smoothness $C^{n,\alpha}$, $n \geqslant 2$, $0 < \alpha < 1$, and with curvature $K > 0$ is immersed in E^3 by two maps $f: D \to E^3$, $f^*: D \to E^3$, that give two surfaces in correct situation. Let M_1 and M_2 be points on one sheet of the hyperboloid of two sheets Γ with equation (22), obtained from the point $M_0(0, 0, 0)$ by maps (23) corresponding to S and S^*, and let $\gamma: [0, 1] \to \Gamma$ be a path on Γ joining M_1 and M_2, $\gamma(0) = M_1$, $\gamma(1) = M_2$, and $\gamma \in C^m$, $0 \leqslant m \leqslant A$. Then there is a disc $\overline{D}_\varepsilon$ in the limits of which the surface S can be applied to S^* by bendings S_t, $0 \leqslant t \leqslant 1$, of smoothness $\hat{C}^{n,\alpha}_{n;m}$ with $S_0 = S$, $S_1 = S^*$, and with $h_t(M_0) = \gamma(t)$.*

In other words, S can be bent into S^* by the family S_t so that the image of the point $(0, 0, 0) \in S_t$ on the hyperboloid Γ is moved on the earlier specified curve $\gamma(t)$ with ends at M_1 and M_2.

A conjecture about the bendability of surfaces of positive curvature was made by Liebmann in 1920 in the following form: if from an ovaloid (a closed regular surface of positive curvature) we remove an arbitrarily small domain, then the remaining part of the ovaloid will admit non-trivial bendings. In the case of a general convex surface this conjecture was proved by A.D. Aleksandrov in 1946, of which we give a few details below (see Remark 5 on p. 211). Let us return to the smooth case. In 1955 Hellwig (Hellwig (1955)) proved that if the boundary of a simply-connected surface of positive curvature satisfies certain conditions of a geometric character, then in the class $C^{4,\alpha}$ the corresponding system (20)–(21) admits a continuous family of solutions that includes the required surface (in fact in Hellwig (1955) bendings of class $\hat{C}^{4;\alpha}_{4;A}$ were constructed); surfaces with such a boundary can also be constructed for the domains mentioned in Liebmann's conjecture, and for a neighbourhood of a given point, and that gives a positive solution of both problems – Liebmann's conjecture and the problem of bendability "in the small". The general case – bendability of a simply-connected surface of positive curvature with a sufficiently smooth boundary – was considered by Fomenko (Fomenko (1965)) for surfaces and deformations of smoothness $C^{3,\alpha}$; in Klimentov (1982) it was shown that any two such identically oriented isometric surfaces can be applied to each other by bendings of class $\hat{C}^{3;\alpha}_{2;A}$; the embeddedness of isometric surfaces is not assumed or asserted either at the initial moment or in the course of their bending.

The specification in Theorem 6.2 of the path $\gamma(t)$ on which the bending of S "goes" is equivalent to the specification of the so-called "bending" function of the surface at a given point, Klimentov (1982), Fomenko (1965). Therefore in the smoothness class $C^{n,\alpha}$, $n \geqslant 3$, Theorem 6.2 can be regarded as proved in these papers. For the smoothness that we have indicated the proof is obtained by the construction by successive approximations of the corresponding family of local solutions of the system (20')–(21).

For a proper understanding of Theorem 6.2 we need to bear in mind that the specification of a point $M \in \Gamma$ as the image of $M_0 \in S$ under the map (23) does not determine the surface S uniquely, so it is impossible to identify Γ either with $\mathfrak{M}$, the set of all local correctly situated isometric immersions (2) in E^3, or with $\hat{\mathfrak{M}}_\varepsilon$, the set of isometric immersions in E^3 of a disc of fixed radius $\varepsilon > 0$ with metric (2) (and one sheet of Γ is not identified with the configuration space of the surface S under its bending). An approach to the description of the topological structure of the space $\hat{\mathfrak{M}}_\varepsilon$ is based on the following simple heuristic argument – when $K > 0$ the system (20)–(21) in the first (linear) approximation is reduced to the generalized Cauchy-Riemann system, each solution of which determines some holomorphic function, and conversely each holomorphic function determines some solution of the system (Vekua (1959)); consequently, in the non-linear (more precisely, quasilinear) case we can expect that between holomorphic functions and solutions of the system (20)–(21) there is some mutually defining relation. In fact, following the general idea of Vekua, in Fomenko (1962) (proof in Fomenko (1965)) formulae were derived that give a one-to-one correspondence between surfaces from one component of $\widehat{\mathfrak{M}}_\varepsilon$ and the set of functions holomorphic in a disc. Klimentov (Klimentov (1982)) gave a critical analysis of the proof from Fomenko (1965) and in turn established a theorem that confirms the general arguments about connections between $\widehat{\mathfrak{M}}_\varepsilon$ and the set of functions holomorphic in a disc. It turns out that in the smoothness class $C^{n,\alpha}$, $n \geqslant 3$, the set $\widehat{\mathfrak{M}}_\varepsilon$ can be represented as an analytic submanifold of the Banach space $C^{n,\alpha}(\overline{D}_\varepsilon)$ modelled in the Banach space of functions of class $C^{n,\alpha}(\overline{D}_\varepsilon)$ holomorphic in D.

Thus, for the case of surfaces of positive curvature the topological structure of the set $\widehat{\mathfrak{M}}_\varepsilon$ is known: it consists of two infinite-dimensional connected components, and any two surfaces of one component can be applied to each other by a bending that is analytic in the parameter.

Remark 5. Along with analytic methods for investigating immersions and bendings of convex surfaces other methods have been developed, which go back to the general approach of A.D. Aleksandrov to the theory of convex surfaces (Aleksandrov (1948)). In particular, on the basis of Aleksandrov's theorems about pasting together and realizing convex metrics it was found that a convex surface is locally not uniquely determined by its metric; see Aleksandrov (1948) and also Efimov (1948a), § 16, and Pogorelov (1969), Ch. 2. Liebman's conjecture is proved here in the following version: if from a general convex closed surface we remove a domain of positive integral curvature, then the remaining part admits non-trivial isometric transformations; see Pogorelov (1969), Ch. 2. Local bendability of a convex surface follows from its local warpability and from the fact that any two isometric and identically oriented general convex surfaces can be applied to each other under the assumption only that the rotation of the boundary (or the geodesic curvature in the smooth case) is of constant sign, see Milka (1973); the bendings can be realized in the class of convex surfaces, but the smoothness of the bending with respect to the parameter is not guaranteed.

Remark 6. Equation (19) in the case of metrics of positive curvature is an equation of elliptic type, so from general properties of solutions of elliptic equations it follows that if two of its solutions, defined in some disc D_ε, coincide in a smaller disc, then they also coincide in D_ε. Hence for surfaces of positive curvature the answer to question 4) in § 2 is: for them $\mathfrak{M}_\varepsilon = \mathfrak{M}_\varepsilon^0$.

6.3. Surfaces of Negative Curvature. Applying the substitution

$$X = \frac{M}{\sqrt{-\varDelta K}}, \qquad Y = \frac{L - N}{2\sqrt{-\varDelta K}}, \qquad Z = \frac{L + N}{2\sqrt{-\varDelta K}}, \qquad \varDelta = EG - F^2,$$

we find that a map of the form (23) transforms S into some domain on a hyperboloid of one sheet $\varGamma$

$$Z^2 = X^2 + Y^2 - 1. \tag{24}$$

The immersibility in E^3 of a metric (2) of negative curvature was proved in Hartman and Wintner (1952) under the following assumptions of smoothness (maybe they are best possible): ds^2 is in C^{n+1} and S is in C^n, $n \geqslant 2$. Application of the methods of their paper enables us to prove the following analogue of Theorems 6.2 and 6.3.

Theorem 6.4. a) *Let M be a point of a hyperboloid of one sheet Γ with equation (24). Then for the metric (2) of smoothness C^{n+1}, $n \geqslant 2$, and with curvature $K > 0$ there is a neighbourhood of the point $(0, 0)$ that admits an isometric immersion in E^3 in the form of a surface S of smoothness C^n for which the map h in (23) transforms the point $M_0(0, 0, 0) \in S$ into the point $M \in \Gamma$.*

b) *Let M_1 and M_2 be two points on Γ joined by a curve $\gamma \colon [0, 1] \to \Gamma$ of class C^m, $0 \leqslant m \leqslant \infty$, or A, and let S_1 and S_2 be two immersions of the metric (2) in E^3 (smoothness as in a)) with an implied correspondence of the points M_i on Γ with the point M_0 on S_i, $i = 1, 2$. Then there is a neighbourhood $U(0, 0) \subset D$ on which S_1 and S_2 can be included in a family of bendings $S(t)$ of smoothness $\hat{C}^n_{n;m}$, with $S(0) = S_1$, $S(1) = S_2$, and the point $M_0 \in S_t$ is transformed by the map h into the point $\gamma(t)$ on Γ. In particular, a surface is bendable onto a mirror image of itself.*

Thus, for the case of negative curvature the set $\mathfrak{M}$ consists of one connected component. Here we should also bear in mind that the choice of the point $M \in \Gamma$ does not uniquely determine the corresponding immersion (2) in E^3; the arbitrariness in the immersion, according to Hartman and Wintner (1952), is determined by the choice of four functions of one argument, with some restrictions at the point $(0, 0)$.

Remark 7. The methods of constructing solutions of equation (19) in Hartman and Wintner (1952) in the case of metrics of negative curvature are such that two of its solutions that coincide in some disc do not necessarily coincide on their extension beyond the disc. Hence for surfaces of negative curvature the sets $\mathfrak{M}_\varepsilon$ and $\mathfrak{M}_\varepsilon^0$ are distinct.

6.4. Neighbourhood of a Parabolic Point. By Theorem 6.1 any surface of class C^4 in a neighbourhood of its parabolic point is bendable and there is applicability (in a wide sense) of two isometric surfaces with parabolic points corresponding to each other. But even in the analytic class of this theorem the question of the structure of the set $\mathfrak{M}$ has not been solved, since firstly the metric (2) with $K(0, 0) = 0$ can be immersed both in the form of a surface with a parabolic point and as a surface with a flat point, and Theorem 6.1 does not give the applicability of such surfaces to each other; secondly there remains open the question of the applicability of mirror reflected surfaces with a parabolic point. We discuss the first question in the section on surfaces with a flat point, and with respect to the second question we have the following theorem.

Theorem 6.5. *A neighbourhood of a parabolic point with stable (zero) order of flattening is not applicable to a mirror reflection of itself* (Hopf and Schilt (1938)).

Criteria for stability are known (Theorem 5.1), for example the surface for Example 2 in §5 is not bendable to its mirror image. This example is interesting in that the curvature of the surface when $a \prec 0$ is negative, that is, Theorem 6.4 about the connectedness of $\mathfrak{M}$ is false if the condition $K < 0$ is violated at the same point M_0. Next it is obvious that a neighbourhood of a parabolic point cannot be bent into its mirror image in the case when in any neighbourhood of it there are points with positive curvature. Hence for the applicability to the mirror image it is necessary that in a neighbourhood of a parabolic point we have $K < 0$ with compulsory presence of the curve $K = 0$, and in the course of such a bending there must inevitably be a situation where the surface has a flattening at the point M_0. Examples of the transition of a parabolic point to a flat point in the course of bending are known (Efimov (1948a) or Hopf and Schilt (1938)), and from them we can obtain an example of the applicability of a surface to its mirror reflection in a piecewise analytic (in the parameter) class of bendings.

Thus, if we restrict ourselves to regular immersions of smoothness C^4 and remain in the class of surfaces with a parabolic point, we find that the corresponding subset $\mathfrak{M}' \subset \mathfrak{M}$ consists of two connected components.

Using the concept of regularity of a point (that is, the presence of a canonical neighbourhood), we can obtain a further refinement of the structure of the set $\mathfrak{M}$; for the details see Efimov (1948b). From the results obtained there we recall only the fact that the index of a parabolic point is unstable under certain conditions on the character of variation of the sign of the curvature of the surface.

We note that for a long time there were no publications on immersions and bendings in the smoothness class C^n, $n \leqslant \infty$, with $K(0, 0) = 0$, but in the recently published works Hong and Zuily (1987), Lin (1985), Lin (1986), and Nakamura and Maeda (1986) local immersibility in E^3 has been obtained in the form of a surface with a parabolic point of metrics (2) with the following conditions: 1) metrics with finite smoothness and with curvature $K \geqslant 0$: if $ds^2 \in C^n$, $n \geqslant 10$, then there is a local immersion in the form of a convex surface $S \in C^{n-6}$ (Lin (1985)); 2) metrics with infinite smoothness and with $K \geqslant 0$: if $ds^2 \in C^\infty$ and K has finite order of zero at the point $(0, 0)$ or the curve $K = 0$ consists of finitely many curves of C^1-smoothness, then there is a convex immersion of $S \in C^\infty$ (Hong and Zuily (1987)); 3) metrics with curvature of variable sign: if $K(0, 0) = 0$ and grad $K(0, 0) \neq 0$, then there is an immersion of $S \in C^{n-3}$ on condition that $K \in C^n$, $n \geqslant 6$ (Lin (1986)); 4) if the curvature $K \in C^\infty$ is zero at the point $(0, 0)$, and in a neighbourhood of it the Hessian of K (as a form) is negative definite (hence $K < 0$), then such a metric admits a local immersion in E^3 in the form of a surface of smoothness C^∞ (Nakamura and Maeda (1985)). The proofs are based on the construction of solutions of the corresponding Darboux equation by special iterations that converge for sufficiently small values of some numerical parameter, so all the resulting surfaces are bendable. But so far there is no complete description of all local immersions and investigation of the connectedness of the set $\mathfrak{M}$ for one of the classes of metrics mentioned above.

§7. Bendability of Surfaces with a Flat Point

The diversity in the picture of bendability planned for surfaces in a neighbourhood of a parabolic point becomes in truth an indescribable realm of the manifold of different possibilities under transition to surfaces with a flat point.

7.1. Non-Applicable Isometric Surfaces. In the cases considered up to now any two isometric surfaces were applicable to each other (at least in the wide sense of the word). But meanwhile Voss in 1895 raised the question formulated in §2 as No. 5: can an isometry always be realized by application? It turns out that for surfaces with a flat point these two concepts – isometry and applicability – are distinct, generally speaking. Namely, the following theorem is true.

Theorem 7.1. *There are non-trivially isometric surfaces S and S^* of class C^A with points $M \in S$ and $M^* \in S^*$ corresponding by an isometry, where M is a flat point on S, such that no neighbourhood of the point M on S can be applied by a bending of class $\hat{C}^2_{1;0}$ to a neighbourhood of the point M^* on S^* or on $\bar{S}^*$, where $\bar{S}^*$ is the mirror image of S^*.*

In the class of bendings of smoothness $\hat{C}^2_{2;0}$ this theorem was proved in Schilt (1937), and in the form mentioned here in Efimov (1948b). The idea of the proof is the same – for the surface S we take an analytic surface with a flat point and with negative curvature close to a flat point, and as S^* we consider an analytic immersion of the metric of the surface S in the form of a surface without a flat point (such an immersion always exists by Theorem 6.1). The surface S can be chosen so that the index of the point M on S is equal to -2 (for example we can take the surface $z = xy^2 + yx^2$); by (17) the index of the corresponding point M^* on S^* is equal to -1, so by the property of stability of the index of a point (Theorem 5.1 d) a bending of S to S^* is impossible. Such an argument goes through if for S we take a surface with stable order of flattening (in particular, a convex surface with positive curvature round a flat point), but here we need, generally speaking, additional investigations on the refinement of the order of smoothness of the bendings under consideration.

7.2. On the Realization of Metrics by Surfaces with a Flat Point. As we see, in all the examples constructed for Theorem 7.1 of surfaces that are isometric but not applicable to each other, one of the surfaces does not have a flat point. In this connection there naturally arises the following question – is this fact regular or is it fortuitous? Apparently in some sense this is a general situation – in any case in the analytic class. Many results refer to the use of this assumption (see Höesli (1950), Hopf and Schilt (1938)).

Theorem 7.2. a) *If the curvature of the metric (2) in a deleted neighbourhood of the point $(0, 0)$ is of constant sign, with Taylor expansion*

$$K = K^{(k)}(x, y) + R(x, y) \qquad (K^{(k)} \text{ is of constant sign}),$$

then any realization of such a metric in the form of a surface with a flat point has a flattening of order $q = k/2$, that is, the surface has the form

$$z = f^{(n)}(x, y) + R(x, y), \qquad n = 2 + k/2.$$

b) *There are metrics (2) with $K(0, 0) = 0$ that do not admit a realization in E^3 in the form of a surface with a flat point.*

c) *There are metrics that have a unique realization in E^3 in the form of a surface with a flat point; to this we can add that there are metrics for which this unique realization with a flattening turns out to be non-bendable.*

A conjecture of Höesli (see Höesli (1950)) is that parts b) and c) of this theorem together cover the "majority" of analytic metrics, that is, for the majority of metrics their analytic realizations with a flat point either do not exist at all or such a realization reduces to a pair of mirror symmetric surfaces. This conjecture was formulated in 1951, but up to now it has not been proved; on the other hand, we know examples of metrics that have two or more realizations with different (non-zero) orders of flattening, see also Höesli (1950).

Such complexity and actual lack of solution of the question about the description of the whole set $\mathfrak{M}$ forces us to restrict ourselves to the investigation of bendability or non-bendability of a given specific surface with flattening. Here the most complete results were obtained by Efimov in a series of papers summarized in the monograph Efimov (1948b) (a survey in Efimov (1958)), and new results have appeared very recently, see Sabitov (1986).

7.3. Non-Bendable Surfaces with a Flattening.

Suppose that the surface $S: z = f(x, y)$ has a representation of the form (16). Following Efimov (1948b), we introduce an integer parameter N that can be defined as follows. Let us consider the equation

$$f_{yy}^{(n)} F_{xx}^{(m)} - 2f_{xy}^{(n)} F_{xy}^{(m)} + f_{xx}^{(n)} F_{yy}^{(m)} = 0, \tag{25}$$

where $F^{(m)}$ is a desired form of degree $m > n$; then $N > n$ denotes the upper bound of those numbers h for which equation (25) has only a zero solution for $m \leqslant n + h$. Under the condition of preserving the initial term $f^{(n)}$ in (16) in the course of bending, the number N has the following meaning: the surface S can be bent only so that in expansions of the form (16) in the process of bending we preserve all terms of degree from n up to $n + N$ and there do not appear degrees less than n; in other words, the osculating paraboloid of order $n + N$ is preserved. Thus, if $N = \infty$, then there are no bendings at all. Hence it remains to find the condition that ensures the invariance of the initial term under bendings, which turns out to be satisfied under the conditions of the following theorem (Efimov (1948b)).

Theorem 7.3. *If the form $f^{(n)}(x, y)$ in (16) does not have multiple real zero directions and $N = \infty$, then the analytic surface S with equation of the form (16) is non-bendable in the class of deformations of smoothness $C_{n;0}^A$.*

Consequently, the main difficulty now reduces to the verification of the condition $N = \infty$. If $f^{(n)}(x, y)$ has the form

$$f^{(n)}(x, y) = a_0 x^n + a_1 x^{n-1} y + \cdots + a_n y^n,$$

then the solubility of each equation (25) when $m > n$ is equivalent to the solubility of some linear system of equations that leads to the corresponding conditions on the rank of the matrix of the system. These conditions in total are expressed as an infinite system of non-identical algebraic equations for the coefficients a_0, $\ldots, a_n$ of the form $f^{(n)}$. Hence, if with each form $f^{(n)}$ we associate the point in E^{n+1} with coordinates $(a_0, \ldots, a_n)$, then it turns out that when $n \geqslant 5$ for almost all points in E^{n+1} we shall have $N = \infty$ for the corresponding forms $f^{(n)}$. This result (see Efimov (1948b), Makarova (1953), Tartakovskij (1953)) shows that almost all surfaces with a flattening are non-bendable. As to specific examples of non-bendability, according to Tartakovskij (1953) for all $n \geqslant 5$ we can propose such a surface

$$f^{(n)}(x, y) = \mu x^n + \lambda x^{n-2} y^2 + \mu x^{n-4} y^4 + y^n, \tag{26}$$

where (λ, μ) is a so-called transcendental point, at which no polynomial $P(x, y)$ with integer coefficients vanishes (in reality the set of admissible pairs (λ, μ) is greater, for example, in the case $n = 5$ we can take even a non-transcendental point (λ, μ), where λ is a transcendental and μ an integer number). We observe, however, that even in small degrees the proof that $N = \infty$ requires large calculations and a complicated technique of analysis, so the discovery of effective methods of verifying the property $N = \infty$ remains a very interesting problem; without a solution of this problem the verification of the non-bendability of a specific surface of the form (16) with the use of Theorem 7.3 is impossible.

One of the main features of the proof of Theorem 7.3 is the establishment of the stability of the order of flattening of the surface under consideration. If we use the criterion of stability of the index of a flat point (see Theorem 5.1), then the fact of the existence of locally non-bendable surfaces can be established in the class of deformations of smoothness $\hat{C}_{1;0}^A$ (Efimov (1948b), § 56).

Thus, for flattenings of order $q \geqslant 3$ almost all analytic surfaces are non-bendable. We shall see below that non-bendable surfaces exist for flattenings of order $q = 2$, and the case $q = 1$ still remains in question.

For the proof of Theorem 7.3 and its analogues the assumption of analyticity is essential, since the main arguments are conducted with the use of expansions in Taylor series. The condition of analyticity has not been removed up to now, and in general the investigation of the non-linear equation (19) in the singular case (degeneracy or change of type) remains an unsolved problem of geometry and differential equations (recently in the papers Hong and Zuily (1987), Lin (1985), Lin (1986) new approaches have been suggested in the cases of degeneracy of elliptic type or change of type along curves).

7.4. Bendable Surfaces with a Flattening. The conditions of Theorem 7.3 are sufficient for non-bendability, but not at all necessary. Therefore surfaces that do not satisfy these conditions can nevertheless be quite non-bendable, and hence there naturally arises the question: do bendable surfaces exist at all? It turns out that such surfaces do exist. With the use of an equation of the form (19′) the following theorem was proved in Dorfman (1957).

Theorem 7.4. *Suppose that an analytic surface S: $z = f(x, y)$ satisfies one of the conditions $f_{xx} = 0$ or $f_{xy} = 0$. Then it admits analytic bendings that preserve the osculating paraboloid of any preassigned order.*

We note that in the conditions of Theorem 7.4 the surface has curvature that vanishes on at least one of the coordinate axes, $x = 0$ or $y = 0$. This suggests the proposition that under the condition that the curvature is of constant sign a neighbourhood of a flat point will always be non-bendable.

7.5. Surfaces of Revolution with Flattening at a Pole. Consideration of surfaces of revolution gives some confirmation to the proposition just stated. Here the proof of non-bendability goes along another path, namely through the use of connections between bendings and infinitesimal bendings (Corollary 3.2). We shall see (§ 8) that almost all surfaces of revolution have rigidity of the 1st or 2nd order, so they are non-bendable in the class of deformations that are analytic in the parameter.

Non-bendability of a given surface does not mean, however, that surfaces isometric to it are necessarily situated far from it: there can exist surfaces that are non-bendable but limiting for the set of isometric surfaces. In the non-local case such examples were constructed by Shor (Shor (1962)), but for general (non-smooth) convex surfaces. Let us consider the local case. Suppose that the meridian L of a surface of revolution S has at points of some denumerable set X (we can consider the more general case of an infinite nowhere dense set) tangents perpendicular to the axis of revolution, with contact of infinite order (Fig. 4). Let $M \in L$ be one such point with abscissa $x_M \in X$. The tangent plane Π_M to S at this point is perpendicular to the axis of revolution and touches S along a parallel. Making a mirror reflection with respect to the plane Π_M of the part of

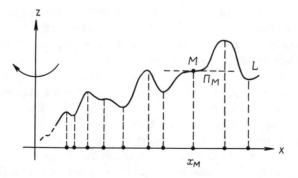

Fig. 4

the surface S with $x_M > x$, we obtain a new surface S_M isometric to S. For this surface S_M we can again consider a new surface obtained by a mirror reflection of part of it with respect to some tangent plane. Repeating this operation with an arbitrary choice of points of X, we obtain a set of C^∞-smooth isometric surfaces of the power of the continuum, among which there are surfaces arbitrarily close to S, which, however, is non-bendable in the class $C^1_{1;A}$ for a definite structure of the set X (Sabitov (1973)). Consequently, giving up analyticity in the smoothness of the surface at once essentially complicates the investigation of isometric transformations of a surface.

Remark 8. This example shows that for surfaces with curvature of variable sign the answer to question 4 in §2 is negative: the spaces $\mathfrak{M}_\varepsilon$ and $\mathfrak{M}_\varepsilon^0$ are different for them.

§8. Infinitesimal Bendings of Surfaces "in the small"

8.1. Equations of Infinitesimal Bendings. Suppose that a surface $S: z = f(x, y) \in C^1$ has a field of infinitesimal bending $U \in C^1$. Then the components ξ, η, ζ of the field U must satisfy the system

$$\xi_x + p\zeta_x = 0,$$
$$\xi_y + \eta_x + p\zeta_y + q\zeta_x = 0,$$
$$\eta_y + q\zeta_y = 0, \qquad (p = f_x, q = f_y).$$

Eliminating ξ and η from here, for $f, \zeta \in C^2$ we obtain the 2nd order equation

$$f_{yy}\zeta_{xx} - 2f_{xy}\zeta_{xy} + f_{xx}\zeta_{yy} = 0. \tag{27}$$

Equation (27) is fundamental in the following sense: when a solution $\zeta(x, y)$ of equation (27) is known, the components ξ and η are determined by quadratures and, in particular, if f and ζ belong to the class C^n, $n \geqslant 2$, then $\xi, \eta \in C^n$ also.

8.2. Rigidity "in the small" of Analytic Surfaces. The existence of surfaces that are rigid "in the small" was first established in a paper by Efimov (Efimov (1948c)). The idea of investigation here goes back to the method used in the proof of existence of non-bendable surfaces: in the analytic case for surfaces of the form $z = f^{(n)}(x, y)$ equation (27) reduces to an infinite totality of equations among which there are equations of the form (25), but beginning with $m = 2$. The condition of solubility of each equation (25) leads to some (its own for each m) algebraic equation for the coefficients of the form $f^{(n)}(x, y)$ and in this way we prove that if for some form $f_0^{(n)}$ none of these equations is satisfied in an identical way, then almost all surfaces $z = f^{(n)}(x, y) + \cdots$ are rigid. In Efimov (1948c) this was established for $n = 9$, so almost all surfaces $z = f^{(9)}(x, y) + \cdots$ are rigid in the analytic class of infinitesimal bendings. Starting from example (26) for $n = 5$, we can show that the corresponding equation (25) does not have solutions for

$m = 2, 3, 4, 5$. Hence the surface

$$z = x^5 + \lambda x^3 y^2 + xy^4 + y^5, \qquad \lambda \text{.is a transcendental number,}$$

is rigid. On the other hand, it is known that the surfaces $z = f^{(3)}(x, y)$ and $z = f^{(4)}(x, y)$ are non-rigid (Berri (1952)), so the smallest degree of a rigid surface of the form $z = f^{(n)}(x, y)$ is equal to 5. But with respect to the smallest degree of rigid surfaces of the form $z = f^{(n)}(x, y) + \cdots$ it is difficult to state any propositions; we can only affirm that $n \geqslant 3$, since in the parabolic case an analytic surface is bendable in the analytic class and by Theorem 3.1 it is non-rigid.

8.3. Analytic Surfaces of Revolution with Flattening at a Pole. The study of infinitesimal bendings has been advanced mostly for surfaces of revolution, since for them the finding of infinitesimal bendings can be reduced to the solution of ordinary equations (Cohn-Vossen (1929)). This is done as follows.

Suppose that the surface $z = \varphi(\sqrt{x^2 + y^2})$ is obtained by rotation about the z-axis of the curve $z = \varphi(\rho), 0 \leqslant \rho < \varepsilon, \varphi'(0) = 0$. Introducing the moving vector $e(\theta) = i \cos \theta + j \sin \theta$ and seeking the field of infinitesimal bendings in the form $\alpha(\rho, \theta)k + \beta(\rho, \theta)e + \gamma(\rho, \theta)e'$, we obtain the system

$$\alpha'_m = \frac{m^2}{\rho} \alpha_m + \frac{m^2 - 1}{\rho \varphi'} \beta_m,$$

$$\beta'_m = -\frac{m^2 \varphi'}{\rho} \alpha_m - \frac{m^2 - 1}{\rho} \beta_m, \qquad \beta_m = -im\gamma_m, \tag{28}$$

where α_m, β_m and γ_m are the Fourier coefficients of the functions α, β and γ. In the smoothness class C^2 and higher, we can reduce the system (28) to one equation

$$\rho \varphi' \alpha''_m + \rho \varphi'' \alpha'_m - m^2 \varphi'' \alpha_m = 0. \tag{29}$$

Non-trivial infinitesimal bendings are obtained for the values $m \geqslant 2$.

In the case of an analytic surface, when

$$\varphi(\rho) = a_0 \rho^{2k} + a_1 \rho^{2k+2} + \cdots, \qquad a_0 \neq 0$$

the well-known Fuchs theory can be applied to equation (29), and we find that equation (29) has solutions of the form

$$\alpha_m = \rho^{\nu_m} \sum_{j=0}^{\infty} A_j \rho^{2j}, \qquad A_0 \neq 0,$$

$$\nu_m = 1 - k + \sqrt{m^2(2k - 1) + (k - 1)^2}. \tag{30}$$

The requirement that the function $\alpha(\rho, \theta) = \alpha_m(\rho)e^{im\theta}$ is analytic leads to the condition $\nu_m = m + 2l$, where l is a natural number, which in turn gives a connection between $x = \nu_m + k - 1$ and $y = m$ in the form of the following Diophantine equation (the so-called Pell equation)

$$x^2 - (2k - 1)y^2 = (k - 1)^2.$$

It turns out that this equation always has a solution with $y \geqslant 2$ (except for the case $k = 5$). Therefore we find that all surfaces of revolution, except those having a flattening of order $q = 8$, are non-rigid (Sabitov (1986)).

8.4. Rigid and Non-Bendable "in the large" Surfaces of Revolution. In the analytic case a peculiarity of infinitesimal bendings of surfaces of revolution is that for a given $k \neq 5$ not all harmonics $\alpha_m(\rho)e^{im\theta}$ exist (when $2k - 1 = n^2$ there are generally only finitely many of them), which gives the possibility of obtaining criteria for rigidity "in the large". Namely, suppose that on a closed surface of revolution homeomorphic to a sphere the two poles have flattening of order $2k_1 - 2$ and $2k_2 - 2$ respectively; we call the orders of the flattenings *consistent* if the neighbourhoods of the poles have infinitesimal bendings with a common number of non-trivial harmonics, and *inconsistent* otherwise.

Lemma 8.1. *There are surfaces of revolution with inconsistent orders of flattening at the poles* (Sabitov (1986)).

As a corollary we find that such surfaces of revolution are rigid "in the large". More interesting, however, is another corollary.

Corollary 8.2. *Surfaces of revolution with inconsistent orders of flattening at the poles are non-bendable in the class of deformations of smoothness* $C_{1;A}^{\infty}$.

The point is that up to now only two classes of closed surfaces non-bendable "in the large" have been known – these are ovaloids or closed convex surfaces and the so-called surfaces of type T, on which the region with curvature $K > 0$ as a whole has integral curvature 4π (for example, a circular torus; for the general definition and the analytic case see Aleksandrov (1938); for the case of class C^3 see Nirenberg (1963)). Hence in the analytic class by Corollary 8.2 we can determine closed and non-convex surfaces that are non-bendable "in the large".

8.5. Non-Analytic Surfaces. In contrast to the non-linear theory of bendings, in the linear problem in a number of cases we can get rid of the requirement that the surface and the field of (infinitesimal) bendings of it are analytic.

8.5.1. Surfaces of Positive Curvature with an Isolated Flat Point. In the fundamental well-studied case such surfaces have the form

$$z = r^n f(x, y), \qquad r^2 = x^2 + y^2, \qquad n > 2, \tag{31}$$

where $f(x, y)$ is a sufficiently smooth function and $f(0, 0) \neq 0$. Since the previous methods are unsuitable here, we need to describe at least "in two words" the process of solving the problem. The main features are the following: firstly, by the choice of the special so-called *adjoint isothermal* coordinates (ξ, η) the determination of infinitesimal bendings reduces to the solution of the system

$$\frac{\partial U}{\partial \bar{\zeta}} - \frac{b(\zeta)}{2\zeta}\bar{U} = 0, \qquad \zeta = \xi + i\eta, \tag{32}$$

where

$$U(\zeta) = \zeta^2 K^{1/4}(\zeta)(\delta M + i\delta L), \qquad \frac{\partial}{\partial \bar{\zeta}} = \frac{1}{2}\left(\frac{\partial}{\partial \xi} + i\frac{\partial}{\partial \eta}\right),$$

$b(\zeta)$ is a function that can be explicitly written out, in particular

$$b(0) = \frac{n - 2}{2\sqrt{n - 1}},$$

and $\delta L = \delta N$ and δM are variations of the coefficients of the second fundamental form of the surface S in adjoint isothermal coordinates (in which $L = N$, $M = 0$); secondly, for a solution of the system (32) one first investigates separately the case of the "model" surface (31) when $f = 1$, for which the system (32) has the form

$$\frac{\partial U}{\partial \bar{\zeta}} - \frac{n - 2}{4\sqrt{n - 1}}\frac{\bar{U}}{\zeta} = 0 \qquad (33)$$

and finally, by taking account of the connection between solutions of the systems (32) and (33) we obtain information about the rigidity or non-rigidity of the surface S. This scheme of investigation, established in a series of papers by Usmanov (a survey of them can be found in Usmanov (1984)), in its first part goes back to Vekua (Vekua (1959)), who studied in this way infinitesimal bendings of surfaces of strictly positive curvature; but in the presence of flattening both the introduction of adjoint isothermal coordinates itself and the solution of the system (32) require "works" with degenerate elliptic systems. The second part, the investigation of the "model" surface $z = (x^2 + y^2)^{n/2}$, was carried out in Efimov and Usmanov (1973). In this paper they were the first to trace the possible connection between the order of flattening of a surface and the "degree" of its non-rigidity in deformations of a given smoothness class (but we must elaborate that the smoothness of not all components of the field of infinitesimal bendings has been studied, but only the vertical component of it, which, generally speaking, without taking account of the smoothness of the surface itself, is insufficient for the corresponding conclusions of the smoothness of the whole field). It turned out that with an increase in the required smoothness the bending field decreases asymptotically to zero, and as a result not in the class C^A but in C^∞ does there arise the appearance of rigidity of the surface. Let us report the precise formulations of certain theorems.

For each $k \geq 3$ we denote by N_k the finite set of rational numbers n determined by the formula

$$n = 2 + 4q\frac{k + q}{k^2 - k - 2q} \qquad \left(q = 1, 2, \ldots, \frac{k^2 - k}{2} - 1\right).$$

Let $N = \bigcup_{k=3}^{\infty} N_k$. We split all the surfaces $z = r^n$ into two classes: S_0 if $n \in N$, and S_1 if $n \notin N$.

Theorem 8.3. a) *All surfaces S_1 are locally non-rigid in any smoothness class $C^l, l < \infty$.* b) *Surfaces from S_0 are locally non-rigid in the class C^A.* c) *For surfaces* from S_1 an infinitesimal bending of the class C^p, $p \geqslant 2$, is a small quantity characterized by the uniform estimate

$$\zeta(x, y) = O(r^{\nu_{1+k_{n,p}}}),$$

where

$$\nu_{1+k_{n,p}} = 1 - \frac{n}{2} + \sqrt{\left(\frac{n-2}{2}\right)^2 + (1 + k_{n,p})^2(n-1)},$$

and the integers $k_{n,p}$ are determined from the inequalities

$$\sqrt{\frac{p(p+n-2)}{n-1}} - 1 < k_{n,p} \leqslant \sqrt{\frac{p(p+n-2)}{n-1}}.$$

Corollary 8.4. *Surfaces from S_1 are locally rigid with respect to infinitesimal bendings of smoothness C^∞.*

As we have already said, the general surface (31) is investigated with the use of an explicitly determined homeomorphism between the solutions of the systems (32) and (33). In particular, if in (31) the function $f \in C^\infty$ and $n \in N$, then such surfaces are locally rigid in the class C^∞ (Usmanov (1984)).

Surfaces of a more general form than in (31) have not actually been studied, and any advance here requires preliminary substantial progress in the investigation of generalized Cauchy-Riemann systems with a singular coefficient.

8.5.2. Local Rigidity in the C^1-Class. Generally speaking, the result of 8.5.1 on C^∞-rigidity of the surfaces (31) has not been represented quite correctly, since we require here for the bending field a smoothness greater than the smoothness of the surface itself (in this respect the result from 8.3 on the C^∞-rigidity of the analytic surface $z = (x^2 + y^2)^5$ is correct). This remark shows that in the nonanalytic case, imposing some requirement on the smoothness of the bending field, we need to take care of the consistency of this requirement with the given smoothness of the surface. In non-singular cases this consistency follows from the corresponding theorems on smoothness of solutions of the elliptic (when $K > 0$) or hyperbolic (when $K < 0$) equation (27) (it is true that here we need to make an important remark: if in the general elliptic case the problem is solved more or less completely, on each convex surface any field U of infinitesimal bendings of it belongs to the Lipschitz class $C^{0,1}$ (Aleksandrov (1937)), and the surface is non-rigid (Pogorelov (1969)), and under the additional condition $S \in C^{n,\alpha}$, $n \geqslant 2$, $0 < \alpha < 1$, $K > 0$, the field $U \in C^{n,\alpha}$ (Pogorelov (1969)), then in the hyperbolic case – even with the strict condition $K < 0$ – we can draw some conclusions about the existence and degree of regularity of the field of infinitesimal bendings only under the a priori assumption $U \in C^2$). But if $K = 0$ at the point under consideration, then even here there is no regular theory, and depending on additional properties of the surface we can encounter very different possibilities. In particular, we have the following theorem.

Theorem 8.5. *For any n, $1 \leqslant n \leqslant \infty$, there is a surface of class C^n that is C^1-rigid in a neighbourhood of a marked flat point (Sabitov (1973)).*

A typical surface of this kind is the surface of revolution

$$z = e^{-1/r} \sin \frac{1}{r}, \qquad r = \sqrt{x^2 + y^2}.$$

It is useful to compare this theorem with the fact that any surface is bendable in the class of C^1-smooth deformations: the linear problem in the C^1-class – the determination of infinitesimal bendings of the 1st order – is sometimes "worse" to solve than the non-linear one! Hence it follows, in particular, that the bendings of surfaces from Theorem 8.5 in the class C^1 cannot depend analytically on the parameter of deformation.

8.5.3. Flattenings of Infinite Order. In Theorem 8.5 by virtue of the condensation of the zeros of K to the point $(0,0)$ we have a flattening of infinite order at the pole. Another investigated case of a flattening of infinite order is a surface of revolution with meridian $z = \varphi(r)$ satisfying the condition

$$\int_0^\varepsilon \frac{\varphi(t)\, dt}{\varphi'(t)t^2} < \infty, \tag{34}$$

where

$$\varphi(r) \in C^n[0, \varepsilon), \qquad 1 \leqslant n \leqslant \infty;$$

$$\varphi'(0) = 0, \qquad \varphi'(r) \neq 0, \qquad r \neq 0.$$

For these surfaces we have $\varphi(r) < \exp(-C/r^\alpha)$ and for them we can state necessary and sufficient conditions for non-rigidity in the classes C^1 and C^2 (see Sabitov (1979b)), and it turns out that if the flattening is very strong, then these conditions of non-rigidity are satisfied. As a result we have the following qualitative picture: a power flattening leads to the possibility of C^∞-rigidity, but leaves non-rigidity in finite smoothness classes; a flattening of infinite order can give rigidity even in the class C^1 (at the same time, in the convex case there is always C^1-non-rigidity, but there may be C^2-rigidity); finally, for very strong flattenings of infinite order there again arises non-rigidity in finite smoothness classes – the influence of the flat point "weakens", and a neighbourhood of it becomes close to a domain on the plane.

8.6. Infinitesimal Bendings of the 2nd Order. Although for surfaces "in the large" there are significantly many papers on infinitesimal bendings of the 2nd order of them, in the local formulation the study of this question has just begun. Non-rigidity of the 2nd order of surfaces in a neighbourhood of a point with $K \neq 0$ follows from Theorems 6.2, 6.3 and 3.1. In a neighbourhood of a parabolic point and a point with a flattening there are only isolated results about infinitesimal bendings of the 2nd order. In Ivanova-Karatopraklieva and Sabitov (1989) for surfaces of revolution we studied the connection between the order of flattening of a pole and the numbers of non-trivial harmonics of infinitesimal bendings of the 1st order extendable to infinitesimal bendings of the 2nd order, and in Sabitov (1979a) there is given a theorem on C^1-rigidity of the 2nd order of a neighbourhood of a pole of a surface of revolution with meridian of the form (34) (however, the proof of the theorem in the form stated in Sabitov (1979a) turned out to be erroneous, but the theorem remains true under the additional assumption that the surface is convex; the analysis of the general case is hindered by the lack of knowledge of a uniform estimate with respect to small ρ as $n \to \infty$ of the solution of the following system with small parameter:

$$\varepsilon y'(\rho) = A(\rho)y + \frac{1}{\rho^2\varphi'(\rho)}z, \qquad A(\rho) = \frac{1}{\rho} - \frac{\varphi}{\rho^2\varphi'(\rho)}$$

$$\varepsilon z'(\rho) = -A^2(\rho)\rho^2\varphi'(\rho)y - A(\rho)z, \qquad \varepsilon = \frac{1}{n^2 - 1} \to 0;$$

here the characteristic polynomial has one double root $\lambda = 0$, to which there corresponds only one eigenvector of the matrix of the system, and this is just a singular case in the general theory of equations with a small parameter, which requires additional assumptions (Vasil'eva and Butuzov (1978)).

A very complete investigation of infinitesimal bendings of the 2nd order can also be carried out for a neighbourhood of a pole of an analytic surface of revolution; here, in particular, it turns out that for the majority of orders of flattening the infinitesimal bendings of the 1st order are not extendable to analytic infinitesimal bendings of the 2nd order, and to determine surfaces with non-rigidity of the 2nd order we can state an algorithm based on the solution of some Diophantine system of equations. The main interest in assertions about the rigidity of the 2nd order is connected, of course, with a deduction about the non-bendability of such surfaces.

Among the questions of general character not solved here we mention the absence of some interpretation of infinitesimal bendings of the 2nd order from the point of view of the theory of shells.

Works on local infinitesimal bendings of the third and higher orders have begun to appear only very recently; see Ivanova-Karatopraklieva (1987/88) and (1990).

8.7. Bendings of Troughs. There is still one class of problems that are re- garded as intermediate problems between bendings of surfaces "in the small" and "in the large" – these are bendings of so-called troughs. These surfaces are thought of as a ring-shaped band containing inside it a planar closed curve with $K = 0$, along which the tangent plane to the surface is stationary. The band can be arbitrarily narrow – in it there is "smallness" of the surface. For such surfaces, particularly for troughs of revolution, there exists a series of complete results, giving criteria for their non-rigidity of the 1st order; with respect to infinitesimal bendings of the 2nd order the troughs in all the cases studied turn out to be rigid; for citations see Sabitov (1979b).

§9. Supplement. Bendings and Infinitesimal Bendings of Polyhedra

9.1. Introduction. The questions of bendings and infinitesimal bendings of polyhedra are an inexhaustible theme of research by many geometers, and interest in them has intensified recently, which in the last instance is apparently explained by new perspectives discovered with the possibilities of wide application of computers for the solution of theoretical and applied problems in this domain (for example, the journal Structural Topology, published by Montreal University, specially declared questions of bendings of polyhedra to be one of the main important directions of its general themes devoted to the study of spatial forms from the viewpoint of mathematics, architecture and the mechanics of engineering constructions). The results obtained and the new ideas in the theory of bendings of polyhedra would be sufficient for a special detailed survey, so we are forced to restrict ourselves to a rather shortened account of the main achievements and analysis of their possible development.[1]

First of all, in the theory of bendings of polyhedra there are two differences from the case of smooth surfaces: firstly, here there is no local theory in the proper sense of the word, that is, there are no properties that depend on the smallness of the neighbourhood under consideration, secondly, there is no outlet to differential equations – in the analytic scheme the theory of bendings of polyhedra is concerned with algebraic equations, and the theory of infinitesimal bendings (of the 1st order) is even concerned with linear equations. In addition, here, of course, far more extensively than in the smooth theory, direct descriptive-geometric and combinatorial considerations are employed.

By a polyhedron in E^3 we shall understand a piecewise linear (continuous in the large and linear on faces) map $f: \mathcal{K} \to E^3$ of some locally finite simplicial complex $\mathcal{K}$, whose body is homeomorphic to a domain on a two-dimensional manifold; we can regard a piecewise Euclidean metric on $\mathcal{K}$ as specified or induced from E^3 by the map f. As usual, the map f can be identified with its image, not forgetting, however, that if there are self-intersections or self-coverings in the image, then points of the image that have different preimages in $\mathcal{K}$ are assumed to be different. If in the definition of a polyhedron we start from a complex $\mathcal{K}$ with possible non-triangular cells, then this case can be reduced to a polyhedron with triangular faces by splitting the faces into triangles by means of diagonals. This does not change the metric on $\mathcal{K}$, since in the theory of bendings of polyhedra it is assumed that the faces are rigid, so the distances between the vertices of the polyhedron on one face are preserved in the course of bending. But, generally speaking, the properties of bendability can change, namely, if a new polyhedron (with additional edges) is non-bendable or rigid (the exact definitions will be given a little later), then the original polyhedron with non-triangular faces will be the same; if the resulting polyhedron with triangular

---

[1] In fact there are now several surveys on this topic; see, for example, Connelly (1992) and Ivanova-Karatopraklieva and Sabitov (1992).

faces is bendable or non-rigid, then the original polyhedron can turn out to be non-bendable or rigid. In other words, the addition of new edges to the faces "weakens" the polyhedron (there arises the additional possibility of curving a face along a "new" edge).

9.2. Polyhedral Metrics and Their Isometric Immersions. If a metric specified on $\mathcal{K}$ is Euclidean on each face, then it is called *polyhedral* (for general questions of introducing a piecewise linear metric on $\mathcal{K}$, see Aleksandrov (1950), Gluck, Krigelman and Singer (1974)). In this case it is required that the map $f: \mathcal{K} \to E^3$ is isometric. If such a map exists, then by the terminology of Aleksandrov (1950) about a polyhedron we say that it is specified by its *development* $\mathcal{K}$ (in the general case the faces of the development are not at all required to be faces of the polyhedron, and the faces are not required to be triangles, Aleksandrov (1950)). Intuitively the specification of the development means that we have triangular faces of the polyhedron of a natural size with an indication of the rule for identifying edges or pasting together triangular faces along common edges.

Hence, if a metric is specified in advance on $\mathcal{K}$, then before we talk about bendings of the polyhedron we first need to verify the existence of an isometric immersion of $\mathcal{K}$ in E^3. Not every development can be isometrically immersed in E^3; for example, if a development R consists of two equal triangles with pairwise identified edges (it is true that this development R is not formally a simplicial complex, but it can easily be converted into a complex by adding an interior vertex to one of the copies of a triangle), then an isometric map of R in E^3 gives a doubly covered triangle, which is not an immersion. We can indicate developments that cannot generally be isometrically mapped into E^3. An example of such a development is shown in Fig. 5. On it we indicate a development of four triangles ABC, ABO_1, BCO_2, CAO_3 with identification of the edges AO_1 and AO_3, BO_1 and BO_2, CO_2 and CO_3 and of all the vertices O_1, O_2, O_3 into one vertex.

We note, however, that if we permit an additional triangulation of the faces of the development, then under a sufficient refinement of the triangulation any (orientable) development admits an isometric embedding in E^3, see Burago and Zalgaller (1960). For example, a twice covered regular triangle can be isometrically embedded in E^3 in the form of an octahedron if to each copy of the triangle

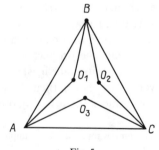

Fig. 5

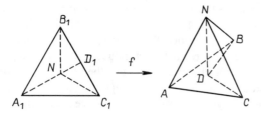

Fig. 6

we add a vertex (Fig. 6) – the centre of the triangle – a point N in the triangle $A_1 B_1 C_1$, a point S in the other copy $A_2 B_2 C_2$, and then embed the four triangles $A_1 N B_1$, $B_1 N D_1$, $D_1 N C_1$ and $C_1 N A_1$ ($B_1 D_1 = D_1 C_1$) in the form of the lateral surface of a pyramid with planar base $ABDC$ (the other triangle is isometrically embedded in E^3 as the mirror symmetric image of this pyramid). This result of Burago and Zalgaller is analogous to the theorem of Kuiper mentioned in §4 about the isometric embeddability in E^3 of any two-dimensional Riemannian manifold: as in the smooth case, the embeddability of a polyhedral metric in E^3 is achieved by means of a fine "corrugation" of the surface. From the viewpoint of bendings such polyhedral immersions have not been studied in practice. For example, can we find such an immersion of a given development with an additional triangulation so that it is bendable? It is assumed that this is improbable, but it would be interesting to investigate isometric transformations of such polyhedra, in particular, to construct the chain of an isometric transition from a convex polyhedron to its mirror image.

Of the general results about immersions of polyhedral metrics the best known is Aleksandrov's theorem about the realization of any complete convex polyhedral metric in E^3 in the form of a convex polyhedron, Aleksandrov (1950). Henceforth we assume that the metric on $\mathscr{K}$ is induced from E^3 by some map $f: \mathscr{K} \to E^3$, so we shall not be concerned with special questions of an isometric immersion in E^3 of a priori specified polyhedral metrics on $\mathscr{K}$.

9.3. Bendings of Polyhedra. Configuration Spaces of Polyhedra.

Suppose that a map $f: \mathscr{K} \to E^3$ determines a polyhedron M; then a polyhedral metric is induced on $\mathscr{K}$. According to the general scheme of §1, we consider the set $\mathfrak{M}$ of all polyhedra obtained by isometric maps of $\mathscr{K}$ in E^3 (a priori we do not require that the map is an embedding or an immersion). In contrast to the smooth case, $\mathfrak{M}$ is always a finite-dimensional set. In fact, since the faces of the complex $\mathscr{K}$ are triangular, the map of $\mathscr{K}$ in E^3 is completely determined if we know the images of the vertices. In turn, if we know the vertices $p_1, \ldots, p_V$ (where V is the number of vertices) of the polyhedron, then we can associate with it a point $m \in E^{3V}$ with coordinates $(x_1, y_1, z_1, \ldots, x_V, y_V, z_V)$, where x_i, y_i, z_i are the coordinates of the point p_i, $1 \leqslant i \leqslant V$, in E^3. Conversely, with each point m of E^{3V} we can obviously associate V points $p_1, \ldots, p_V$ of E^3. Consequently, with the set $\mathfrak{M}$ there is associated in a one-to-one way some set in E^{3V}. Intuitively the set

$\mathfrak{M}$ can be interpreted as the set of positions of the polyhedron M in E^3 under all possible isometric maps of $\mathscr{K}$ in E^3, so $\mathfrak{M}$ is naturally called the configuration space of the polyhedron M. Since $\mathfrak{M}$ is represented by some set of points in E^{3V}, the topology on $\mathfrak{M}$ can be regarded as induced from the Euclidean topology in E^{3V}. Now, as in the general case, we define a bending of the polyhedron M as a path in a connected component $\mathfrak{N} \subset \mathfrak{M}$, beginning at $M \in \mathfrak{N}$. In order to exclude from consideration polyhedra congruent to M, we shall assume that one face on M is fixed; then M as a rigid body admits only a mirror reflection with respect to the plane of the fixed face, so any bending of the polyhedron M, if it exists, is necessarily non-trivial.

The metric of a polyhedron with triangular faces is completely determined by the lengths of the edges. Let l_{ij} be the length of the edge joining the vertices with numbers i and j. In this case all isometric polyhedra of $\mathfrak{M}$ have vertices $p_1, \ldots, p_V$ satisfying the conditions

$$|p_i - p_j| = l_{ij}, \qquad (i, j) \in \tilde{I}, \tag{35}$$

where $\tilde{I}$ is the set of pairs (i, j) of numbers of the vertices joined by edges. Consequently, in E^{3V}, by the association described above, to the set $\mathfrak{M}$ there corresponds some algebraic variety. Assuming, as agreed, that one face is fixed, we see that the algebraic variety is bounded. Therefore we immediately obtain the following description of the set $\mathfrak{M}$: it consists of finitely many compact connected components. In addition, connected components of a bounded algebraic variety are separated from each other by some positive distance, and so it follows that if two isometric polyhedra are sufficiently close to each other, then they will belong to one connected component of $\mathfrak{M}$, and so they are applicable to each other. Thus, for polyhedra we have a new version of the answer to question 5 in § 2: if isometric polyhedra are sufficiently close to each other, then the isometry between them reduces to a bending.

Another consequence of the structure of $\mathfrak{M}$ as an algebraic variety is that if two polyhedra are applicable to each other, then the application can be realized by an analytic path; in other words, if there is a continuous path in $\mathfrak{M}$, then there is an analytic path with the same ends, Gluck (1975). Therefore bendings of polyhedra can always be sought in the class of bendings that are analytic in the parameter.

Next, for polyhedra we can give an answer to the question formulated in Remark 3 on p. 187: an algebraic variety does not have a boundary, so a polyhedron M cannot be a boundary point in $\mathfrak{M}$ of its connected component $\mathfrak{N}$: in $\mathfrak{N}$ there is a path passing *through* M. Intuitively this means that a polyhedron in the process of bending cannot arrive at the position from which a bending later goes only by the reverse motion with a repetition of the positions already passed through.

Let us illustrate all that we have said above by an example of bendings of a closed planar quadrangle (although here all the dimensions are smaller by one, nevertheless the idea of algebraic investigation is the same). Suppose that the edges of the quadrangle have lengths a, b, c, d. There is always an edge l such that the two adjacent edges have length no smaller than that of l. Suppose that the length of l is $b = 1$, and that the lengths of the adjacent edges are equal to a and c, $a \geqslant c \geqslant 1$. We restrict ourselves to a consideration of the following cases (we regard the edge l as fixed):

1) $a = b = c = 1$. Then $\mathfrak{M}$, depending on the values of d, has the following topological structure:

$\quad\quad A_{11}$) $d = 0 \quad \mathfrak{M} = S_0$ (two points)

$\quad\quad A_{12}$) $0 < d < 1 \quad \mathfrak{M} = S^1 \cup S^1$ (two circles)

$\quad\quad A_{13}$) $d = 1 \quad \mathfrak{M}$ has the form as in Fig. 7

$\quad\quad A_{14}$) $1 < d < 3 \quad \mathfrak{M} = S^1$

$\quad\quad A_{15}$) $d = 3 \quad \mathfrak{M}$ is one point; the case $d > 3$ is impossible.

2) $b = 1, a = c > 1$. Then

$\quad\quad A_{21}$) $d = 0 \quad \mathfrak{M} = S^0$

$\quad\quad A_{22}$) $0 < d < 1 \quad \mathfrak{M} = S^1 \cup S^1$

$\quad\quad A_{23}$) $d = 1 \quad \mathfrak{M}$ has the form as in Fig. 8

$\quad\quad A_{24}$) $1 < d < 2a - 1 \quad \mathfrak{M} = S^1$

$\quad\quad A_{25}$) $d = 2a - 1 \quad \mathfrak{M} = S^1 \vee S^1$

$\quad\quad A_{26}$) $2a - 1 < d < 2a + 1 \quad \mathfrak{M} = S^1$

$\quad\quad A_{27}$) $d = 2a + 1 \quad \mathfrak{M}$ is one point; the case $d > 2a + 1$ is impossible.

Fig. 7

Fig. 8

We can give a similar description of $\mathfrak{M}$ in the remaining cases, and also generally in the case of bendings on the plane of an arbitrary closed n-gon. As to configuration spaces of polyhedra, the study of them is still only at an initial stage: configuration spaces of octahedra are known, Bushmelev and Sabitov (1990), and of so-called degenerate suspensions, Sabitov (1983). Generally it seems that problems of the theory of bendings of polyhedra can conventionally be split into two large groups: the establishment of the fact of bendability or non-bendability of a given polyhedron M, and the description of the configuration space $\mathfrak{M}$ for a bendable polyhedron M.

Remark 9. Regarding the l_{ij} in (35) as variables, we see that the set of all polyhedra $f: \mathscr{K} \to E^3$ together with their metric is an algebraic variety in a Euclidean space of dimension $q = 3V + I$, where $I = \operatorname{card} \tilde{I}$; the complement to this variety in $E^q \setminus E^{3V}$ determines metrics on $\mathscr{K}$ that cannot be isometrically mapped into E^3.

9.4. Infinitesimal Bendings of Polyhedra and Their Connection with Bendings.
In the geometrical scheme the definition of infinitesimal bendings of polyhedra is analogous to the smooth case. Let p_i^0, $1 \leqslant i \leqslant V$, be the vertices of the polyhedron M, and $z_i^{(j)}$ the vectors applied to the points p_i, $1 \leqslant k \leqslant j \leqslant n$. Let us consider a deformation of the polyhedron under which the vertices p_i go over to the position

$$p_i(t) = p_i^0 + t^k z_i^{(k)} + \cdots + t^n z_i^{(n)}, \qquad 1 \leqslant i \leqslant V. \tag{36}$$

If the lengths of the edges of the polyhedron are changed to the order $o(t^n)$, $t \to 0$, then the deformation (36) is called an infinitesimal bending of order (k, n). Analytically this is expressed by the relation

$$|p_i(t) - p_j(t)| = |p_i^0 - p_j^0| + o(t^n), \qquad t \to 0.$$

$$(i, j) \in \tilde{I}.$$

Since in the case of triangular faces the lengths of the edges determine a metric on the whole polyhedron, all the distances on the polyhedron are also changed to the order $o(t^n)$.

Non-triviality of the infinitesimal bending (36) is ensured if at the three points p_1^0, p_2^0, p_3^0 that determine a fixed face of the polyhedron M the deformation (36) leaves these points fixed up to $o(t^n)$, that is, if $z_i^{(k)} = \cdots = z_i^{(n)} = 0$, $1 \leqslant i \leqslant 3$, and there exists a point p_j where $z_j^{(k)} \neq 0$, $j > 3$.

The equations of infinitesimal bendings of order (k, n) have the following form (on the assumption that in the representation (36) instead of $z_i^{(l)}$ we write $2z_i^{(l)}$, $k \leqslant l \leqslant n$, $1 \leqslant i \leqslant V$):

$$(p_i - p_j)(z_i^{(l)} - z_j^{(l)}) + \sum_{m=k}^{l-k} (z_i^{(m)} - z_j^{(m)})(z_i^{(l-m)} - z_j^{(l-m)}) = 0,$$

$$l = k, \ldots, n \tag{37}$$

(supposing that $z_i^{(m)} = 0$, $0 \leqslant m \leqslant k - 1$). In particular, infinitesimal bendings of the 1st order are obtained as a solution of the linear system

$$(p_i - p_j)(z_i - z_j) = 0, \qquad (i, j) \in \tilde{I}. \tag{38}$$

In this system for a non-trivial infinitesimal bending there are $3V - 9$ unknowns and $3V - 3\chi - 3$ equations, where χ is the Euler characteristic of the complex $\mathscr{K}$.

Since bendings of polyhedra can always be assumed to be analytic in the parameter, for bendings we have the representation (36) with $n = \infty$, so the equations of the bendings are given by the system (37) with $n = \infty$ and $k \geqslant 1$. As in Theorem 3.1 we thus see that a bendable polyhedron admits infinitesimal bendings of order $(1, 1)$ and $(1, 2)$. But the connections between bendings and infinitesimal bendings of polyhedra are considerably richer than in the smooth case. Let us consider this question in more detail.

If we recall that a polyhedron with V vertices is represented by a point P in E^{3V}, then the trajectory of the polyhedron in the process of this bending is represented by some curve γ in E^{3V}. We know that non-trivial bendings of polyhedra take place in a set of non-rigid polyhedra, which is an algebraic variety T whose equation is obtained from the condition rank $L < 3V - 9$, where L is the matrix of the system (38). At each point $s \in T$ there is at least one $3V$-dimensional non-zero vector $z(S) = (z_1, \ldots, z_V)$ with $z_1 = z_2 = z_3 = 0$. A bendable polyhedron belongs to T and in the process of bending it remains on T. Let $\gamma: P = P(t)$ be its trajectory; then $P'(t)$ is a tangent vector to γ and simultaneously the vector of an infinitesimal bending of the polyhedron $P(t)$. Consequently, a bending of a polyhedron takes place in the set of those non-rigid polyhedra for which the field of its infinitesimal bending is tangent to the manifold T. Hence we have two conclusions: 1) for infinitesimal bendings of the 1st order to be extendable to bendings of a polyhedron it is necessary that the vector $z(S)$ should belong to the tangent plane of the manifold T: 2) the trajectory of a bending of a polyhedron is the vector curve on T of the field of vectors $z(S)$. In turn, the first requirement distinguishes on T a submanifold T_1 along which $z(S)$ is tangent to T; obviously, the trajectory lies on T_1 and we again find that the field $z(S)$ must be tangent not only to T, but also to T_1. Extending this process, in the end in finitely many steps we distinguish submanifolds $T_m \subset T_{m-1} \subset \cdots \subset T_1 \subset T$ such that T_m consists entirely of trajectories of bendings of the polyhedron, that is, there is a finite algorithm for testing the bendability of the polyhedron.

What we have said above intersects with the fact that non-bendable polyhedra can admit infinitesimal bendings not greater than some order or, in other words, if a polyhedron is non-bendable, then for any given $k \geqslant 1$ it is rigid with respect to infinitesimal bendings of order (k, n) for all sufficiently large n. This connection between bendings and infinitesimal bendings of polyhedra was noticed, according to Connelly (1980), by Gromov and is based on the following algebraic theorem of Artin (Artin (1969), see also Artin (1968)): for the system of polynomial equations $f(x, y) = 0$, where $f = (f_1, \ldots, f_m)$, $x = (x_1, \ldots, x_n)$ and $y = (y_1, \ldots, y_N)$, there is an integer β, depending on n and N, the total degree d

of the polynomials f and a non-negative integer α, such that if

$$f(x, y) = 0 \qquad (\text{mod } x^\beta), \qquad \beta = \beta(n, N, d, \alpha), \tag{39}$$

for some polynomial $\bar{y}(x)$, then the system has a solution in the form of a convergent power series $y(x)$ which coincides with $\bar{y}(x)$ up to the terms x^α. Hence, as a consequence, if the system $f(x, y)$ has a solution $\bar{y}(x)$ in the form of a formal power series in powers of x, then it has a solution in the form of a convergent power series coinciding with $\bar{y}(x)$ up to the power x^α specified in advance.

The system of polynomial equations that needs to be solved in order to find a bending of polyhedra is obtained from a representation of the condition for a deformation to be isometric in the form

$$\Delta p_{ij} \Delta z_{ij} + \Delta z_{ij}^2 = 0, \qquad (i, j) \in \tilde{I}, \tag{40}$$

where $\Delta p_{ij} = p_i - p_j, \Delta z_{ij} = z_i - z_j$.

This system consists of I 2nd order equations for $3V$ unknowns, and the application of Artin's theorem to it leads to the assertion stated above that if a polyhedron admits infinitesimal bendings of order (k, n) with sufficiently large n, then it is bendable. Moreover, Artin's theorem in principle enables us to solve the question of the extendability of a given infinitesimal bending of order (k, n), $k \geqslant 1$, to a bending; we only need to be able to calculate Artin's number $\beta = \beta(1, 3V, 2, n)$ for the system (40) from (39), and then to verify whether the system (40) admits a polynomial solution z approximate in the sense of (39) that coincides up to degree n with the given field of infinitesimal bendings of order (k, n).

We conclude this subsection with the following remarks.

Remark 10. The consideration of analogues of Darboux surfaces for infinitesimal bendings (of the 1st order) of polyhedra does not occur in the literature. There exists an analogue of a field of rotations, but the construction in it of a diagram of rotations at once runs into specific difficulties.

Remark 11. As in the smooth case, for infinitesimal bendings of polyhedra we have projective invariance. Recently some papers were devoted to this question with a critical discussion[1].

Remark 12. We can represent a deformation of a polyhedron under an infinitesimal bending of order (k, n) in E^{3V} as some path. Choosing as parameter the length of an arc (if $k = 1$) or a power of it (if $k > 1$), we can obtain a standard representation of all equivalent fields of infinitesimal bendings of a given order.

9.5. Uniquely Determined Polyhedra. If the configuration space $\mathfrak{M}$ of a polyhedron M consists of one or two points, we say that M is uniquely determined by its metric. The case when $\mathfrak{M}$ consists of one point can only be in the degenerate situations when the whole polyhedron M is situated on a plane and its metric

---

[1] On this see, for example, Mathematical Reviews **87h**: 52021 a, b, c.

is such that the "exit" of M into space is impossible. A tetrahedron, for example, has a configuration space of two points. Since unique determination in the general class of polyhedra is apparently a very rare property, it is natural to look for such features in definite classes of polyhedra. One such class is known – the class of convex polyhedra. A famous theorem of Cauchy asserts that two closed convex polyhedra uniformly composed of equal faces are congruent, that is, they can be superimposed by a motion with a possible addition of a mirror reflection. In our conditions we can formulate Cauchy's theorem as the unique determination of a convex polyhedron in the class of convex polyhedra. This theorem with different versions of the proofs and with various comments is mentioned in many books and surveys (for wider generalizations of it, see Aleksandrov (1950); of the later works see, for example, Milka (1986), Berger (1977), Gluck (1975), and Kuiper (1979)), so we restrict ourselves to an analysis of its connection with the question of bendability of a convex polyhedron.

If a polyhedron M is strictly convex, then its non-bendability follows from Cauchy's theorem. But if M is not strictly convex, that is, there are vertices on it around which the total angle is 2π, then in the course of bending the convexity may be broken, so Cauchy's theorem on the coincidence of M with the polyhedron M' obtained in the course of the bending is inapplicable. Aleksandrov showed in Aleksandrov (1950) that if the polyhedron M' is obtained by a triangulation of a strictly convex polyhedron M (not necessarily with triangular faces) by the addition of vertices only on the edges of M, then such a polyhedron M' is non-bendable; one of the proofs can be obtained as a consequence of rigidity of the first order of such a polyhedron. If there are "false" vertices on the polyhedron, around which all the faces lie on one plane, then such a polyhedron will not be rigid. But it turns out that it remains rigid with respect to infinitesimal bendings of the 2nd order, and in such a case it is again non-bendable. This result of Connelly (Connelly (1980)) concludes the long history of the question of bendings and infinitesimal bendings of closed convex polyhedra in the same sense as in due course the question of unique determination of closed convex polyhedra was "closed down" by a theorem of Olovyanishnikov, which asserts that a convex polyhedron is uniquely determined in the class of convex surfaces (a proof of this theorem can be found in Efimov (1948a)).

9.6. Non-Bendable Polyhedra. The first serious way out from the closed world of convex polyhedra to the "open cosmos" of general polyhedra was the investigation of Connelly (Connelly (1974)) of non-bendable *suspensions*. Not only the result but also the method of this paper is interesting. The combinatorial structure of a suspension is perceived from its other name – bipyramid: over a closed broken line L of n links there are constructed lateral surfaces of two pyramids, one with a vertex at some point N, the other with a vertex at another point S. The broken line L is called the equator of the suspension, and the points N and S are called the poles of the suspension. The combinatorial image of a suspension or the corresponding simplicial complex $\mathcal{K}$ can be represented as two convex

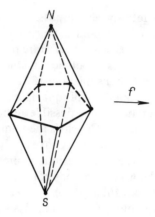

f

Fig. 9

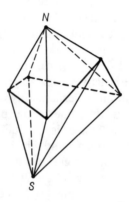

Fig. 10

n-gonal pyramids with a common planar base L and with vertices N and S on different sides of the plane of the base (Fig. 9).

Under a continuous map $f: \mathcal{K} \to E^3$, linear on each face, the image $f(\mathcal{K})$ can be a non-convex polyhedron in E^3, even with self-intersections and self-coverings with possibly coincident N and S (Fig. 10). The case $N \neq S$ was considered in Connelly (1974). He showed that if the spatial distance $|NS|$ is fixed, then such a suspension cannot be bent (for the exclusion of certain degenerate cases, see Sabitov (1983)). If we assume that in the course of bending the distance $|NS|$ continuously changes, it turns out that we can then construct some algebraic function $F(x)$ that is identically zero in a neighbourhood of the point $x = x_0$, where x_0 is the distance $|NS|$ for the original position of the suspension. Then extending $F(x)$ to the complex z-plane and studying the behaviour of $F(z)$ on different sheets of its Riemann surface, we can obtain information about the necessary geometric properties of a bendable suspension. In particular, it turns out that if the suspension bounds a non-zero volume in some sense (for example, it is embedded in E^3), then such a suspension is non-bendable.

Thus, the idea of using the methods of the theory of functions of a complex variable in the theory of bendings of polyhedra consists in the following: we need to introduce certain parameters $x_1, \dots, x_m$ (for example, the spatial distances between certain vertices not belonging to one edge), the specification of which certainly uniquely determines the position of the polyhedron, together with the given lengths of its edges, then establish certain algebraic dependences $F_j(x_1, \dots, x_m) = 0, j = 1, \dots, n$, that are identically satisfied for values $x_1, \dots, x_m$ sufficiently close to their original values, and then go over to complex variables $z_1, \dots, z_m$ and consider the equations $F_j(z_1, \dots, z_m) = 0$ for values of the arguments that do not have direct geometrical sense. If we succeed in showing that the fulfilment of the equalities $F_j(z_1, \dots, z_m) = 0$ on $\mathbb{C}^m$ leads to a contradiction with some geometrical properties of the polyhedron, then a polyhedron with these properties cannot be bent with the variables $x_1, \dots, x_m$, that is, it does not change its form by the sense of the choice of parameters $x_1, \dots, x_m$.

Another class of non-convex non-bendable polyhedra consists of *pyramids*, combinatorially definable as polyhedra with n vertices, among which there is one vertex from which $n - 1$ edges start out; of course, it is assumed that this polyhedron, like a topological space, is homeomorphic to a closed (without boundary) manifold. What is interesting here is not only the fact of non-bendability (and it is true for all immersed pyramids), but also that there are pyramids of any topological form, including non-orientable ones; see Sabitov (1989).

Thus, we draw up the following scheme: embedded polyhedra with n vertices among which there is one with degree $n - 1$ (the *degree of a vertex* is the number of edges issuing from it) are non-bendable – these are pyramids; next, polyhedra with n vertices among which there are two with degree $n - 2$ are also non-bendable – these are generalized suspensions (that is, not necessarily homeomorphic to a sphere). Now it is natural to ask whether this series can be extended in a general way: are there embedded polyhedra with n vertices among which there are k vertices with degree $n - k$, non-bendable for any k, $3 \leqslant k \leqslant n - 3$, on condition that there is no vertex with degree greater than $n - k$ (without the last condition there cannot be non-bendability: for example, in the bendable polyhedron of Klaus Stefen (see 9.7 below) with 9 vertices there are 4 vertices with degree $9 - 4 = 5$, moreover in it there is one vertex with degree $6 > 5$).

The non-bendable polyhedra studied in this section are not all uniquely determined, however. Sometimes we have succeeded in constructing polyhedra that are isometric but not equal to them, as follows. There is a general method for constructing isometric surfaces starting from a known non-rigid surface. Let z_i be the field of infinitesimal bendings of the 1st order of a polyhedron with vertices p_i, $1 \leqslant i \leqslant V$. Then the two polyhedra M_1 and M_2 with vertices $p_i + tz_i$ and $p_i - tz_i$ will be non-trivially isometric. Since the existence of non-rigid suspensions embedded in E^3 is known (Aleksandrov and Vladimirova (1962)), we find that for arbitrarily small $t > 0$ there are two polyhedra situated in the t-neighbourhood of each other and non-trivially isometric to each other. On the other hand, each of these suspensions M_1 and M_2 is non-bendable, so by 9.4 in a sufficiently small neighbourhood of each of them there are no other suspensions isometric to them.

9.7. Bendable Polyhedra. The difficulty and lack of study of problems of bendings associated with non-convex polyhedra are visible at least from the fact that after Cauchy's result for more than 150 years it was not known whether it can be extended to all embedded (or even immersed) polyhedra or not. An example of a non-immersed bendable polyhedron was constructed by Bricard (Bricard (1897)), also long after Cauchy.

The first attempt to "estimate" the degree of definability of a polyhedron by some a priori specified parameters of it was undertaken by Legendre (Legendre (1806), Note VIII). He showed that the number of parameters necessary for a determination of the position of the vertices of a polyhedron (of the type of a sphere) with known combinatorial structure is equal to the number of its edges. It was natural to assume that for such parameters we can take the lengths of the edges of the polyhedron, but Legendre himself disproved this assumption. As an example he considered a quadrangular prism: knowledge of the lengths of all its 12 edges does not enable us to restore the prism uniquely, since its bases can be chosen to be different in form, but with the same lengths of edges.

We observe that Legendre's idea of counterexample is suitable only for polyhedra with non-triangular faces, since in the case of triangular faces knowledge of the lengths of the edges completely determines the form of each face. Bricard showed that in the case of triangular faces the lengths of the edges do not always uniquely determine the position of the vertices of the polyhedron. He considered octahedra and obtained a complete description of types of bendable octahedra. In all bendable polyhedra known at present elements of Bricard's bendable octahedra have been used, so to complete the picture we describe their metric characteristics and structure in E^3.

An octahedron can be regarded as a suspension with a four-link equator, taking as poles any pair of opposite vertices. In the figures we shall denote the vertices of an octahedron in its combinatorial scheme and the images of its vertices under the construction of a bendable octahedron in E^3 by the same letters.

1st type of bendable octahedra. Opposite edges of the equator of an octahedron have equal lengths: $A_1 B_1 = A_2 B_2$, $A_1 B_2 = A_2 B_1$, and a bendable octahedron in E^3 has the following structure. The top "half" $N A_1 B_1 A_2 B_2$ of the octahedron is realized in E^3 in an arbitrary way; the equator in E^3 always has an axis of symmetry l that passes through the midpoints of the segments $A_1 A_2$ and $B_1 B_2$ (if the equator is planar, then l passes through the point of intersection of the diagonals of the parallelogram perpendicular to its plane); as the image of S we take the point symmetrical to N with respect to the line l (Fig. 11). Then the edges of the octahedron have the distribution of lengths shown in Fig. 12;

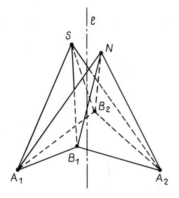

Fig. 11

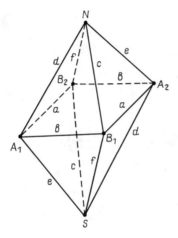

Fig. 12

bendings of the octahedron repeat the bendings of the quadrangular pyramid $NA_1B_1A_2B_2$ in E^3.

2nd type. Adjacent edges of the equator of the octahedron are equal: $A_1B_1 = A_1B_2$, $A_2B_1 = A_2B_2$, and a bendable octahedron in E^3 has the following struc- ture. The top part $NA_1B_1A_2B_2$ of the octahedron is realized in E^3 in an arbitrary way; the equator in E^3 always has a plane of symmetry P, which passes through the line A_1A_2 and bisects the dihedral angle between the half-planes $A_1A_2B_1$ and $A_1A_2B_2$; for the image of S we take the point symmetrical to N with respect to the plane P (Fig. 13). The edges of such an octahedron have the distribution of lengths as in Fig. 14, and the bendings of the octahedron repeat the bendings of the quadrangular pyramid $NA_1B_1A_2B_2$.

3rd type. This type is the most difficult to describe. It is characterized by the following condition: all three equators of the octahedron are such that in them,

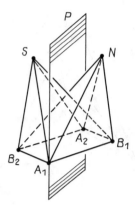

Fig. 13

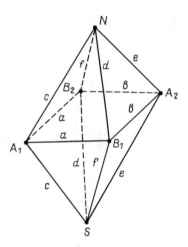

Fig. 14

in their planar situation, we can inscribe circles with a common centre, touching
either the sides of the equators themselves or their extensions (Lebesgue (1967)).
This means that on each equator the sums of pairs of either opposite sides or
adjacent sides are equal, but this is not enough: it is necessary that in all six
tetrahedral angles of the octahedron the opposite planar angles are either equal
or mutually complementary with respect to π. The structure of the corresponding bendable octahedron in E^3 is as follows: the top part $NA_1B_1A_2B_2$ of
the octahedron is realized in E^3 in such a way that there is pairwise symmetry
of the lines NA_1 and NA_2, NB_1 and NB_2 with respect to some plane σ passing
through the common line of intersection of the bisecting planes of the angles of
the quadrangle $A_1B_1A_2B_2$ (for a given plane σ the point N is uniquely determined); the image of S is determined as the point obtained by the intersection of
two lines, one of which is symmetrical to NA_1 with respect to the plane bisecting

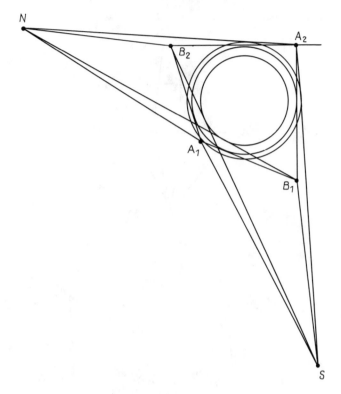

Fig. 15

between A_1B_1 and A_1B_2 and the other is symmetrical to NA_2 with respect to the plane bisecting between A_2B_1 and A_2B_2. The bendings of the octahedron repeat the bendings of the pyramid $NA_1B_1A_2B_2$. An important difference between bendings of the octahedron of the 3rd type and those of the other two types is that, in the process of bending, an octahedron of the 3rd type twice takes up a position where it all lies on one plane; for a representation of it we most often use the planar position of it, see Fig. 15.

We observe that we can realize each model in the form of a convex octahedron, so there are convex octahedra that are non-bendable, but are nevertheless not uniquely determined in the class of general polyhedra.

With respect to the metrical structure two or even all three types under a definite choice of the lengths of edges can be isometric. This means that some octahedra of different types can have coincident configuration spaces. The classification of bendable octahedra according to the form of their configuration space is carried out in Bushmelev and Sabitov (1990).

The Bricard octahedra are not embedded or even immersed in E^3. The first rather complicated example of an embedded bendable polyhedron was constructed by Connelly in 1977, then in 1978 Stefen found a bendable polyhedron

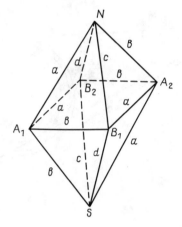

Fig. 16

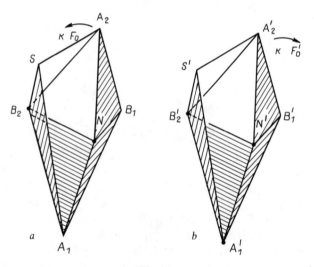

Fig. 17

embedded in E^3 that has 9 vertices in all. Descriptions of these polyhedra can be found, for example, in Berger (1977), Connelly (1978), and Kuiper (1979). Here we describe only the Stefen polyhedron. Let us consider the 1st type of Bricard octahedron with the dimensions shown in Fig. 16. We construct in E^3 the Bricard octahedron, removing from it the faces SA_2B_1 and SA_1B_1; we obtain, with the corresponding situation of vertices, a polyhedron F with boundary $A_1SA_2B_1$ that is embedded in E^3 (Fig. 17a). It is bent, as part of a bendable octahedron, and in the course of bending the distance SB_1 does not change. Therefore we can assume that the "empty" triangle A_1SB_1 is fixed, and in this case the movements of the vertex A_2 under the bending reduce to rotation about the fixed line SB_1. We now take another copy of the same polyhedron: let us denote it by F' (Fig. 17b). There are positions F_0 and F'_0 respectively of the polyhedra F and F', obtainable in the course of their bending, such that by pasting the sides $A_1 S$ and $A_1 B_1$ of F_0 to the corresponding sides $A'_1 S'$ and $A'_1 B'_1$ of F'_0 we obtain a polyhedron M with boundary $A_2 S A'_2 B_1$ that is embedded in E^3. With rotations of the point A_2 around the line SB_1 we can associate rotations of A'_2 about the same line such that the distance $A_2 A'_2$ is not changed, so the polyhedron M is bent "in the large". It remains to paste the boundary of M by the "cap" of the two triangles $A_2 S A'_2$ and $A_2 A'_2 B_1$ and we obtain a closed embedded bendable polyhedron.

It is natural to ask whether there are bendable polyhedra embedded or immersed in E^3 with fewer than 9 vertices. It has been shown that among polyhedra with fewer than 8 vertices there are no such polyhedra, see Maksimov (1987); we can show that there are none of them among polyhedra with 8 vertices, but for the present only one of torus type (since any polyhedron of genus $\chi < 0$ has more than 8 vertices there remains only the case of a polyhedron of sphere type).

As we see, the discovery of bendable polyhedra is a very non-trivial problem. This is no accident, since there are in a certain sense very few of them. Namely, if we return to the representation of polyhedra in the form of points in the space E^{3V}, then it turns out that bendable polyhedra can fill only a set of measure zero in E^{3V}. For polyhedra of the type of a sphere this assertion in an implicit form has been known for a long time (Aleksandrov (1950), Ch. 2, § 6 and Poznyak (1960)). In an explicitly expressed and specially proved form this result was obtained in Gluck (1975) with the use of the following general idea: from the rigidity of the 1st order there follows non-bendability, and non-rigid polyhedra fill only a hypersurface in E^{3V} determined by a certain algebraic equation. One would think that for polyhedra of higher topological type the property of scarce bendability should be obtained "more easily", since they have more edges for the same number of vertices, but nevertheless the proof of this fact was obtained very recently (Fogelsanger (1987)) on the basis of the following idea: triangulation of a polyhedron is transformed into a simpler form, for which it is proved that almost all its realizations in E^3 are non-bendable, and on going over from one triangulation to another the property of non-bendability in the general position is preserved. Another proof can be obtained by establishing that the algebraic varieties considered in 9.3 and 9.4 and corresponding to bendable polyhedra do not coincide with the whole space, and so the set of bendable polyhedra fills no more than some algebraic hypersurface in E^{3V}.

9.8. Conjecture on the Invariance of the Volume of a Bendable Polyhedron. For all known examples of bendable polyhedra the following remarkable property holds: the volumes bounded by these polyhedra remain constant in the course of their bending (for the Stefen polyhedron this is immediately obvious from the description of its bendings in the previous section). Intuitively this means that these bendable polyhedra are not suitable for the manufacture of their bellows – if in such polyhedra we remove one face (that is, we make a triangular performation) and deform them isometrically, then the air from such bellows will not go outside, since the pressure inside will remain constant (by the way, this fact must find some technical application, of course). The generally accepted conjecture, which should obviously be called the Connelly-Sullivan conjecture, is that any closed bendable polyhedron in the course of bending bounds a volume of the same quantity. By the intuitive interpretation mentioned above this conjecture is figuratively called the "bellows" conjecture.

Of course, it is tempting to assume that this conjecture is due to Euclid (Definition 10 in Book XI of Euclid's "Elements" states: equal and similar solid figures are such as are contained by similar planes equal in number and magnitude), but even Cauchy's theorem cannot be interpreted as a repetition or confirmation of Euclid's idea, since there are reasons for supposing that Euclid had in mind only totally specific polyhedra; on this see Milka (1986).

It is easy to construct examples of isometric polyhedra that bound different volumes. On the other hand, we know that the configuration space of a given polyhedron as a whole consists of finitely many connected components. Therefore the Connelly-Sullivan conjecture can be formulated in a different way as follows: all polyhedra isometric to a given one can have only finitely many values of generalized volumes (generalized volumes are formally calculated from the same formula $\frac{1}{3}\Sigma(p_i p_j p_k)$, where p_i are the vertices and the summation is over all faces, which gives volumes of embedded polyhedra). Since the volume of a polyhedron changes continuously under a bending, from the fact that the number of its values is finite it will follow that it remains constant under a bending.

One possible way of proving the conjecture that the volume is invariant under a bending was to establish that the volume is stationary under an infinitesimal bending. However, it turned out that there are non-rigid polyhedra whose volume under an infinitesimal bending is not stationary; see V.A. Aleksandrov (1989). Since, starting from such polyhedra, we can construct two arbitrarily close isometric polyhedra (see 9.6), we thus find that for any $\varepsilon > 0$ we can determine isometric polyhedra situated in an ε-neighbourhood of each other and having different volumes. If the Connelly-Sullivan conjecture is true, then the fact that the volume of a polyhedron is non-stationary under an infinitesimal bending of it is an indication that this field of infinitesimal bendings cannot be extended to a bending.

The assumption that an infinitesimal bending of the 1st order can be extended to an infinitesimal bending of the 2nd order also does not guarantee that the volume is stationary. This is not difficult to verify by the example of non-rigid octahedra considered in Gluck (1975). Therefore to obtain a proof that the volume of a polyhedron is invariant under a bending through the study of the behaviour of the volume under infinitesimal bendings requires conversions to infinitesimal bendings of high orders.

On the contrary, for smooth surfaces there is an encouraging result: for all non-rigid surfaces of revolution their volume remains stationary under an infinitesimal bending; see V.A. Aleksandrov (1989). Although up to now there are no examples of regular closed bendable surfaces, nevertheless information on the behaviour of their volume under a priori assumed bendings can be useful.

§ 10. Concluding Remarks

Thus, from the general survey of the circle of problems posed in § 2 we can make the deduction that these problems have a more or less developed apparatus for research only in the case of power orders of flattening, and for an arbitrary form of flattenings only individual results are known; therefore for the problem 1′) from § 2 no approaches have so far been developed.

Let us say a few words about problem 7). In the analytic case an immersion of the metric (2) in E^3 in the form of a surface with points of general position is obtained as a solution of a certain Cauchy problem, so the corresponding connected components in $\mathfrak{M}$ for such surfaces are infinite-dimensional. There is infinite-dimensionality – without the requirement of analyticity – also in the case $K \neq 0$; see Theorems 6.3 and 6.4 and the discussion of them. But if the surface S has flattening at a marked point, then there are still no results about the dimension of the configuration space (set of positions) $\mathfrak{N}_S \subset \mathfrak{M}$ of the surface S under all possible bendings of it. Generally we can say that the question of parametrization of the configuration space of a surface arising in the course of bendings of it is not new in the literature: in this connection see § 9, in which we show the finite-dimensionality of the configuration space of a polyhedron – a result that goes back to Legendre in a certain sense. In modern works the question of parametrization (with the precise statement of parametrization itself) has been repeatedly discussed also for the investigation of bendings of surfaces of positive curvature with different boundary conditions. But in these works the configuration space $\mathfrak{N}$ is an ordinary finite-dimensional linear space, so the question of possible topological or other singularities in the structure of $\mathfrak{N}$ has not arisen. Apparently the first person who drew attention in the literature to the interest in the study of the topological structure of the configuration space of bendable polyhedra was Gluck (Gluck (1975)). Since then a number of works on this theme have appeared, concerned only with polyhedra, it is true. In the smooth case, following Berger, Bryant and Griffiths (1981), the question can be put as follows: suppose that two isometric surfaces in correct situation have identical osculating paraboloids of some order n; do they then coincide? Their coincidence implies that $\mathfrak{M}$ has dimension that does not exceed the number of coefficients of osculating paraboloids of the corresponding order. Such a situation occurs in the multidimensional case; see below. However, we should bear in mind that there are analytic surfaces that are bendable with the preservation of the osculating paraboloid of any preassigned order; see Dorfman (1957).

In the multidimensional case, in the local theory of bendings of surfaces the first classical result was a theorem of Beez in 1878 on the local unique determination of a hypersurface M^n in R^{n+1}, $n \geqslant 3$, on condition that the subspace of degeneracy of its second fundamental form is not very extensive (again an analogue of the absence of flattening of the surface!). For the general type of degeneracies and flattenings in the analytic class of surfaces and deformations for surfaces, analogues of Efimov's theorems about their non-bendability and rigidity have been established in the "majority" of cases; see Lashchenko (1987) and (1989). For convex hypersurfaces tests for rigidity and unique determination have been obtained even in non-smooth cases; see Sen'kin (1978). In the case of codimension greater than one the result of Beez was strengthened by Allendoerfer in 1939 up to codimension $r \leqslant \left[\dfrac{n}{3} \right]$, again for the neighbourhood of a point lying in general position in some sense; a detailed account of the history and results of Beez, Allendoerfer and others can be found in Spivak (1979); for a new account of the proof of Allendoerfer's theorem see Chern and Osserman (1981). The main observation here is that under certain conditions in algebro-differential equations of an immersion of M^n in E^{n+r} the algebraic part of the system turns out to be a determining part: any solution of it satisfies the differential equations of Codazzi's system.

Of recent research in this direction we must name above all the article Berger, Bryant and Griffiths (1981) with details in Griffiths and Jensen (1987). In these papers, depending on the codimension for immersions "in general position" it is established that either they are non-bendable or the arbitrariness in the immersion is determined by a certain number of numerical parameters or functions of several arguments. As an answer to question 7), more interesting are the cases of codimension $r \leqslant (n-1)(n-2)/2$: then the immersion of M^n "in general position" in E^{n+r} is either uniquely determined (when $r \leqslant n$, $n \geqslant 8$; or $r \leqslant 6$, $n = 7$; $r \leqslant 4$, $n = 6, 5$; $r \leqslant 3$, $n = 4$) or the configuration space is finite-dimensional and under a fixing of the osculating paraboloid of some order the surface is not bent.

For some criteria for local bendability of multidimensional surfaces see Jacobowitz (1972a, b), (1982a, b) and Yanenko (1952), (1954). There are still only a few papers devoted to the study of local infinitesimal bendings of multidimensional surfaces; different criteria for rigidity, depending on the codimension and local structure of the surface, can be found in Markov (1980, 1987), where he gives quite a detailed bibliography. For a global (that is, not including local coordinates) definition of infinitesimal bendings of the 1st order of surfaces in Riemannian spaces see Viotsekhovskij (1977) and Goldstein and Ryan (1975).

Comments on the References

Most of the references are cited in the text during the exposition. The history and results of the theory of infinitesimal bendings in the 19th century can be found in Darboux (1896), for example. Among the recent survey works, the closest to our theme are Chapters 11 and 12 of Spivak (1979) (we note that terms that distinguish the concepts of non-bendability and unique determination were introduced here in the English language literature possibly for the first time). Unsolved (up to now!)

problems can be found in Efimov (1948a). The article Ivanova-Karatopraklieva (1988) is devoted to a survey of works on infinitesimal bendings of surfaces of mixed curvature; this is a field in which there are still not many results, since even the statements of the problems require new approaches. For methods of constructing surfaces with a precisely defined number of fields of infinitesimal bendings of the first order see Reshetnyak (1962) and Trotsenko (1980).

For polyhedra, in §9 we have not touched on an interesting series of questions that connect bendings of a polyhedron with bendings of the framework of its edges, nor with problems of calculating the stresses in the framework caused by a different distribution of forces in the edges; for this see for example Whiteley (1984, 1987a, 1987b) and especially Connelly (1992). In these and other works cited in the text polyhedra are also considered in multidimensional spaces.

Survey articles on immersions are Gromov and Rokhlin (1970), Poznyak (1973), and Poznyak and Sokolov (1977); a new survey on bendings of surfaces is given in Ivanova-Karatopraklieva and Sabitov (1991) and (1992).

References*

Aleksandrov, A.D. (1937): Infinitesimal bendings of nonregular surfaces. Mat. Sb., Nov. Ser. *1*, 307–322 (Russian), Zbl.14,413

Aleksandrov, A.D. (1938): On a class of closed surfaces. Mat. Sb., Nov. Ser. *4*, 69–77 (Russian), Zbl.20,261

Aleksandrov, A.D. (1948): The Intrinsic Geometry of Convex Surfaces. Gostekhizdat, Moscow. German transl.: Die Innere Geometrie der Konvexen Flächen. Akademie-Verlag, Berlin, 1955, Zbl.38,352

Aleksandrov, A.D. (1950): Convex Polyhedra. Gostekhizdat, Moscow. German transl.: Konvexe Polyeder. Akademie-Verlag, Berlin, 1958, Zbl.41,509

Aleksandrov, A.D. and Vladimirova, S.M. (1962): Bending of a polyhedron with rigid faces. Vestn. Leningr. Univ. *17*, No. 13 (Ser. Mat. Mekh. Astron. No. 3), 138–141 (Russian), Zbl.135,407

Aleksandrov, V.A. (1989): Remarks on a conjecture of Sabitov on the rigidity of the volume under an infinitesimal bending of a surface. Sib. Mat. Zh. *30*, No. 5, 16–24. Engl. transl.: Sib. Math. J. *30*, No. 5, 678–684 (1989), Zbl.687.53007

Artin, M. (1968): On the solutions of analytic equations. Invent. Math. *5*, 277–291, Zbl.172,53

Artin, M. (1969): Algebraic approximation of structures over complete local rings. Inst. Hautes Etud. Sci., Publ. Math. *36*, 23–58, Zbl.181,488

Belousova, V.P. (1961); Infinitesimal deformations of surfaces. Vestn. Kiev. Univ. Ser. Mat. Mekh., *4*, 49–57 (Ukrainian).

Berger, E., Bryant, R., Griffiths, P. (1981): Some isometric embedding and rigidity results for Riemannian manifolds. Proc. Natl. Acad. Sci. USA *78*, 4657–4660, Zbl.468.53040

Berger, M. (1977): Géométrie. Vols. 1–5. Cedic, Paris, Zbl.423.51001

Berri, R.Ya. (1952): An integral invariant of binary forms of the fourth degree. Usp. Mat. Nauk 7, No. 3, 125–130 (Russian), Zbl.49,153

Bol, G. (1955): Projektive und affinen Eigenschaften des Darbouxschen Flächenkranzes, Abh. Math. Semin. Univ. Hamburg *20*, 64–96, Zbl.65,145

Borisov, Yu.F. (1965): $C^{1,\alpha}$-isometric immersions of Riemannian spaces. Dokl. Akad. Nauk SSSR *163*, 11–13. Engl. transl.: Sov. Math. Dokl. *6*, 869–871 (1965), Zbl.135,403

Boudet, R. (1961): Sur quelques propriétés géométriques des transformations infinitésimales des surfaces. Thesis. Fac. Sci. Univ. Aix-Marseille, 78 pp.

---

*For the convenience of the reader, references to reviews in Zentralblatt für Mathematik (Zbl.), compiled using the MATH database, and Jahrbuch über die Fortschritte der Mathematik (Jbuch) have, as far as possible, been included in this bibliography.

Bricard, R. (1897): Mémoire sur la théorie de l'octaèdre articulé. J. Math. Pures Appl. 5, 113–148, Jbuch 28, 624

Burago, Yu.D., Zalgaller, V.A. (1960): Polyhedral embedding of a development Vestn. Leningr. Univ. 15, No. 7, 66–80 (Russian), Zbl.98,354

Bushmelev A.V., Sabitov, I.Kh. (1990): Configuration spaces of Bricard octahedra. Ukr. Geom. Sb. 33, 36–41 (Russian)

Calabi, E., Hartman, P. (1970): On the smoothness of isometries. Duke Math. J. 37, 741–751, Zbl.203,543

Chern, S.S., Osserman, R. (1981): Remarks on the Riemannian metric of a minimal submanifold. Lect. Notes Math. 894, 49–90, Zbl.477.53056

Cohn-Vossen, S.E. (1929): Unstarre geschlossene Flächen. Math. Ann. 102, 10–29, Jbuch 55, 1016

Cohn-Vossen, S.E. (1936): Bendability of surfaces in the large. Usp. Mat. Nauk 1, 33–76. Zbl.16,225 (Reprinted in the book: Cohn-Vossen, S.E., Some questions of differential geometry in the large. Fizmatgiz, Moscow, 1954 (Russian))

Connelly, R. (1974): An attack on rigidity. I, II. Preprint, Cornell Univ., appeared in: Bull. Am. Math. Soc. 81, 566–569 (1975), Zbl.315.50003

Connelly, R. (1978): A flexible sphere. Math. Intell. 1, 130–131, Zbl.404.57018

Connelly, R. (1980): The rigidity of certain cabled frameworks and the second order rigidity of arbitrarily triangulated convex surfaces. Adv. Math. 37, 272–299, Zbl.446.51012

Connelly, R. (1992): Rigidity, in: Handbook of convex geometry (P. Gruber and J. Wills, eds.) (to appear)

Darboux, G. (1896): Léçons sur la théorie générale des surfaces. Part 4, 2nd. ed. Gauthier-Villars, Paris, Jbuch 25, 1159

Dorfman, A.G. (1957): Solution of the equation of bending for some classes of surfaces. Usp. Mat. Nauk 12, No. 2, 147–150 (Russian), Zbl.85,366

Efimov, N.V. (1948a): Qualitative questions of the theory of deformations of surfaces. Usp. Mat. Nauk 3, No. 2, 47—158. Engl. transl.: Am. Math. Soc. Transl. 6, 274–323, Zbl.30,69

Efimov, N.V. (1948b): Qualitative questions of the theory of deformations of surfaces "in the small". Tr. Mat. Inst. Steklova 30, 1–128 (Russian), Zbl.41,488

Efimov, N.V. (1948c): On rigidity in the small. Dokl. Akad. Nauk SSSR 60, 761–764 (Russian), Zbl.39,382

Efimov, N.V. (1952): Some theorems about rigidity and non-bendability. Usp. Mat. Nauk 7, No. 5, 215–224 (Russian), Zbl.47,150

Efimov, N.V. (1958): A survey of some results on qualitative questions of the theory of surfaces. Proc. 3th All-Union Congr. of Math. 1956, Acad. Sci. Moscow, Vol. 3, 401–407, Zbl.87,361

Efimov, N.V., Usmanov, Z.D. (1973): Infinitesimal bendings of surfaces with a flat point. Dokl. Akad. Nauk SSSR 208, 28–31. Engl. transl.: Sov. Math. Dokl. 14, 22–25 (1973), Zbl.289.53005

Fogelsanger, A. (1987): The generic rigidity of minimal cycles. Preprint, Cornell Univ. 60 pp.

Fomenko, V.T. (1962): Investigation of solutions of the basic equations of the theory of surfaces. Dokl. Akad. Nauk SSSR 144, 69–71. Engl. transl.: Sov. Math. Dokl. 3, 686–689 (1962), Zbl.117,383

Fomenko, V.T. (1965): Bending of surfaces with preservation of congruence points, Mat. Sb., Nov. Ser. 66, 127–141 (Russian), Zbl.192,272

Gluck, H. (1975): Almost all simple connected closed surfaces are rigid. Lect. Notes Math. 438, 225–239, Zbl.315.50002

Gluck, H., Krigelman, K., Singer, D. (1974): The converse to the Gauss-Bonnet theorem in PL. J. Differ. Geom. 9, 601–616, Zbl.294,57014

Goldstein, R.A., Ryan, P.J. (1975): Infinitesimal rigidity of submanifolds. J. Differ. Geom. 10, 49–60, Zbl.302.53029

Griffiths, P.A., Jensen, G.R. (1987): Differential systems and isometric embeddings. Ann. of Math. Stud., No. 114, Zbl.637.53001

Gromov, M.L., Rokhlin, V.A. (1970): Immersions and embeddings of Riemannian manifolds. Usp. Mat. Nauk 25, No. 5, 3–62. Engl. transl.: Russ. Math. Surv. 25, No. 5, 1–57 (1970), Zbl.202,210

Gulliver, R.D., Osserman, R., Royden, H.L. (1973): A theory of branched immersions of surfaces. Am. J. Math. 95, 750–812, Zbl.295.53002

Hartman, P., Wintner, A. (1951): Gaussian curvature and local embedding. Am. J. Math. 73, 876–884, Zbl.44,184

Hartman, P., Wintner, A. (1952): On hyperbolic partial differential equations. Am. J. Math. 74, 834–876, Zbl.48,333

Hellwig, G. (1955): Über die Verbiegbarkeit von Flächenstucken mit positiver Gausscher Krümmung. Arch. Math. 6, 243–249, Zbl.64,158

Höesli, R. (1950): Spezielle Flächen mit Flächpunkten und ihre lokale Verbiegbarkeit. Compos. Math. 8, 113–141, Zbl.38,335

Hong, J., Zuily, C. (1987): Existence of C^∞ local solutions for the Monge-Ampère equation. Invent. Math. 89, 645–661, Zbl.648.35016

Hopf, H., Schilt, H. (1938): Über Isometrie und stetige Verbiegung von Flächen. Math. Ann. 116, 58–75, Zbl.19,20

Isanov, T.G. (1977): On the extension of infinitesimal bendings. Dokl. Akad. Nauk SSSR 234, 1257–1260. Engl. transl.: Sov. Math. Dokl. 18, 842–846 (1977), Zbl.376.53002

Ivanova-Karatopraklieva, I. (1982–1983): Properties of the fundamental field infinitesimal bendings of a surface of revolution. God. Sofij. Univ. Fak. Mat. Mekh. 76, 21–40 (Bulgarian), Zbl.637.53005

Ivanova-Karatopraklieva, I. (1987–1988): Infinitesimal third-order bendings of surfaces of revolution with flattening at the pole. Ann. Univ. Sofia Fac. Math. Mec. 81 (to appear), (Russian)

Ivanova-Karatopraklieva, I. (1988): Infinitesimal bendings of surfaces of mixed curvature. Proc. XVII Spring Conf. of the Union of Bulgarian Math. pp. 49–56

Ivanova-Karatopraklieva, I. (1990): Infinitesimal bendings of higher order of rotational surfaces. Compt. Rend. de l'Acad. Bulgare des Sci. 43, No. 12

Ivanova-Karatopraklieva, I., Sabitov, I.Kh. (1989): Infinitesimal second-order bendings of surfaces of revolution with flattening at the pole. Mat. Zametki 45, 28–35. Engl. transl.: Math. Notes 45, No. 112, 19–24 (1989), Zbl.662.53003

Ivanova-Karatopraklieva, I., Sabitov, I.Kh. (1991): Bendings of surfaces. I. Problems of Geometry 23, 131–184 (Russian)

Ivanova-Karatopraklieva, I., Sabitov, I.Kh. (1992): Bendings of surfaces. II. Problems of Geometry 24 (to appear) (Russian)

Jacobowitz, H. (1972a): Extending isometric embeddings. J. Differ. Geom. 9, 291–307, Zbl.283.53025

Jacobowitz, H. (1972b): Implicit function theorems and isometric embeddings. Ann. Math., II. Ser. 95, 191–225, Zbl.214,129

Jacobowitz, H. (1982a): Local analytic isometric deformations. Indiana Univ. Math. J. 31, No. 1, 47–55, Zbl.502.53004

Jacobowitz, H. (1982b): Local isometric embeddings. Semin. Differ. Geom., Ann. Math. Stud. 102, 381–393, Zbl.481.53018

Kann, E. (1970): A new method for infinitesimal rigidity of surfaces with $K \geqslant 0$. J. Differ. Geom. 4, 5–12, Zbl.194,525

Klimentov, S.B. (1982): On the structure of the set of solutions of the basic equations of the theory of surfaces. Ukr. Geom Sb. 25, 69–82 (Russian), Zbl.509.53021

Klimentov, S.B. (1984): On the extension of higher-order infinitesimal bendings of a simply-connected surface of positive curvature. Mat. Zametki 36, 393–403. Engl. transl.: Math. Notes 36, 695–700 (1984), Zbl.581.53002

Kuiper, N. (1955): On C^1-isometric imbeddings. I, II. Nederl. Akad. Wetensch. Proc. Ser. A 58 (Indagationes Math. 17) 545–556; 683–689, Zbl.67,396

Kuiper, N. (1979): Sphères polyédriques flexibles dans E^3, d'après Robert Connelly. Lect. Notes Math. 710, 147–168, Zbl.435.53043

Kuznetsov, V.A. (1987): The structure of a neighbourhood of an isolated zero of the Lipschitz-Killing curvature on an m-dimensional surface in E^n. Proc. All-Union Conf. on Geometry "in the large". Inst. Math. Sib. Division of the USSR Academy of Sciences, Novosibirsk, p. 64

Lashchenko, D.V. (1987): On the rigidity "in the small" of certain classes of hyper surfaces. (Deposited at VINITI, No. 3258, pp. 1–21)

Lashchenko, D.V. (1989): On the rigidity "in the small" of certain classes of surfaces. (Deposited at VINITI, No. 2121, pp. 1–23)

Lebesgue, H. (1902): Intégrale, longueur, aire. Ann. Math. Pura Appl. (3) 7, 231–359, Jbuch 33, 307

Lebesgue, H. (1967): Octaèdres articulés de Bricard. Enseign. Math., II. Ser. 13, 175–185, Zbl.155,493

Legendre, A. (1806): Eléments de géométrie. 6th ed., Paris

Lin, C.S. (1985): The local isometric embedding in R^3 of two-dimensional Riemannian manifolds with non-negative curvature. J. Differ. Geom. 21, 213–230, Zbl.584.53002

Lin, C.S. (1986): The local isometric embedding in R^3 of two-dimensional Riemannian manifolds with Gaussian curvature changing sign cleanly. Commun. Pure Appl. Math. 39, 867–887, Zbl.612.53013

Makarova, Z.T. (1953): Investigation of an integral invariant of binary forms of degree $n > 4$. Mat. Sb., Nov. Ser. 33, 233–240 (Russian), Zbl.52,17

Maksimov, I.G. (1987): Investigation of the bendability of polyhedra with few vertices. Proc. All-Union Conf. on Geometry "in the large". Inst. Math. Siberian Division of the USSR Academy of Sciences, Novosibirsk, p. 75 (Russian)

Markov, P.E. (1980): Infinitesimal bendings of some multidimensional surfaces, Mat. Zametki 27, 469–479. Engl. transl.: Math. Notes 27, 232–237 (1980), Zbl.436.53052

Markov, P.E. (1987): Infinitesimal higher-order bendings of multidimensional surfaces in spaces of constant curvature, Mat. Sb., Nov. Ser. 133, 64–85. Engl. transl.: Math. USSR, Sb. 61, No. 1, 65–85 (1988), Zbl.629.53020

Milka, A.D. (1973): Continuous bendings of convex surfaces, Ukr. Geom. Sb. 13, 129–141 (Russian) Zbl.288.53045

Milka, A.D. (1986): What is geometry "in the large"?, Nov. Zh. Nauk. Tekhn. Ser. "Mat. Kibernet." 3, 3–31 (Russian)

Nakamura, G., Maeda, Y. (1985): Local isometric embedding of 2-dimensional Riemannian manifolds into R^3 with nonpositive Gaussian curvature, Proc. Japan Acad., Ser. A 61, 211–212

Nash, J. (1954): C^1-isometric embeddings, Ann. Math., II. Ser. 60, 383–396, Zbl.58,377

Nirenberg, L. (1963): Rigidity of a class of closed surfaces, in: Nonlinear problems, Proc. Symp. Madison 1962, 177–193, Zbl.111,344

Pogorelov, A.V. (1967): Geometrical methods in the nonlinear theory of elastic shells, Nauka, Moscow (Russian) Zbl.168,456

Pogorelov, A.V. (1969): Extrinsic geometry of convex surfaces, Nauka, Moscow. Engl. transl.: Am. Math. Soc., Providence, RI, 1973, Zbl.311.53067

Pogorelov, A.V. (1986): Bendings of Surfaces and Stability of the Shells. Nauka, Moscow: Engl. transl.: Providence, RI (1988), Zbl.616.73051

Poznyak, E.G. (1959): A relation between non-rigidity of the first and second order for surfaces of revolution, Usp. Mat. Nauk 14, No. 6, 179–184 (Russian), Zbl.97,370

Poznyak, E.G. (1960): Nonrigid closed polyhedra, Vestn. Mosk. Univ. Ser. I. 15, No. 3, 14–19 (Russian), Zbl.98,354

Poznyak, E.G. (1973): Isometric immersions of two-dimensional Riemannian metrics in Euclidean spaces, Usp. Mat. Nauk 28, No. 4, 47–76. Engl. transl.: Russ. Math. Surv. 28, No. 4, 47–77 (1973), Zbl.283.53001

Poznyak, E.G., Sokolov, D.D. (1977): Isometric immersions of Riemannian spaces in Euclidean spaces. Itogi Nauki Tekh., Ser. Algebra, Topologija, Geom. 15, 173–211. Engl. transl.: J. Sov. Math. 14, 1407–1428 (1980), Zbl.448.53040

Reshetnyak, Yu.G. (1962): Nonrigid surfaces of revolution. Sib. Mat. Zh. 3, 591–604 (Russian), Zbl. 119,372

Reshetnyak, Yu.G. (1982): Stability theorems in geometry and analysis. Nauka, Novosibirsk (Russian), Zbl.523.53025

Sabitov, I.Kh. (1965a): The local structure of Darboux surfaces. Dokl. Akad. Nauk SSSR 162, 1001–1004. Engl. transl.: Sov. Math. Dokl. 6, 804–807 (1965), Zbl.131,193

Sabitov, I.Kh. (1965b): Some results on infinitesimal bendings of surfaces "in the small" and "in the large". Dokl. Akad. Nauk SSSR 162, 1256–1258. Engl. transl.: Sov. Math. Dokl. 6, 862–864 (1965), Zbl.131,193

Sabitov, I.Kh. (1967): A minimal surface as the rotation graph of a sphere. Mat. Zametki 2, 645–656. Engl. transl.: Math. Notes 2, 881–887 (1968), Zbl.162,247

Sabitov, I.Kh. (1973): Rigidity of "corrugated" surfaces of revolution. Mat. Zametki *14*, 517–522. Engl. transl.: Math. Notes *14*, 854–857 (1973), Zbl.283.53004

Sabitov, I.Kh. (1979a): Infinitesimal bendings of surfaces of revolution with exponential flattening at the pole. Proc. 7th All-Union Conf. on modern problems of geometry. Beloruss. State Univ., Minsk, p. 240

Sabitov, I.Kh. (1979b): Trough, Mat. Entsiklopediya. Vol. 2, p. 409. Engl. transl. in: Encycl. Math., Reidel, Dordrecht, 1988

Sabitov, I.Kh. (1983): Description of bendings of degenerate suspensions. Mat. Zametki *33*, 901–914. Engl. transl.: Math. Notes *33*, 462–468 (1983), Zbl.528.51012

Sabitov, I.Kh. (1986): Investigation of the rigidity and non-bendability of analytic surfaces of revolution with flattening at the pole. Vestn. Mosk. Univ., Ser. I 1986, No. 5, 29–36. Engl. transl.: Mosc. Univ. Math. Bull. 41, No. 5, 33–41 (1986), Zbl.633.53006

Sabitov, I.Kh. (1987): Some results and problems of the local theory of bendings. Proc. All-Union Conf. on geometry in the large, Inst. Math. Siberian Division of the USSR Academy of Sciences, Novosibirsk, p. 108 (Russian)

Sabitov, I.Kh. (1988): Isometric immersions and embeddings of locally Euclidean metrics in R^2. Tr. Semin. Vektorn. Tenzorn. Anal. *23*, 147–156 (Russian)

Sabitov, I.Kh. (1989): New classes of unbendable polyhedra. Proc. All-Union Conf. on geometry and analysis, Inst. Math. Siberian Division of the USSR Academy of Sciences, Novosibirsk, p. 72 (Russian)

Sauer, R. (1935): Infinitesimale Verbiegungen zueinander projektiver Flächen. Math. Ann. *111*, 71–82, Zbl.10,374

Sauer, R. (1948): Projektive Transformationen des Darboux'schen Flächenkranzes. Arch. Math. *1*, 89–93, Zbl.31,269

Schilt, H. (1937): Über die isolierten Nullstellen der Flächenkrümmung und einige Verbiegkeitssätze. Compos. Math. *5*, 239–283, Zbl.18,169

Sen'kin, E.P. (1978): Bending of convex surfaces. Itogi Nauki Tekh., Ser. Probl. Geom. *10*, 193–222 (Russian). Engl. transl.: J. Sov. Math. *14*, 1287–1305 (1980), Zbl.423.53047

Shor, L.A. (1962): An example of a discontinuum of nontrivially isometric convex surfaces. Usp. Mat. Nauk *17*, No. 5, 157–160 (Russian), Zbl.168,424

Sinyukov, N.S. (1986): On the development of modern differential geometry in Odessa State University in recent years. Izv. Vyssh. Uchebn. Zaved., Mat., No. 1, 69–74. Engl. transl.: Soviet Math. *30*, No. 1, 92–99 (1986), Zbl.602.01031

Spivak, M. (1979): A comprehensive introduction to differential geometry. Vol. 5, 2nd. ed., Publish or Perish, Berkeley, CA, Zbl.306.53003

Tartakovskij, V.A. (1953): The *N*-invariant of N.V. Efimov in the theory of bending of surfaces. Mat. Sb., Nov. Ser. *32*, 225–248 (Russian), Zbl.50,160

Trotsenko, D.A. (1980): Non-rigid analytic surfaces of revolution. Sib. Mat. Zh. *21*, No. 5, 100–108. Engl. transl.: Sib. Math. J. *21*, 718–724 (1980), Zbl.467.53001

Usmanov, Z.D. (1984): Infinitesimal bendings of surfaces of positive curvature with a flat point, in: Differential Geometry, Warsaw 1979, Banach Cent. Publ. *12*, 241–272 (Russian), Zbl.559.53002

Vasil'eva, A.B., Butuzov, V.F. (1978): Singularly perturbed equations in critical cases. Moscow Univ. Press. Moscow (Russian)

Vekua, I.N. (1959): Generalized analytic functions. Fizmatgiz, Moscow. Engl. transl.: Pergamon Press, London-Paris-Frankfurt; Addison-Wesley, Reading, MA, 1962, Zbl.92,297

Vekua, I.N. (1982): Some general methods of constructing different versions of the theory of shells. Nauka, Moscow, Zbl.598.73100. Engl. transl.: Pitman, Boston etc., 1985

Vincensini, P. (1962): Sur les propriétés géométriques des transformations infinitésimales des surfaces et ses relations avec la théorie de congruences des sphères. Ann. Sci. Éc. Norm. Super., III. Ser. *79*, 299–319, Zbl.108,345

Vojtsekhovskij, M.I. (1977): Infinitesimal bending. Mat. Entsiklopediya, Vol. 1, p. 435. Engl. transl. in: Encycl. Math., Reidel, Dordrecht, 1988

Vojtsekhovskij, M.I. (1979): Darboux surfaces. Mat. Entsiklopediya Vol. 2, p. 16. Engl. transl. in: Encycl. Math., Reidel, Dordrecht, 1988

Whiteley, W. (1984): Infinitesimally rigid polyhedra. I. Statics of frameworks. Trans. Am. Math. Soc. *285*, 431–465, Zbl.518.52010

Whiteley, W. (1987a): Applications of the geometry of rigid structures. Proc. Conf. on computer-aided geometric reasoning (Sophia-Antipolis, 1987), Vol. II,.INRIA, Rocquencourt, pp. 217–251

Whiteley, W. (1987b): Rigidity and polarity. I. Statics of sheet structures. Geom. Dedicata *22*, 329–362, Zbl.618.51006

Yanenko, N.N. (1952): Some necessary conditions for bendable surfaces V_m in $(m + q)$-dimensional Eudlidean space. Tr. Semin. Vektorn. Tenzorn. Anal. *9*, 236–287 (Russian), Zbl.48,389

Yanenko, N.N. (1954): On the theory of embedding of surfaces into multidimensional Eudlidean space. Trudy Mosk. Mat. O.-va *3*, 89–180 (Russian), Zbl.58,156

Zalgaller, V.A. (1962): Possible singularities of smooth surfaces. Vestn. Leningr. Univ. 17, No. 7 (Ser. Mat. Mekh. Astron. No. 2), 71–77 (Russian), Zbl.122,170

Author Index

Subject Index

Encyclopaedia of Mathematical Sciences
Editor-in-chief: R. V. Gamkrelidze

Geometry

Volume 28: **R. V. Gamkrelidze** (Ed.)
Geometry I
Basic Ideas and Concepts of Differential Geometry
1991. VII, 264 pp. 62 figs. ISBN 3-540-51999-8

Volume 29: **E. B. Vinberg** (Ed.)
Geometry II
Geometry of Spaces of Constant Curvature
1993. Approx. 260 pp. ISBN 3-540-52000-7

Volume 48: **Yu. D. Burago, V. A. Zalgaller** (Eds.)
Geometry III
Theory of Surfaces
1992. Approx. 270 pp. 80 figs.
ISBN 3-540-53377-X

Volume 70: **Yu. G. Reshetnyak** (Ed.)
Geometry IV
Nonregular Riemannian Geometry
1993. Approx. 270 pp. ISBN 3-540-54701-0

Algebraic Geometry

Volume 23: **I. R. Shafarevich** (Ed.)
Algebraic Geometry I
Algebraic Curves. Algebraic Manifolds and Schemes
1993. Approx. 300 pp. ISBN 3-540-51995-5

Volume 35: **I. R. Shafarevich** (Ed.)
Algebraic Geometry II
Cohomological Methods in Algebra. Geometric Applications to Algebraic Surfaces
1995. Approx. 270 pp. ISBN 3-540-54680-4

Volume 36: **A. N. Parshin, I. R. Shafarevich** (Eds.)
Algebraic Geometry III
1995. Approx. 270 pp. ISBN 3-540-54681-2

Volume 55: **A. N. Parshin, I. R. Shafarevich** (Eds.)
Algebraic Geometry IV
Linear Algebraic Groups. Invariant Theory
1993. Approx. 310 pp. ISBN 3-540-54682-0

Number Theory

Volume 49: **A. N. Parshin, I. R. Shafarevich** (Eds.)
Number Theory I
Fundamental Problems, Ideas and Theories
1993. Approx. 340 pp. 16 figs.
ISBN 3-540-53384-2

Volume 62: **A. N. Parshin, I. R. Shafarevich** (Eds.)
Number Theory II
Algebraic Number Theory
1992. VI, 269 pp.
ISBN 3-540-53386-9

Volume 60: **S. Lang**
Number Theory III
Diophantine Geometry
1991. XIII, 296 pp. 1 fig.
ISBN 3-540-53004-5

Encyclopaedia of Mathematical Sciences
Editor-in-chief: R. V. Gamkrelidze

Algebra

Volume 11: **A. I. Kostrikin, I. R. Shafarevich** (Eds.)
Algebra I
Basic Notions of Algebra
1990. V, 258 pp. 45 figs. ISBN 3-540-17006-5

Volume 18: **A. I. Kostrikin, I. R. Shafarevich** (Eds.)
Algebra II
Noncommutative Rings. Identities
1991. VII, 234 pp. 10 figs. ISBN 3-540-18177-6

Volume 37: **A. I. Kostrikin, I. R. Shafarevich** (Eds.)
Algebra IV
Infinite Groups. Linear Groups
1993. Approx. 220 pp.
ISBN 3-540-53372-9

Volume 57: **A. I. Kostrikin, I. R. Shafarevich**
Algebra VI
Combinatorial and Asymptotic Methods of Algebra
1994. Approx. 260 pp.
ISBN 3-540-54699-5

Volume 58: **A. N. Parshin, I. R. Shafarevich**
Algebra VII
Combinatorial Group Theory. Applications to Geometry
1993. Approx. 260 pp. 38 figs.
ISBN 3-540-54700-2

Volume 73: **A. I. Kostrikin, I. R. Shafarevich** (Eds.)
Algebra VIII
Representations of Finite-Dimensional Algebras
1992. Approx. 188 pp. 98 figs.
ISBN 3-540-53732-5

Topology

Volume 24: **S. P. Novikov, V. A. Rokhlin** (Eds.)
Topology II
Homotopies and Homologies
1994. Approx. 235 pp. ISBN 3-540-51996-3

Volume 17: **A. V. Arkhangel'skij, L. S. Pontryagin** (Eds.)
General Topology I
Basic Concepts and Constructions. Dimension Theory
1990. VII, 202 pp. 15 figs. ISBN 3-540-18178-4

Volume 50: **A. V. Arkhangel'skij** (Ed.)
General Topology II
Compactness. Homologies of General Spaces
1994. Approx. 270 pp. ISBN 3-540-54695-2

Volume 51: **A. V. Arkhangel'skij** (Ed.)
General Topology III
1994. Approx. 240 pp. ISBN 3-540-54698-7